ACS SYMPOSIUM SERIES 423

# Barrier Polymers and Structures

**William J. Koros,** EDITOR
*The University of Texas at Austin*

Developed from a symposium sponsored by
the Division of Polymer Chemistry, Inc.
at the 197th National Meeting
of the American Chemical Society,
Dallas, Texas,
April 9–14, 1989

American Chemical Society, Washington, DC 1990

**Library of Congress Cataloging-in-Publication Data**

Barrier Polymers and Structures
  William J. Koros, editor

  p.    cm.—(ACS Symposium Series, 0097–6156; 423).

"Developed from a symposium sponsored by The Division of
Polymer Chemistry, Inc. at the 197th Meeting of the American
Chemical Society, Dallas, Texas, April 9–14, 1989."

  Includes bibliographical references.

  ISBN 0–8412–1762–9

  1. Food—Packaging—Congresses. 2. Plastics in packaging—
Congresses.

  I. Koros, William J., 1947–    . II. Polymer Chemistry, Inc.
III. American Chemical Society. Meeting (197th : 1989 : Dallas,
Tex.) IV. Series.

TP374.B37   1990
668.4'9—dc20                              90–275
                                              CIP

# ACS Symposium Series

## M. Joan Comstock, *Series Editor*

### *1990 ACS Books Advisory Board*

# Foreword

The ACS SYMPOSIUM SERIES was founded in 1974 to provide a medium for publishing symposia quickly in book form. The format of the Series parallels that of the continuing ADVANCES IN CHEMISTRY SERIES except that, in order to save time, the papers are not typeset but are reproduced as they are submitted by the authors in camera-ready form. Papers are reviewed under the supervision of the Editors with the assistance of the Series Advisory Board and are selected to maintain the integrity of the symposia; however, verbatim reproductions of previously published papers are not accepted. Both reviews and reports of research are acceptable, because symposia may embrace both types of presentation.

# Contents

# Preface

Barrier resins have revolutionized the packaging industry and, therefore, have been the focus of intense investigation. Numerous interesting developments related to barrier applications have been reported in journals and at various technical meetings. Typically, however, the information is spread in a dilute fashion through a large number of outlets. The time seemed right to publish a collection of both fundamental and practical principles involved in making and using barrier polymers and structures. The project took the form of this book, which provides a comprehensive treatment of the state of science and technology in the area of barrier polymers and barrier structures.

The topics covered will be of direct interest to industrial scientists making or using barrier packaging. Moreover, government regulators who work with the packaging industry should find the book useful in indicating trends in materials and processes in the industry. Academic researchers working in fundamentals of sorption and transport processes will also find several good updates in these areas.

## Acknowledgments

The National Science Foundation and the ACS are acknowledged for partial support of my time in the preparation of the overview chapter and the coordination of the book. It has been a sincere pleasure working with Cheryl Shanks and Beth Pratt-Dewey of the ACS on this project; their good nature and efficiency were examples for me. Thanks are extended to the session chairs and participants in the symposium on which the book is based. Finally, I express my deepest gratitude to the authors whose work is presented in this volume and the many excellent reviewers who helped perfect the papers with their thoughtful comments.

## Dedication

This book is dedicated to Vivian T. Stannett, whose pioneering contributions to understanding the fundamentals of barrier polymers continue to inspire and motivate scientists in this field.

WILLIAM J. KOROS
The University of Texas at Austin
Austin, TX 78712

January 25, 1990

# Chapter 1

# Barrier Polymers and Structures: Overview

**William J. Koros**

**Department of Chemical Engineering, The University of Texas at Austin, Austin, TX 78712**

This introductory chapter provides an overview of the
papers presented at the symposium on Barrier Polymers
and Barrier Structures that was sponsored by the Polymer
Chemistry Division of the American Chemical Society at
the Spring 1989 meeting.  A total of nineteen papers
from the symposium are included in this volume.   Topics
covered include barrier transport fundamentals, advanced
composite structures, reactive surface treatments and
the effects of orientation on barrier properties.
Relationships between polymer molecular structure and
barrier efficacy are also treated in detail.   Time and
history dependent phenomena associated with retorting of
barrier laminates are discussed from the standpoint of
theoretical modeling and experimental characterization
of  the barrier layers.   The effects of concentration
dependent diffusion, flavor scalping and nonFickian
transport phenomena are also discussed.   The coverage,
therefore, is broad while providing sufficient depth to
provide a state-of-the-art update on the major technical
issues facing the barrier packaging field.

The development of efficient packaging materials and containers has
allowed the evolution of the modern market system based on
concentrated production facilities supplying individuals residing
far from the ultimate source of products.  The goal of the packaging
industry has been to provide increasingly more cost effective means
of preserving the quality of materials with as close to their
as-produced natures as possible.
  Modern packaging is a sophisticated technology rooted deeply
in fundamental polymer science.  Nevertheless, the sheer size and
competitiveness of the packaging industry also makes it an extremely
practical field with an eye to applying polymer science to achieve
the bottom line of cost and barrier efficiency. This book seeks to

0097–6156/90/0423–0001$06.25/0

reflect the dual nature of this technology by presenting fundamental
principles along with complementing discussions of applications of
these principles by leaders in the field.

Many excellent reviews of the fundamental processes by which
small molecules penetrate between the segments comprising a
polymeric film are available (1-10). These reviews, coupled with
past symposia publications related to barriers and the allied field
of membranes (11-18) should allow a newcomer to the field to rapidly
gain the needed background to become an active participant in this
dynamic technology.

The chapters in the present book reflect the current research
and development directions of active university and industrial
participants in the barrier field. The overview provided by this
first chapter gives a framework for appreciating the more detailed
subsequent discussion of topics that are at the cutting edge of this
evolving technology. Therefore, it is anticipated that this book
will be of interest to both the current expert and the newcomer to
the field.

## Fundamentals

In addition to providing a container to prevent scattering and bulk
phase mixing of components, modern packages control the exchange
of components between the package contents and the external
environment. For instance, the protection from attack by oxygen is
among the most common function served in food packaging. Even this
general requirement, however, has manifold aspects. The degrees of
sensitivity of different materials to environmental effects are
clearly very different and the oxygen barriers required to
successfully store these products differ accordingly. Similar
considerations apply to water permeation, since foods and many
pharmaceuticals show varying degrees of stability in a dry state as
compared to that in the presence of water. Table I (19)
illustrates typical ranges of sensitivity to oxygen and water vapor
for a spectrum of common foods. The table also illustrates the
manifold requirements needed in terms of oil and flavor and/or aroma
component barriers that packaging must provide for the different
foods.

As packaging capabilities increase, the complexity of the
applications that can be treated also increases considerably,
thereby explaining the ever expanding markets for packaging
materials. An interesting example in which complex requirements
must be met involves the storage of blood platelets (20). Blood
platelets are living cells that both consume oxygen to live and
generate carbon dioxide as a metabolic byproduct. The generation of
carbon dioxide presents problems to viability, since it tends to
cause undesirable changes in the natural pH unless it can escape.
An added requirement enters because the aqueous solution containing
the platelets must not lose a significant amount of water by
permeation. This case, therefore, illustrates the need for an
advanced "controlled atmosphere" package, or "smart package" that is
able to allow relatively free exchange of oxygen and carbon dioxide
with the external environment while essentially preventing outward
permeation losses of water. Similar considerations apply

Table I: Permeation protection required for various foods and beverages for a one year shelf life at 25°C †

| Food or Beverage | Estimated maximum tolerable oxygen gain, ppm | Estimated maximum water gain or loss wt. percent | High oil barrier req'd? | High volatile organics barrier req'd? |
|---|---|---|---|---|
| Canned milk, meats | 1 to 5 | -3 | yes | - |
| Baby foods | 1 to 5 | -3 | yes | yes |
| Beer, ale, wine | 1 to 5* | -3 | - | yes |
| Instant coffee | 1 to 5 | +2 | yes | yes |
| Canned vegetables, soups, spaghetti | 1 to 5 | -3 | - | - |
| Canned fruits | 5 to 15 | -3 | | yes |
| Nuts, snacks | 5 to 15 | +5 | yes | - |
| Dried foods | 5 to 15 | +1 | - | - |
| Fruit juices, drinks | 10 to 40 | -3 | - | yes |
| Carbonated soft drinks | 10 to 40* | -3 | - | yes |
| Oils, shortenings | 50 to 200 | +10 | yes | - |
| Salad dressings | 50 to 200 | +10 | yes | yes |
| Jams, pickles, vinegars | 50 to 200 | -3 | - | yes |
| Liquors | 50 to 200 | -3 | - | yes |
| Peanut butter | 50 to 200 | +10 | yes | - |

* Less than 20% loss of $CO_2$ is also required.

† (Data taken from ref. 19.)

to the packaging of fruits and vegetables that require inward
permeation of carbon dioxide and respiration of oxygen without loss
of water (21).

Modern packages typically regulate the contact between the
contents and the environment by a process  known as permeation.
The first good description of the permeation process in polymers
dates back to 1831 when Mitchell (22) noted that  natural rubber
membranes allowed the passage of carbon dioxide faster than
hydrogen under equivalent conditions.  Mathematically, one can
describe the permeation process in terms of Eq(1) using a
*permeability coefficient* of component i, $P_i$,

$$P_i = \frac{\text{(steady state flux of i)}}{\Delta p_i / \ell} \tag{1}$$

The permeability is defined in terms of the steady state
permeation rate of component i per unit area divided by the
normalized partial pressure difference of component  i , $\Delta p_i$, acting
across the membrane of thickness, $\ell$ (see Fig. 1) (23,24).

Accurate description of barrier films and complex barrier
structures, of course, requires information about the composition
and partial pressure dependence of penetrant permeabilities in each
of the constituent materials in the barrier structure.  As
illustrated in Fig. 2 (a-d), depending upon the penetrant and
polymer considered, the permeability may be a function of the
partial pressure of the penetrant in contact with the barrier layer
(15).  For gases at low and intermediate pressures, behaviors shown
in Fig. 2a-c  are most common.  The constant permeability in Fig.2a
is seen for many fixed gases in rubbery polymers, while the response
in Fig. 2b is typical of a simple plasticizing response for a more
soluble penetrant in a rubbery polymer.   Polyethylene and
polypropylene containers are expected to show upwardly inflecting
permeability responses like that in Fig. 2b as the penetrant
activity in a vapor or liquid phase increases for strongly
interacting flavor or aroma components such as d-limonene which are
present in fruit juices.

The permeability vs pressure response in Fig. 2c is found for
most gases in glassy polymers (24-27).  In general, the magnitude of
the decline in permeability with pressure depends upon the glass
transition temperature of the polymer and the critical temperature
of the gas.  This type of response is related to intersegmental
packing defects present in glasses due to their hindered mobilities,
and is discussed in Chapter 2 (28) under the heading of "dual mode"
sorption theory.

At sufficiently high penetrant partial pressures in glassy
polymers, the onset of plasticization produces the upturn in
permeability seen in Fig. 2d.  While this behavior is shown for
the acetone-ethyl cellulose system, it is also typical of many
other systems. In fact, if one extends the pressure range of the
measurements in Fig. 2c to 900 psia, an  upturn in permeability like
that seen in Fig. 2d is observed (29).  This response is, therefore,
a combination of the dual mode response in Fig. 2c and the
plasticization response in Fig. 2b at high sorbed concentrations.

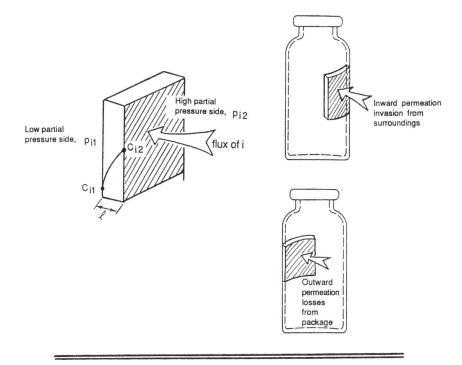

Figure 1: Summary of solution-diffusion model relationships. Typically the diffusion coefficient, $D_i$, decreases as the penetrant molecular weight increases. On the other hand, the solubility coefficient, $S_i$, tends to increase with increasing penetrant molecular weight. As a result, the permeability, $P_i$, may either decrease or increase with increasing penetrant molecular weight, depending upon which factor dominates.

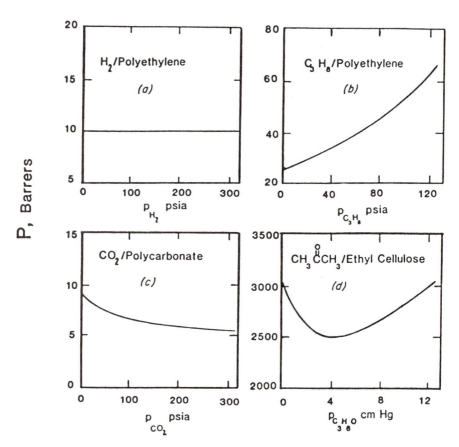

Figure 2: The pressure dependency of various penetrant-polymer systems:
(a) 30°C, $H_2$/polyethylene, (b) 20°C propane/polyethylene, (c) 35°C, carbon
dioxide/ polycarbonate, (d) 40°C, acetone/ ethyl cellulose. (Reprinted with
permission from ref. 15. Copyright 1987   John Wiley and Sons.)

Graham, who was one of the first to consider the
permeabilities of natural rubber films to a wide range of gases,
found responses such as that seen in Fig. 2a.    The description he
formulated in 1866 of the so-called "solution-diffusion" mechanism
still prevails today (30).    He postulated that a penetrant leaves
the external phase in contact with the membrane by dissolving in the
upstream face of the film and then undergoes molecular diffusion to
the downstream face where  it evaporates into the external phase
again.    Mathematically, one can state the solution-diffusion model
in  terms of permeability, solubility and diffusivity coefficients,
as shown in Eq(2).

$$P = D\,S \hspace{4cm} (2)$$

where the diffusion coefficient , D, characterizes the average
ability of the dissolved penetrant to move among the polymer
segments comprising the film.  The solubility coefficient, S, is
thermodynamic in nature and is related to the slope of the
equilibrium sorption isotherm of the penetrant of interest, so a
large value of S implies a large tendency for the penetrant to
dissolve, or "sorb", into the polymer.   In the more complex cases
shown in Fig. 2c-d, the solution-diffusion mechanism still applies;
however, the D and S parameters can be strong functions of the
penetrant activities or partial pressures in contact with the
barrier at its two surfaces.

## Gas Barrier Materials

If one is using a barrier to minimize outward diffusion of a
component, e.g. $CO_2$ , the downstream condition "1" in Fig. 1 would
correspond to the external atmosphere.  On the other hand, for
minimization of inward diffusion, e.g. $O_2$ barriers,  the downstream
condition "1" corresponds to the internal package contents.   In
either case, the penetrant concentration is typically very low at
the downstream face. For such cases where  $C_{i1} \sim 0$ , one can simply
represent S as the secant slope of the equilibrium sorption isotherm
for the particular polymer-penetrant pair (31).
    Clearly, in seeking a good barrier material for excluding a
given penetrant, it is desirable to minimize both D and S as much as
possible without sacrificing cost and processing ease.  Extensive
studies of synthetic rubbery materials were performed to identify
possible replacements for natural rubber(3,32-35).   The development
of butyl and nitrile rubbers with both adequate resilience and
greatly reduced air and even hydrocarbon permeability for tires and
gasket applications was an early success of these structure-
permeability studies.
    Such materials, however, are not appropriate for packaging
films, and regenerated cellulose or cellophane  was the first
practical glassy film to find large scale use in packaging (36).
Cellophane is an excellent gas barrier in the dry state; however, it
is sensitive to moisture and loses barrier capability at high
relative humidities (4,23). Moreover, it is not thermoplastic or
heat sealable, so replacements were clearly desirable.

Polyvinyl chloride (PVC) is an economical thermoplastic film with good barrier properties to gases and water.    Later developments of even higher barrier Saran® type resins based on copolymers of vinylidene chloride and vinyl chloride produced many new packaging opportunities.    These materials are still very popular today (37-39).    Consideration of the structure-permeability relationships for the Saran  materials in Chapter 6(40) indicates that their evolution is still continuing as better understanding of the influence of sequence distribution and even more subtle issues of chain microstructure emerge.

Concerns about health effects caused by low levels of residual monomers in PVC and acrylonitrile resins promoted the consideration of condensation polymers such as poly(ethylene terephthalate) (PET) (41).    Packaging resins based on acrylonitrile are extraordinarily good barriers that were considered for carbonated beverage bottles. The furor over residual monomers in these materials stunted their growth in spite of improved technology to reduce residuals to barely detectable levels.    The step reaction polymer, polyethylene terephthalate(PET) has intrinsically much lower residual monomers in the as-made polymer, and it has become the material of choice for the carbonated beverage market (42).

Ethylene-vinyl alcohol (EVOH) copolymers were discovered to be excellent oxygen barriers and did not require dealing with dangerous monomers, so they became popular soon after their introduction in the US in the mid 1970's (43).    These polymers combine a strong hydrogen bonded amorphous phase like that of cellophane with a partially crystalline nature.    While these materials have incredibly low permeabilities to oxygen in the dry state, like cellophane, they lose their barrier capability at high relative humidities (43).    Chapters 8-11(44-47) discuss approaches to engineer around this sensitivity to humidity through the use of a coextruded laminate structure.    While this approach minimizes direct contact of the barrier layer with high water activities, during steam retorting for sterilization, complex measures are needed to avoid serious long term degradation of the barrier (48,49).    As noted in the case of the older Saran  resins, much remains to be learned regarding the effects of chain microstructure on ultimate film properties for the EVOH family of polymers.

## Directions in the Search for New Barrier Materials

The search for basic principles to guide the development of more effective barrier materials based on condensation polymerization is ongoing.    Advanced polyesters such as poly(ethylene 2,6-naphthalene dicarboxylate) or PEN have been reported to have as much as 5 times lower permeability than conventional PET; however, proprietary considerations have prevented the publication of complete details concerning such materials.

The chapters devoted to structure-permeability properties of the more commercially advanced amorphous nylons (Chapter 5 (50)) and experimental polycarbonates (Chapter 7 (51))  illustrate many of the general principles needed for optimizing the barrier properties of essentially any polymer family. These chapters consider the importance of intersegmental packing and intrasegmental mobility on the ability of a small molecule like oxygen to execute its thermally activated diffusive motions through glassy environments typical of high barrier polymers. The high barrier amorphous nylon resins discussed in Chapter 5 (50) have the desirable property of becoming better, rather than worse oxygen barriers as the humidity increases.

The glassy polymers such as the aromatic polyamides and polycarbonates have significant hindrances to intramolecular mobility. The data for these materials appear to be correlated fairly well in terms of the "specific free volume" discussed by Lee(52). Structural variations that suppress the ability to pack tend to reduce the quality of the barrier while those that improve the ability to pack produce better barriers. The free volume in this case is defined as the difference between the actual polymer molar volume at the temperature of the system and at 0°K. This latter parameter is determined by group contribution methods.

The above approach lumps all volume together, and for glassy materials this oversimplification may account for some of the scatter in such correlations. For glasses, regions of nonuniform packing are thought to contribute to the apparent free volume, besides that which arises from segmental and subsegmental oscillations and vibrations (25). This localized unrelaxed volume due to packing defects is not as likely to contribute to diffusion as is the generally distributed free volume which is contributed by segmental motions (25).

Therefore, while lumping the two contributions together simplifies broad comparisons between different materials, it would be interesting, but tedious to base comparisons on specific free volumes that have been adjusted for the contribution of the packing defects to see if scatter in the correlations is reduced. Fortunately such corrections are of less importance for materials such as the polyamides considered in Chapter 5 with similar glass transition temperatures, so scatter is rather small even without their consideration (50).

Several detailed analyses of the diffusion process in both rubbery polymers and in hindered glasses are offered in Chapter 2 (28). Approximate molecular interpretations have been offered for the parameters in these models (25). Nevertheless, more work is needed to verify any molecular scale connection between such parameters and the structures and motions of the polymer backbone. Spectroscopy and molecular modeling of the differences in segmental motions in a systematically varied family of polymers, e.g, the polyesters, or polyamides, can offer insight in some cases. Unfortunately, the exact segmental motions involved in the diffusive process are only partially understood, so one must be cautious about drawing conclusions based on such studies unless they are supported by actual complementary transport data. Hopefully the structure-property results presented in this book will further stimulate thinking to improve the connection between spectroscopi-cally sensed motions, and diffusion to complement the correlations based on specific free volume in Chapters 5 & 7 (50,51).

The importance of segmental motions in the diffusion process is reflected by the Arrhenius expression for the diffusion coefficient given by Eq(3) (23,24).

$$D = D_0 \exp[ -E_D/RT]$$

(3)

The majority of the activation energy for execution of a diffusion jump is used to produce a transient gap of sufficient size between surrounding segments to allow movement of the penetrant over the length of one diffusional step, $\lambda$. As shown in Fig. (3a), a good correlation exists between the activation energy and the preexponential factor in Eq(3) for elastomers. This correlation is largely independent of the type of rubber and gas type (3). The triangle points that have been added to the plot are data for rubbery semicrystalline PET, and these points appear to fit the curve well in spite of the fact that PET is not an elastomer like the other materials on the plot.

Figure (3b) compares similar data for preexponential factors and activation energies for three glassy polymers: PET, bisphenol-A polycarbonate(PC) and tetramethyl bisphenol-A polycarbonate(TMPC). TMPC has a Tg over 50°C higher than that of PC and almost 120°C above that of PET. Moreover, the sub-Tg transition of TMPC is over 150°C higher than that of PC, indicating that it's motions are highly hindered. In spite of marked differences between the polymers, all three glassy polymers appear to fit reasonably on a single correlation line. If one considered polymers with lower Tg's so the difference between the measured temperature and Tg was small, the scatter in the glassy state correlation would probably increase, reflecting the transition region between the two states(23).

It appears that the slope of the correlation line for the glassy materials is very similar to that for the rubbery materials, but there is an offset in the value of $D_0$ by roughly two orders of magnitude relative to rubbery materials having similar activation energies. The fact that rubbery PET follows the top line, while glassy PET follows the bottom one is impressive evidence for a change in some aspect of the transport process in the glass transition interval. The preexponential factor can be written in terms of the activated state theory to give:

$$D_0 = \kappa \; \lambda^2 \; kT/h \; \exp( \Delta S_D/R) \tag{4}$$

The $\kappa$ factor is essentially equal to 1/6 for an isotropic medium, and it should be the same for both rubbery and glassy polymers. The lower values of $D_0$ in the glassy state, therefore, must be due to either a large reduction in the jump length, or to a significantly lower entropy of activiation associated with formation of the activated state in the glassy material as compared to that in the rubbery state. Most molecular visualizations of the diffusion process suggest that although the jump length may be somewhat smaller in glasses as compared to rubbers, they are probably on the order of one to three collision diameters of the penetrant in both media (53). Therefore, the large difference in the preexponential factors in the glassy and rubbery states is likely to be due to significantly lower entropies of activation in the glass as compared to the rubbery state. Molecular modeling calculations may one day be able to explain systematic differences in $D_0$ for structurally-related materials to permit molecular design of improved barriers. Achieving this goal in the near future is unlikely, so systematic experimental studies coupled with fundamentally based correlations

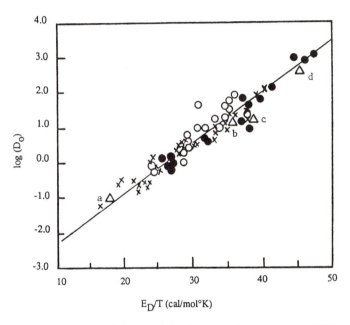

Figure 3a: Illustration of the correlation between the preexponential factor in Eq(3) and the activation energy divided by the absolute temperature at the midpoint of the temperature range over which it was evaluated for a large number of penetrants in different elastomers. Note that all points are relatively well correlated by a single line (3). The triangle points are for He (a), $N_2$(b), $CO_2$(c), and $CH_4$(d) in rubbery PET (23).

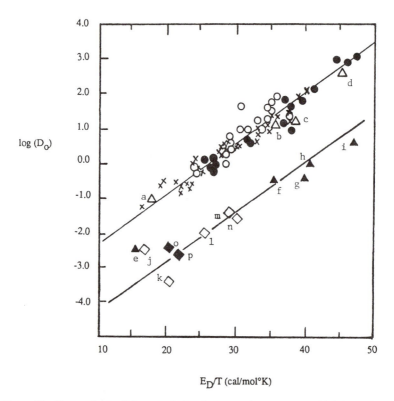

Figure 3b: Comparison of the correlation between the preexponential factor in Eq(3) and the activation energy divided by the absolute temperature at the midpoint of the temperature range over which it was evaluated for a large number of penetrants in different glassy and rubbery polymers.    The open triangle points are the same as in Fig. 3a.  The closed triangle points are for  He (e), $O_2$(f), $CO_2$(g),  $N_2$(h) and $CH_4$ (i) in glassy PET (23).  The open diamonds are for $H_2$ (j), Ar(k), $O_2$(l),  $CO_2$(m) and  Kr (n) in glassy PC (23).  The closed diamonds are for $N_2$(o) and $O_2$(p)  in glassy TMPC (26).

like those discussed in Chapters 5 & 7 provide the most efficient means for discovering new high performance materials (50,51).

## Complex Barrier Responses

The above discussions of the diffusion coefficient are a bit overly simplified, since both the diffusivity and the solubility coefficients in Eq(2) may vary with penetrant activity at a fixed temperature (15, 26, 54-56). Indeed, as suggested by the variety of permeability responses shown in Fig. 3, D and S can be functions of the state of the polymer and the activity of the penetrant as well as the temperature. Chapter 2(28) discusses the differences between rubbery and glassy polymers and surveys the types of theoretical analyses that exist for gas sorption and transport processes. Besides reviewing currently accepted analyses of these two states of amorphous materials, including the well-known dual mode sorption model, this chapter suggests the need for molecular-scale modeling. This theme is also echoed in Chapter 6(40) where preliminary results from such modeling for the high barrier Saran® materials are considered.

Chapter 3(57) discusses the reduced permeability commonly observed when crystallinity is added to either a rubbery or glassy matrix. The reduction in permeability results from reductions in both the solubility and diffusivity parameters (58-62). This chapter also reviews the important effects of combined orientation and crystallinity which are dependent on whether orientation occurs during the formation of the barrier, or subsequent to it in a cold drawing process (62-66). The effects of process-induced orientation of several commercial barrier resins is considered in further detail in Chapter 12(67).

Besides the above conventional effects, Chapter 3 summarizes data suggesting the ability of some gases to sorb and diffuse inside the actual crystals of poly(4-methyl-1-pentene) (68,69). Finally, Chapter 3 considers liquid crystalline polymers, which seem to form a new class of materials in terms of barrier responses(57). The high barrier nature of liquid crystal polymers appears to be largely due to their unusually low solubility coefficients for typical penetrants. This is quite different from the case for most high barriers like EVOH, and polyacrylonitrile that typically function due to the unusually low mobilities of penetrants in their matrices (70).

Chapter 4(71) focuses on the characterization of sorption kinetics in several glassy polymers for a broad spectrum of penetrants ranging from the fixed gases to organic vapors. The sorption kinetics and equilibria of these diverse penetrants are rationalized in terms of the polymer-penetrant interaction parameter and the effective glass transition of the polymer relative to the temperature of measurement. The kinetic response is shown to transition systematically from concentration *independent* diffusion, to concentration *dependent* diffusion , and finally to complex *nonFickian* responses. The nonFickian behavior involves so-called "Case II" and other anomalous situations in which a coupling exists between the diffusion process and mechanical property relaxations in the polymer that are induced by the invasion of the penetrant (72-78).

## Complex Barrier Structures

As shown in Fig. 4, one can use a variety of barrier structures besides that of the simple monolithic container or film to control the exchange between the package and the external environment. Cases b-d involve the addition of permeation resistances in series, while cases e & f involve the introduction of an effectively longer permeation path length by interposing objects with large aspect ratios in the barrier. Barrer (79) has provided a comprehensive analysis of both the laminate and filled systems, including the possibility of finite permeability through the dispersed phase as in Fig.4f. He gives expressions for the steady state permeability and transient time lag associated with the approach to steady state permeation for the case of constant diffusivities and solubility coefficients. Examples of such barrier structures are discussed in the following chapters.

It was noted earlier that although EVOH is an outstanding oxygen barrier in the dry state, complex and undesirable transformations in its barrier properties occur as the relative humidity is increased (44-46). The most significant of these changes are referred to generally as "retort shock" and reflects interactions of water with hydrogen bonds responsible for the excellent oxygen barrier of this resin under dry conditions. Under high temperature retort conditions used for sterilization of food, catastrophic changes can be wrought in the barrier properties of the barrier layer of laminate films (48). This phenomenon, discussed in Chapters 8-11 is affected by the degree of crystallinity, the extent of orientation and the time and temperature of the retort process(44-47). Severe disruptions in the amorphous phase hydrogen bond network and even some of the small, less perfect crystalline domains are believed to occur in extreme retort shock cases. Unmitigated, this effect has serious results, amounting to roughly a 300% loss in oxygen barrier efficacy even a year after the retort process. In Chapter 9, modification of the laminate construction (Fig. 4b), including the incorporation of desicants in some cases, is shown to provide hopeful approaches to mitigating retort shock(45).

The use of other materials as the central layer of the barrier laminate is, of course, feasible. Obvious candidates for this application include the high barrier amorphous polyamides (Chapter 5 (50)) and the liquid crystalline polyesters (Chapter 3(57)) which either develop slightly improved barriers under elevated relative humidity conditions or at least do not lose barrier properties. No reports are yet available concerning the performance of such structures.

Besides the coextruded laminate structure in Fig. 4b, cases c-f are also viable structures for some applications. Chapter 11(47) discusses the addition of inorganic fillers to EVOH copolymer to achieve large increases in barrier properties in some applications. The effects of different loadings of mica flake in several polymers other than EVOH was also recently reported to be effective (80).

Reactive modifications of inexpensive low barrier resins as a means of optimizing cost-benefit properties of barrier structures

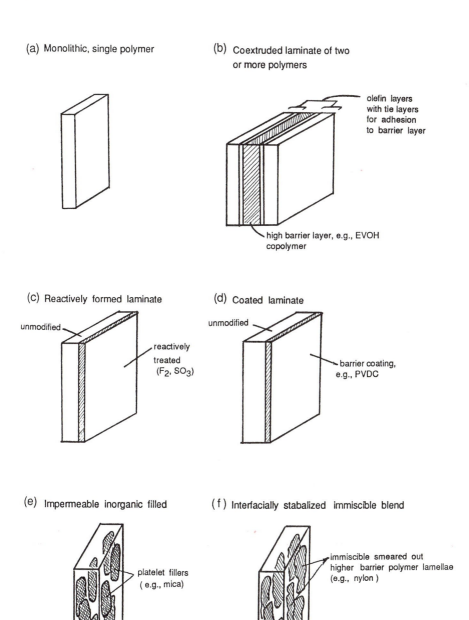

Figure 4: Primary types of barrier structures.

(Fig. 4c) can be compared to lamellar blending of small amounts of immiscible higher barrier polymers (Fig. 4f) by consideration of Chapters 14, 15 and 13, resp.(81-83) The efficacies of the various treatments for the preparation of solvent resistant automotive gasoline tanks and hydrocarbon storage bottles are treated in these chapters.

## Flavor Scalping by Package Walls

Whereas most packaging involves preventing the passage of material between the external environment and the internal package contents, the problem of uptake of components *into* the barrier wall also deserves attention. This problem was first encountered when PET bottles were introduced. It was found that although actual loss of $CO_2$ to the external environment might be small, absorption into the wall could account for measurable decarbonation of the contained beverage (84). More recently, the issue has involved the "flavor scalping" of critical flavor or aroma components due to sorption into the olefin inner liners of fruit juice composite wall containers (85-87). This problem can also be complicated further by destruction of flavor components by invading oxygen in some cases.

The complex biophysical nature of taste and odor sensations makes it difficult to define the precise agents that are most critical in retaining pleasing product properties. It may even be possible that major components such as d-limonene which exist at high levels in fruit juices may not be the most critical compounds that are sensed to be missing in off-flavor products. Nevertheless, a major component like d-limonene with the ability to interact strongly with the olefin inner liners may swell and plasticize the olefin structure sufficiently to allow the rapid penetration of the critical minor components. The critical component might be present in such a low level that it would be unable to swell the sample sufficiently to allow the debilitating loss in flavor that can occur in the presence of the d-limonene.

The detailed consideration of the flavor scalping problem in reference 18 and in Chapters 16-18 (88-90) of the present book provide a good introduction to quantifying and engineering around this complex problem. In addition to optimization of crystallinity and orientation factors to suppress d-limonene uptake in flavor scalping, one might expect that reactive treatments such as those used for rendering gas tanks highly impervious to hydrocarbon uptake could help with such flavor scalping that is driven by d-limonene and other terpene sorption. While this idea appears attractive, a recent analysis has shown that it is unlikely to succeed if the primary flavor loss occurs due to d-limonene sorption (87).

## Rubber-Solvent Systems

Strong interactions of vapors and liquid penetrants with rubbery polymers are discussed in Chapters 19 and 20 (91-92). These systems are extremely important in the continued improvement of gaskets, hoses, and protective apparel. Chapter 20(92) shows that anomalous sorption behavior can be observed in rubbers exposed to high activities of strong swelling solvents. Evidence is presented to show that the anomalous behavior is probably due to a breakdown of nonisothermal conditions instead of inhibited segmental relaxation effects as in the case of glassy materials cited in Chapter 4(71).   Similar effects have been noted for water vapor sorption due to heat transfer limitations coupled with the large heat of vaporization and rapid diffusion rates of water.  The present case, however, is believed to be the first example of the problem involving an organic vapor.  This study is useful reading for anyone considering the use of vapor sorption- desorption studies to estimate penetrant diffusivities in rubbery media where uptake rates are rapid and may appear to produce anomalous sorption responses.

Chapter 19 considers the effects of polymer-penetrant interactions on the sorption of aromatic penetrants into a polyurethane thermoplastic elastomer(91).  A direct liquid immersion approach was used, so the heat transfer problems noted above should not be important.  Nevertheless, non Fickian phenomena are still observed.  Unlike simple elastomers, thermoplastic elastomers achieve their crosslinked natures by formation of microdomains of either crystalline or glassy hard segments.  The anomalous sorption behavior presumably reflects interactions of the solvents with these microdomains (93-94).

## Conclusion

The packaging industry is entering a period of high visibility and high expectations.  The technology discussed in the following chapters gives hints of the available resources in terms of materials and package structures that must be used to meet these challenges.  Current estimates (85) suggest that the dollar value of polymers used to manufacture packaging alone will account for almost $16 Billion in 1990. Gaskets, hoses, protective apparel, and encapsulants or masks for microelectronics (95) add considerably to the size of the entire market that depends upon packaging related technology.

Clearly, an incredible number of issues face the modern packaging engineer.  Solvent attack, oxygen invasion, flavor losses, water losses or gains must be regulated using materials that hopefully will not cost an inordinate fraction of the value of the package contents.  In addition to the treatment of purely technical issues treated an important social consideration involving the acceptable handling of wastes generated by the use of polymeric materials in packaging must be considered.  Indeed, this issue, as much as first cost considerations of packaging approaches will become increasingly important as society awakens to both the limits of our ability to bury our wastes and also the value of polymeric wastes if handled correctly in recycling programs.

## Literature Cited

1.   Crank J. and Park G. S. *Diffusion in Polymers*; Eds.; Academic: New York, 1968.
2.   Rogers, C. E. *In Physics and Chemistry of the Organic Solid State*; Fox, D., Labes, M. M., and Weissberger, A., Eds.; Wiley-Interscience: New York, 1965; Vol II, Chapt.6.

3.  van Amerongen, G. J. Rubber Chem. Technol., 1964, 37, 1065.
4.  Bixler, H. J. and Sweeting, O. J., In Ed., The Science and Technology of Polymer Films, Sweeting, O. J.,Ed.; John Wiley and Sons: New York, 1971; Vol. II.
5.  Hopfenberg, H. B. Permeability of Plastic Films and Coatings to Gases, Vapors and Liquids; Ed.; Plenum: New York, 1974.
6.  Crank, J. The Mathematics of Diffusion; 2nd Ed.; Clarendon Press: Oxford, UK, 1975
7.  Felder, R. M. and Huvard, G. S. Methods of Experimental Physics,1980, 16c, 315.
8.  Stern, S. A. and Frisch, H. L. CRC Crit. Rev. in Solid State and Mat. Sci., 1983, 11(2), 123, CRC Press: Boca Raton, Fl.
9.  Hopfenberg, H. B.and Paul D. R. In Polymer Blends; Paul, D. R. and S. Newman; Eds.; Academic: New York, 1978; Chapt. 10.
10. M. Salame In The Wiley Encyclopedia of Packaging Technology; Bakker, M., Ed.; John Wiley and Sons: New York, 1986; pp. 48-54.
11. Koros, W. J.; Fleming, G. K.; Jordan, S. M.; Kim, T. H. and Hoehn, H. H. Prog. Polym. Sci., 1988, 13, 339 .
12. Vieth, W. R.; Howell, J. M. and Hsieh, J. H. J. Membr. Sci. 1976, 1, 177.
13. Vieth, W. R. Membrane Systems,: Analysis and Design; Hanser Publishers: New York, 1988; Chapt.1-3
14. Stannett, V. T., Koros; W. J., Paul, D. R.; Lonsdale, H. K. and Baker, R. W. Adv. in Polym. Sci., 1979, 32, 71.
15. Koros, W. J. and Chern. R. T.  In Handbook of Separation Process Technology; Rousseau R. W., Ed.; John Wiley and Sons: New York, 1987; Chapter 20.
16. Comyn, J. Polymer Permeability, Ed.; Elsevier Applied Science Publishers: New York, 1985.
17. Duda, J. L. and Vrentas, J. S. In  Encyclopedia of Polymer Science; Kroschwitz, J. I., Ed.; John Wiley and Sons: New York, 1986;  vol. 5, p. 36.
18. Hotchkiss, J. H. Food and Packaging Interactions, Ed.; American Chemical Society: Washington, 1988; Symposium Series No. 365.
19. Plastics Engineering, May 1984, p. 47.
20. PL 732™ Blood Containers, 1983 Travenol Product Guide, Travenol Labs, Inc., Deerfield IL.; p. 11.
21. Mod. Plastics, Aug. 1985, p. 57.
22. Mitchell, J. K. R. Inst. J., 1831, 2, 101.
23. Stannett, V. T.  Chapter  2 in Ref.1.
24. Rogers, C. E.,  Chapter  2 in Ref 16.
25. Chern, R. T.; Koros, W. J.; Sanders, E. S.;  Chen, S. H., and Hopfenberg, H. B. In Industrial Gas Separations; Whyte, T. E. Jr.; Yon, C. M.  and Wagener E. H., Eds; American Chemical Society: Washington,1983; Symposium Series No. 223, p. 47.
26. Muruganandam, N.; Koros, W. J. and D. R. Paul J. Polym. Sci., Polym. Phys. Ed., 1987, 25, 1987.
27. Koros, W. J. and Paul, D. R. J. Polym. Sci., Polym. Phys. Ed., 1978, 16, 2171.
28. Stern, S. A. and S. Trohalaki In  Barrier Polymers and Barrier Structures; Koros, W. J.,  Ed.; American Chemical Society: Washington, D.C., 1990; Chapter 2.

29. Jordan, S. M.; Fleming, G. K., and Koros, W. J. J. Membr. Sci., 1987, 30, 191.
30. Graham, T. Philos. Mag., 1866, 32, 401.
31. Chern, R. T.; Koros, W. J.; Hopfenberg, H. B. and Stannett, V. T. In Material Science Aspects of Synthetic Polymer Membranes; Lloyd, D. R., Ed.; American Chemical Society: Washington, 1984; Symposium Series No. 269, Chapter 2.
32. Barrer, R. M., Trans. Faraday Soc., 1939, 35, 628.
33. Barrer, R. M. and Skirrow, G., J. Polym. Sci., 1948, 3, 549.
34. Prager, S. and Long, F. A., J. Am. Chem. Soc., 1951, 73, 4072.
35. van Amerongen, G. J., J. Polym. Sci., 1950, 5, 307.
36. Briston, J. H. In The Wiley Encyclopedia of Packaging Technology; Bakker, M., Ed.; John Wiley and Sons: New York, 1986; p. 329.
37. DeLassus, P. T., J.Vinyl Technol., 1979, 1, 14.
38. Brown, W. E. and DeLassus, P. T., Polym. Plast. Technol. Engr., 1980, 14(2), 171.
39. Brown, W. E., In The Wiley Encyclopedia of Packaging Technology, Bakker, M., Ed.; John Wiley and Sons: New York, 1986; p. 692.
40. Bicerano, J., Burmester; A. F.; Delassus, P. T. and Wessling, R. A., in Barrier Polymers and Barrier Structures; Koros, W. J., Ed.; American Chemical Society: Washington, D.C., 1990; Chapter 6.
41 McCaul, J. P., In The Wiley Encyclopedia of Packaging Technology, Bakker, M., Ed.; John Wiley and Sons: New York, 1986; p.474.
42. Wyeth, N. C. In High Performance Polymers; Their Origin and Development; Seymour, R. B., and Kirshenbaum, G. S., Eds.; Elsevier: New York, 1986; p. 417.
43. Blackwell, A. L., In High Performance Polymers; Their Origin and Development; Seymour, R. B., and Kirshenbaum, G. S., Eds.; Elsevier: New York, 1986; p. 425.
44. Gerlowski, L. E. In Barrier Polymers and Barrier Structures; Koros, W. J., Ed.; American Chemical Society: Washington, D.C., 1990; Chapter 8.
45. Tsai, B. C. and Wachtel, J. A. In Barrier Polymers and Barrier Structures; Koros, W. J., Ed.; American Chemical Society: Washington, D.C., 1990; Chapter 9.
46. Alger, M. M., Stanley, T. J. and Day, J. In Barrier Polymers and Barrier Structures; Koros, W. J., Ed.; American Chemical Society: Washington, D.C., 1990; Chapter 10.
47. Bissot, T. C. In Barrier Polymers and Barrier Structures; Koros, W. J., Ed.; American Chemical Society: Washington, D.C., 1990; Chapter 11.
48. Tsai, B. C., and Jenkins, B. J., J. Plastic Film & Sheeting, 1988, 4, 63.
49. U. S. Patent No. 4,407,897 issued to American Can Company, Greenwich, Conn.
50. Krizan, T. D., Coburn, J. C. and Blatz, P. S. In Barrier Polymers and Barrier Structures; Koros, W. J., Ed.; American Chemical Society: Washington, D.C., 1990; Chapter 5.
51. Schmidhauser, J. C. and Longley, K. L. In Barrier Polymers and

Barrier Structures; Koros, W. J., Ed.; American Chemical
Society: Washington, D.C., 1990; Chapter 7.

52. Lee, W. M., Polym. Engr. and Sci., 1980, 20(1), 65.

53. Kumins, C. A. and Kwei, T. K.; Chapter 4 in ref. 1.

54. Barrer, R. M., Barrie, J. A. and Slater, J. J. Polym. Sci.,
1957, 23, 315.

55. Stern, S. A. and Saxena, V. J. Membr. Sci., 1980, 7, 47.

56. Saxena, V. and Stern, S. A. J. Membr. Sci., 1982, 12, 65.

57. Weinkauf D. H. and Paul, D. R. In Barrier Polymers and Barrier
Structures; Koros, W. J., Ed.; American Chemical Society:
Washington, D.C., 1990; Chapter 3.

58. Doty, P. M., Aiken, W. H. and Mark, H. Ind. Engr. Chem., 1946,
38, 788.

59. Meyers, A. W., Rogers, C. E., Stannett, V. T. and Szwarz, M.,
Tappi, 1958, 41, 716.

60. Michaels, A. S. and Bixler, H. J. J. Polym. Sci., 1961, 50,
413.

61. Michaels, A. S., Vieth, W. R., and J. A. Barrie, J. Appl.
Phys., 1963, 34, 1 & 13.

62. Klute, C. H. J. Appl. Polym. Sci., 1959, 1, 340.

63. Wang, L. H. and Porter, R. S. J. Polym. Sci., Polym. Phys. Ed.,
1984, 22, 1645.

64. El-Hibri, M. J. and Paul, D. R. J. Appl. Polym. Sci., 1985, 30,
3649.

65. Yasuda, H., Stannett, V. T., Frisch, H. L., and Peterlin, A.
Die Macromol. Chem., 1964, 73, 188.

66. Slee, J. A., Orchard, G. A., Bower, D. I., Ward, I. M., J.
Polym. Sci., Polym. Phys. Ed., 1989, 27, 71.

67. Shastri, R., Dollinger, S. E., Roehrs, and Brown, C. N. In
Barrier Polymers and Barrier Structures; Koros, W. J., Ed.;
American Chemical Society: Washington, D.C., 1990; Chapter 12.

68. Winslow, F. H., ACS Symposium Series No. 95, American Chemical
Society: Washington, D.C., 1979, p. 11.

69. Puleo, A. C., Paul, D. R. and Wong, K. P. Polymer, 1989, 30,
1357.

70. Chiou, J. S. and Paul, D. R., J. Polym. Sci., Polym. Phys. Ed.,
1987, 25, 1699.

71. Berens, A. R., In Barrier Polymers and Barrier Structures;
Koros, W. J., Ed.; American Chemical Society: Washington, D.C.,
1990; Chapter 4.

72. Petropoulos, J. H. J. Membr. Sci., 1984, 17, 233.

73. Durning, C. J., and Russel, W. B. Polymer, 1985, 26, 119.

74. Enscore, D. J., Hopfenberg, H. B. and Stannett, V. T. Polymer,
1977, 18, 793.

75. Hopfenberg, H. B. and Frisch, H. L. J. Polym. Sci., Part B:
Polym. Phys., 1969, 7, 405.

76. Lasky, R. C., Kramer, E. J., and Hui, C. Y. Polymer, 1988, 29,
673.

77. Thomas, N. L., Windle, A. H. Polymer, 1982, 23, 529.

78. Sarti, G. Polymer, 1979, 20, 827.

79. Barrer, R. M.; Chapter 6 in Ref. 1.

80. Cussler, E. L., Hughes, S. E., Ward, W. J. and Aris, R., J.
Membr. Sci., 1988, 38, 161.

81. Walles, W. E.,   In Barrier Polymers and Barrier Structures;
    Koros, W. J., Ed.; American Chemical Society:  Washington,
    D.C., 1990; Chapter 14.
82. Hobbs, J.P., Anand, M., and Campion, B. A.,   In Barrier Polymers
    and Barrier Washington,   D.C., 1990; Chapter 15.
83. Subramanian, P. M.  In Barrier Polymers and Barrier Structures;
    Koros, W. J., Ed.; American Chemical Society: Washington,   D.C.,
    1990; Chapter 13.
84. Fenelon, P. J. Polym. Engr. and Sci., 1973, 13, 440.
85. Hotchkiss, J. H.;   Chapter 1 in Ref. 18.
86. Landois-Garza, J. and Hotchkiss, J. H.; Chapter 4 in Ref. 18.
87. Farrel, C. J., Ind. Engr. Chem. Res., 1988, 27, 1946.
88. Hansen, A. P. and Arora, D. K,   In Barrier Polymers and Barrier
    Structures;  Koros, W. J., Ed.; American Chemical Society:
    Washington,  D.C., 1990; Chapter 16.
89. Miltz, J., Mannheim, C. H., and Harte, B. R.,   In Barrier
    Polymers and Barrier  Structures;  Koros, W. J., Ed.; American
    Chemical Society: Washington,   D.C., 1990; Chapter 17.
90. Strandburg, G., DeLassus, P. T., and Howell, B. A.,   In Barrier
    Polymers and Barrier  Structures;  Koros, W. J., Ed.; American
    Chemical Society:  Washington,  D.C., 1990; Chapter 18.
91. Aithal, U. S., Aminabhavi, T. M. and Cassidy, P. E., In Barrier
    Polymers and Barrier  Structures;  Koros, W. J., Ed.; American
    Chemical Society: Washington,  D.C., 1990; Chapter 19.
92. Waksman, L S., Schneider, N. S., and Sung, N., In Barrier
    Polymers and Barrier Structures;  Koros, W. J., Ed.; American
    Chemical Society:  Washington,  D.C., 1990; Chapter 20.
93. Koberstein, J. T. and R. S. Stein  J. Polym. Sci., Polym. Phys.
    Ed., 1983, 21, 1439.
94. Chiang, K. T., and Sefton, M.V. J. Polym. Sci., Polym. Phys.
    Ed., 1977, 15, 1927.
95. Prasad, S. K., Advanced Materials & Processes; Aug.1986, p. 25.

RECEIVED November 14 , 1989

# Chapter 2

# Fundamentals of Gas Diffusion in Rubbery and Glassy Polymers

## S. A. Stern and S. Trohalaki

### Department of Chemical Engineering and Materials Science, Syracuse University, Syracuse, NY 13244-1190

This paper reviews some of the more important models and mechanisms of gas diffusion in rubbery and glassy polymers in light of recent experimental data.

Diffusion (transport) of gases in polymers is an important, and in some cases, controlling factor in a number of important applications, such as protective coatings, membrane separation processes, and packaging for foods and beverages. Therefore, a better understanding of the mechanisms of gas diffusion in polymers is highly desirable in order to achieve significant improvements in these applications and to develop new ones.

From a formal (macroscopic) viewpoint, the diffusion process can be described in many cases of practical interest by Fick's two laws (1-5). These laws are represented by the following equations for the isothermal diffusion of a substance in or through a ν-dimensional, hyperspherical polymer body of sufficiently large area [ν=1 for a slab or membrane (film), ν=2 for a hollow cylinder, and ν=3 for a spherical shell] (2):

$$J = -\omega_\nu \ r^{(\nu-1)} \ D\frac{\partial c(r,t)}{\partial r} \tag{1}$$

and

$$\frac{\partial c}{\partial t} = \frac{1}{r^{\nu-1}} \frac{\partial}{\partial r} \left( r^{\nu-1} \ D \ \frac{\partial c}{\partial r} \right), \quad R_2 < r < R_2 \tag{2}$$

0097–6156/90/0423–0022$10.50/0

where J is the local diffusion rate of the penetrant gas; $c(r,t)$ is the local penetrant concentration at a position coordinate r and at time t; D is the local mutual diffusion coefficient; and $\omega_1=1$ for a slab, $\omega_2=2\pi$ for a hollow cylinder, and $\omega_3=4\pi$ for a spherical shell. J is taken through the unit area of slab, the unit length of cylinder, and the whole area of the spherical shell. D depends on the nature of the penetrant/polymer system and can be constant or a function of concentration.

Integration of Equation 1 for the desired geometry and boundary conditions yields the total rate of diffusion of the penetrant gas through the polymer. Integration of Equation 2 yields information on the temporal evolution of the penetrant concentration profile in the polymer; Equation 2 must be augmented by the desired initial and boundary conditions of interest. The above relations apply to homogeneous and isotropic polymers.

The equilibrium concentration (solubility), c, of a penetrant gas dissolved in a polymer can be related to the pressure, p, of the penetrant by the isothermal relation:

$$c = S(c)p \tag{3}$$

where $S(c)$ [or $S(p)$] is a solubility coefficient. When the concentration of the penetrant in the polymer is very low, Equation 3 reduces to a form of Henry's law; the solubility coefficient is then independent of c (or p). In such cases, the diffusion coefficient may also become independent of c.

It is often of practical interest to determine the permeation rate of a gas, $J_s$, <u>across</u> a polymer membrane under steady-state conditions. When the membrane is nonporous, gas permeation is commonly described in terms of a "solution-diffusion" mechanism (1-6). The permeation rate can be obtained from Equation 1, since the diffusion of the penetrant gas in the membrane is the rate-determining step in the permeation process. Steady-state permeation is achieved if the constant gas pressures $p_h$ (the "upstream" pressure) and $p_\ell$ ($<p_h$) (the "downstream" pressure) are maintained at the membrane interfaces. For a sheet (plane) membrane of thickness $\delta$, Equation 1 yields for $\nu=1$, $\omega=1$, (1-6):

$$J_s = \bar{P} \ (p_h - p_\ell)/\delta \tag{4}$$

where
$$\bar{P} \equiv \bar{D} \cdot \bar{S} \tag{5}$$

where $\bar{P}$ is a mean permeability coefficient, $\bar{D}$ is a mean diffusion coefficient defined by the relation

$$\bar{D} = \int_{c_\ell}^{c_h} D(c)/(c_h - c_\ell) \ dc \tag{6}$$

and $\bar{S}$ is a function defined by the relation

$$\bar{S} = \frac{(c_h - c_\ell)}{(p_h - p_\ell)} \tag{7}$$

$c_h$ and $c_\ell$ are the equilibrium concentrations of the penetrant dissolved at the membrane interfaces at the pressures $p_h$ and $p_\ell$, respectively. When $p_h \gg p_\ell$, and thus $c_h \gg c_\ell$, as is often the case in membrane separation processes, $\bar{S}$ reduces to

$$\bar{S} = S_h = c/p \big|_h \tag{8}$$

where $S_h$ is the solubility coefficient evaluated at pressure $p_h$.

Equations 4-8 indicate that $\bar{P}$ is a product of a diffusion coefficient (a kinetic factor) and of a solubility coefficient (a thermodynamic factor). A common method of inferring the mechanism of permeation is to determine the dependence of $\bar{P}$ on p (or c) and on the temperature. This requires, in turn, a knowledge of the dependence of D and S on these variables. As is shown below, this dependence is very different above and below the glass-transition of the polymer.

The problem of gas diffusion in, and permeation through, inhomogeneous polymers is more complex, but has been considered by a number of investigators, e.g., refs. (1-3,5,7). When the polymer is highly plasticized by the penetrant, the diffusion coefficient may also become a function of time and of "history", but these non-Fickian cases will not be discussed here (1,3,5,6,8-11).

Considerable effort has been made during the last two decades to develop a "microscopic" description of gas diffusion in polymers, which is more detailed than the simplified continuum viewpoint of Fick's laws. It has been known for a long time that the mechanism of diffusion is very different in "rubbery" and "glassy" polymers, i.e., at temperatures above and below the glass-transition temperature, Tg, of the polymers, respectively. This is due to the fact that glassy polymers are not in a true state of thermodynamic equilibrium, cf. refs. (1,3,5,7-11). Some of the models and theories that have been proposed to describe gas diffusion in rubbery and glassy polymers are discussed below. The models selected for presentation in this review reflect only the authors' present interests.

DIFFUSION MODELS FOR RUBBERY POLYMERS

The various models developed to describe the diffusion of small gas molecules in polymers generally fall into two categories: (1) molecular models analyze specific penetrant and chain motions together with the pertinent intermolecular forces, and, (2) "free-volume" models attempt to elucidate the relationship between the diffusion coefficient and the free volume of the system, without consideration of a microscopic description.

1. Molecular Models

Molecular models commonly assume that fluctuating microcavities or "holes" exist in the polymer matrix and that, at equilibrium, a definite size distribution of such holes is established on a time-average basis. A hole of sufficient size may contain a dissolved penetrant molecule, which can "jump" into a neighboring hole once it acquires sufficient energy. Diffusive motion results only when holes which have become vacant in this manner are occupied by other penetrant molecules. A net diffusive flux arises in a preferred direction in response to a driving force, otherwise molecules will diffuse in random directions since their motion has a Brownian nature. The flux magnitude depends on the concentration of holes which are large enough to accommodate a penetrant molecule.

Molecular models include these characteristics largely to describe the Arrhenius behavior of diffusion coefficients observed experimentally, i.e.

$$D = D_o \exp \left(-E_{app}/RT\right)$$ (9)

where $E_{app}$ is the apparent energy of activation for diffusion, $D_o$ is a constant, R is the universal gas constant, and T is the absolute temperature. A correlation is found between $E_{app}$ and the molecular diameter of the penetrant, but no theoretical expression for D has been obtained with molecular models.

a) The Model of Meares

The model of Meares is only of historical interest because it was the first molecular model for diffusion in polymers. Meares (12) found that the activation energy for diffusion correlates linearly with the square of the penetrant diameter, but not with the diameter cubed. Therefore, he inferred that the elementary diffusion step is not governed by the energy necessary to create a hole that can accommodate a penetrant molecule, but by the energy required to separate polymer chains so that a cylindrical void is produced which allows the penetrant molecule to "jump" from one equilibrium position to another. The length of the cylindrical void is the jump length, $\lambda$, which can be calculated from the relation proposed by Meares:

$$E_d = \frac{\pi}{4} \sigma^2 N_A \lambda \text{ (CED)}$$ (10)

where $E_d$ is the actual activation energy for diffusion; $\sigma$ is the collision diameter of the penetrant; CED is the cohesive energy density of the polymer; and $N_A$ is Avogadro's number. For He, $H_2$, Ne, $O_2$ and Ar in poly(vinyl acetate) at 26°C, Meares found that $\lambda$ varied from 2.6Å for He to 28.9Å for Ar. A relation for the activation entropy was also derived.

b) The Model of Brandt

Brandt (13) used a simple molecular model that takes into account polymer structure in order to estimate $E_d$. The activated state in diffusion involves two polymer chains which bend symmetrically in order to create a passageway for the penetrant molecule (see Figure 1a). Cooperation of neighboring polymer segments is necessary only for penetrants that are too large to pass through the existing space between the chains. The activation energy consists of an intermolecular contribution, $E_b$, due to the repulsion the bent chains experience from their neighbors, and an intramolecular contribution, $E_i$, due to the resistance of the molecular chains to bending. The total energy stored within f degrees of freedom, $E^*$, is then

$$E^* = E_b + E_i + E_{th} \tag{11}$$

where $E_{th}$ is the thermal energy which is not part of the activation energy. By expressing $E_b$, $E_i$, and $E_{th}$ in terms of f and other molecular parameters, Brandt found $E_d(=E_b+E_i)$ to have a minimum for values of f between 12 and 19. Using Barrer's zone theory (14), apparent activation energies appropriate to the diffusion of ethane in polyethylene were calculated and found to be 25 to 70% lower than experimental values. This can be attributed to low estimates of the internal pressure used in the calculations and to other motions, e.g., unsymmetrical ones, that lead to successful diffusion jumps.

Brandt and Anysas (15) tested this model by studying the diffusion of gases in fluorocarbon polymers. Their results, as well as data from other sources, showed that plots of $E_{app}$, the activation energy found experimentally, versus $\sigma^2$ were nonlinear with a downward concavity, in contrast to the data of Meares. $E_{app}$ was assumed to be equivalent to $E_d$. These plots did not converge to the origin, but to an abscissa-intercept such that $E_{app}$ vanished for hypothetical penetrants for which $\sigma^2$ was 5 to 8 Å. While the curvature of $E_{app}$ as a function of $\sigma^2$ was not consistent with any then-current theory, the intercept is consistent with Brandt's model which relates it to the "free cross-section" or the average free distance between polymer-chain surfaces.

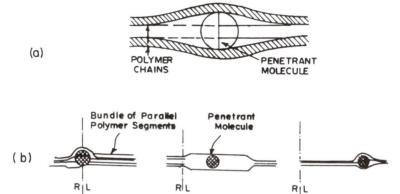

(a)

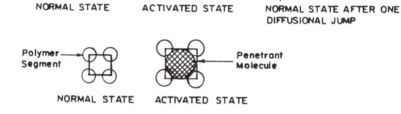

(b)

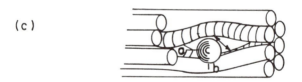

(c)

Figure 1.    a)    Brandt's model of the activated state for diffusion.
                  (Reprinted from J. Phys. Chem., 63,
                  1080, 1959, by W.W. Brandt. Copyright 1959 American
                  Chemical Society.)

             b)    The Activation diffusion process according to
                   DiBenedetto and Paul. (A.T. DiBenedetto and
                   D.R. Paul, J. Polym. Sci., A, 2, 1001, 1964,
                   copyright c 1964 John Wiley & Sons, Inc. Reprinted
                   by permission of John Wiley & Sons, Inc.).

             c)    Proposed diffusion model of Pace and Datyner
                   (R. J. Pace and A. Datyner, J. Polym. Sci., Polym.
                   Phys. Ed., 17, 437, 1979, copyright c 1979 John
                   Wiley & Sons, Inc. Reprinted by permission of John
                   Wiley & Sons, Inc.)

The model of Brandt suggests that an activation energy is not required for diffusion of penetrants that are sufficiently small. Brandt and Anysas also found, contrary to the model of Meares, that values of $E_{app}$ for a given penetrant in different polymers were not simply proportional to the cohesive energy density of the polymer.

### c) The Model of DiBenedetto and Paul

The molecular model of DiBenedetto and Paul (16) was the first to predict the downward concavity of $E_d$ as a function of the square of the penetrant collision diameter observed experimentally. The model is an outgrowth of DiBenedetto's cell theory (17) for amorphous polymers in which the polymer is considered to be a homogeneous, single-component phase containing Avogadro's number of identical n-center segments. Each center, defined as a repeat unit, is subject to a cylindrically symmetric potential field formed by four nearest-neighbors. A dissolved gas molecule is assumed to behave as a three-dimensional harmonic oscillator within a cell or void formed by a bundle of four parallel polymer segments.

A diffusion "jump" is construed to involve coordinated segmental rotations and vibrations that produce a cylindrical void adjacent to the penetrant molecule. Exchanging one vibrational degree of freedom for one degree of translational freedom, the penetrant diffuses into the cylindrical void. The activation energy is then the potential difference between the "normal" dissolved state and the "activated" state in which the void is present (see Figure 1b). Neglecting the interaction of the penetrant with its surroundings, as well as the work required to compress surrounding chains, the activation energy is equated to the potential energy change for the partial breaking of van der Waals' bonds between four n-center polymer segments.

Assuming $E_d$ to be equivalent to $E_{app}$, DiBenedetto and Paul fitted their expression for $E_d$ to experimental data, using an adjustable parameter representing the number of centers per polymer segment involved in a single diffusion step. Values of this parameter range from 5 to 9, in agreement with values estimated from viscous-flow data. Inconsistent with experimental observations is the prediction that $E_d \to 0$ as $\sigma \to 0$. This arises because, even though the passageway in the activated state consists of a free volume plus a volume created by expansion of the chain bundles, only the latter volume is associated with $E_d$.

## 2. Free-Volume Models

The free-volume models reviewed here and in a later section are based on Cohen and Turnbull's theory (18) for diffusion in a hard-sphere liquid. These investigators argue that the total free volume is a sum of two contributions. One arises from molecular vibrations and cannot be redistributed without a large energy change, and the second is in the form of discontinuous voids. Diffusion in such a liquid is not due to a thermal activation process, as it is taken to be in the molecular models, but is assumed to result from a re-distribution of free-volume voids caused by random fluctuations in local density.

## a) The Model of Fujita

Fujita (19) related the thermodynamic diffusion coefficient, $D_T$, to the fractional free volume, $v_f$, by

$$D_T = RTA_d \exp(-B_d/v_f) , \tag{12}$$

where $A_d$ and $B_d$ are characteristic parameters which depend only on the nature of the penetrant/polymer system. For low penetrant concentrations, $v_f$ is given by

$$v_f(T,v) = v_f(T,0) + \gamma(T)v , \tag{13}$$

where $v$ is the volume fraction of the penetrant, $v_f(T,0)$ is the average fractional free volume of the pure polymer at temperature $T$, and $\gamma(T) [= (\partial v_f/\partial v)_T]$ is a measure of the penetrants effectiveness in increasing the free volume.

Combining Equations 12 and 13 and noting that $D_T \to D(v=0)$ as $v \to 0$, cf. Figure 2, Fujita and Kishimoto (20) found that

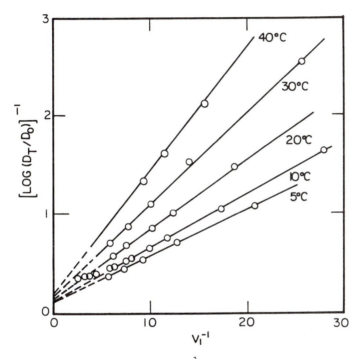

Figure 2. Plot of $[\log(D_T/D(0)]^{-1}$ as a function of reciprocal volume fraction of benzene, $v_1$, in poly(ethyl acrylate). (A. Kishimoto and Y. Enda, J. Polym. Sci., A-1, 1799, 1963, copyright $^c$ 1963 John Wiley & Sons, Inc. Reprinted by permission of John Wiley & Sons, Inc.)

$$\left[ \ln\left(\frac{D_T}{D(0)}\right) \right]^{-1} = \frac{v_f(T,0)}{B_d} + \frac{\left[ v_f^2(T,0) \right]}{B_d \gamma(T) v} \tag{14}$$

and

$$\ln\left(\frac{D(0)}{RT}\right) = \ln A_d - \frac{B_d}{v_f(T,0)} \tag{15}$$

$A_d$, $B_d$, and $\gamma$ can be obtained from a plot of $[\ln(D_T/D(0)]^{-1}$ versus $1/v$ together with a plot of $\ln[D(0)/RT]$ versus $1/v_f(T,0)$, but, because of the intercept from the latter plot is usually very small, substantial errors are incurred. A plot of Equation 14 is illustrated in Figure 2.

The fractional free volume of pure polymer as a function of temperature, $v_f(T,0)$, çan be found from

$$v_f(T,0) = v_{fs}(T_s,0) + \alpha(T-T_s) \tag{16}$$

applicable to polymers only in the temperature range Tg $\leq$ T $\leq$ Tg + 100 K. The fractional free volume, $v_{fs}(T_s,0)$ of pure polymer is a reference state at temperature $T_s$, and $\alpha$, a parameter characteristic of the polymer, can be evaluated from the dependence of steady-flow viscosity on temperature (21,22). Alternatively, Tg can be identified with $T_s$ so that $v_{fs}(Tg,0)$ and $\alpha$ assume values of 0.025 and 0.048°C$^{-1}$, respectively, as suggested by Williams, Landel, and Ferry (23). This value of $\alpha$ is in reasonable agreement with the difference between the coefficients of thermal expansion of the polymer above and below Tg.

Fujita's model is valid for penetrant/polymer systems with diffusion coefficients that exhibit a strong concentration dependence, such as organic vapors in amorphous polymers, (20,22,24-27), but fails to describe the difference between water in poly(vinyl acetate) and in poly(methyl acrylate) (28). This may be due to the hydrogen-bonding nature of water rather than to a failure of the model. Fujita viewed his theory as inappropriate for small penetrant molecules, whose diffusion is largely independent of concentration, because the critical hole size for such penetrants is smaller than that required for viscous flow of the mixture. This discrepancy was corrected by Frisch, Klempner, and Kwei (29), who proposed a reference volume that need not coincide with that for viscosity. However, even without this correction, Fujita's model has been shown to adequately describe the absorption kinetics of small molecules in rubbery polymers, e.g., of $CH_4$, $C_2H_4$, $C_3H_8$, and $CO_2$ in polyethylene (30).

Stern, Frisch, and coworkers have extended Fujita's free-volume model to the permeation of light gases (31-33) (see Figure 3) and binary gas mixtures (34,35) (see Figure 4) through polymer membranes. The extended model was found to describe satisfactorily the dependence of permeability coefficients on pressure and temperature for a variety of light gases in polyethylene, as well the dependence on composition for several binary mixtures in the same polymer. The validity of the extended model is limited to total penetrant concentrations of up to 20-25 mol-%.

The effets of crystallinity and of inert fillers on diffusion have been treated in the context of free-volume theory by Kreituss and Frisch (36,37).

b)  Other Free-Volume Models

Other free-volume models have been discussed by Frisch and Stern (5), by Kumins and Kwei (38), and by Rogers and Machin (39). Free-volume models which are applicable to both rubbery and glassy polymers are described in a following section of this review.

DIFFUSION MODELS FOR GLASSY POLYMERS

1.  Effect of Glassy Polymer State

The mechanisms of gas diffusion are very different at temperatures above and below the glass-transition temperature, $T_g$, of the polymers, i.e., when the polymers are in their "rubbery" or "glassy" state, respectively (1,3-8). The difference in these mechanisms is reflected in the significant differences observed in the dependence of the diffusion coefficient, as well as of the permeability and solubility coefficients, on the penetrant gas pressure or concentration in polymers and on the temperature.

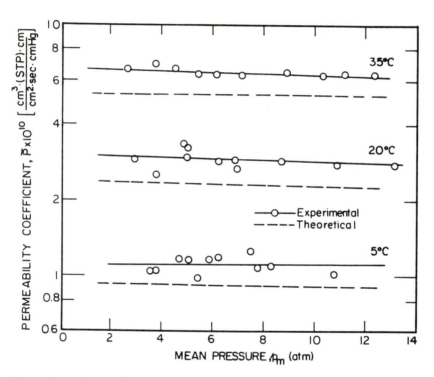

Figure 3a.  Comparison of experimental permeability coefficients
with values predicted by Stern, Frisch, and coworkers'
extension of Fujita's free-volume model for Ar in
polyethylene. (S. A. Stern, S. R. Sampat, and S. S.
Kulkarni, J. Polym. Sci.: Part B: Polym. Phys., 24,
2149, 1986, copyright ᶜ 1986 John Wiley & Sons, Inc.
Reprinted by permission of John Wiley & Sons, Inc.)

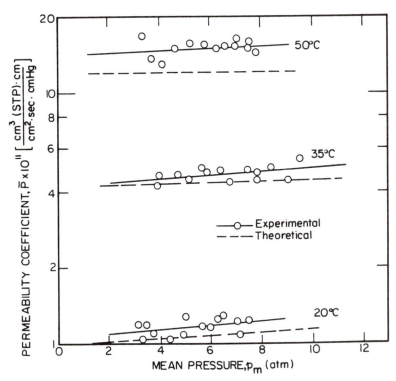

Figure 3b.  Comparison of experimental permeability coefficients with values predicted by Stern, Frisch, and coworkers' extension of Fujita's free-volume model for $SF_6$ in polyethylene. (S.A. Stern, S.R. Sampat, and S.S. Kulkarni, J. Polym. Sci.: Part B: Polym. Phys., 24, 2149, 1986, copyright © 1986 John Wiley & Sons, Inc. Reprinted by permission of John Wiley & Sons, Inc.)

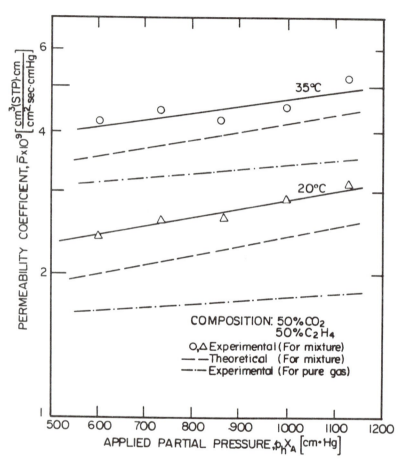

Figure 4.    Permeation of a 50-mole% $CO_2$ - 50-mole% $C_2H_4$ mixture
             through polyethylene.  Dependence of permeability
             coefficients for $CO_2$ on applied partial pressure of
             $CO_2$.  (S. A. Stern, G. R. Mauze, and H. L. Frisch,
             J. Polym. Sci., Polym. Phys. Ed., 21, 1275, 1983,
             copyright ᶜ 1983 John Wiley & Sons, Inc.  Reprinted by
             permission of John Wiley & Sons, Inc.)

For example, the diffusion coefficients for light gases in rubbery polymers are often independent of concentration, or increase slightly with increasing concentration (or pressure). These gases commonly have low critical temperatures and their solubility in polymers is very low (e.g., within the Henry's law limit). By contrast, the diffusion coefficients for the same gases in glassy polymers are highly nonlinear functions of the penetrant concentration (or pressure). More specifically, the diffusion coefficients increase significantly with increasing concentration and reach a constant value at sufficiently high concentrations. The increase in the diffusion coefficients becomes more marked as the temperature is lowered below Tg. The diffusion coefficients considered here are effective (i.e., experimentally determined) quantities. The permeability and solubility coefficients for light gases in glassy polymers also are strongly nonlinear functions of pressure (or concentration).

The differences in the transport and solution behavior of gases in rubbery and glassy polymers are due to the fact that, as mentioned previously, the latter are not in a state of true thermodynamic equilibrium ($\underline{1},\underline{3}-\underline{8}$). Rubbery polymers have very short relaxation times and respond very rapidly to stresses that tend to change their physical conditions. Thus, a change in a temperature causes an immediate adjustment to a new equilibrium state (e.g., a new volume). A similar adjustment occurs when small penetrant molecules are absorbed by a rubbery polymer at constant temperature and pressure: absorption (solution) equilibrium is very rapidly established. Furthermore, there appears to exist a unique mode of penetrant absorption and diffusion in rubbery polymers ($\underline{7}$).

By comparison, glassy polymers have (on the average) very long relaxation times. Hence, "in the presence of a penetrant, the motions of whole polymer chains or of portions thereof are not sufficiently rapid to completely homogenize the penetrant's environment. Penetrant (molecules) can thus potentially sit in holes or irregular cavities with very different intrinsic diffusional mobilities" ($\underline{8}$). In other words, there could exist more than one mode of penetrant absorption and diffusion in glassy polymers.

2. The Dual-Mode Sorption Model

A phenomenological theory known as the "dual-mode sorption" model offers a satisfactory description of the dependence of diffusion coefficients, as well as of solubility and permeability coefficients, on penetrant concentration (or pressure) in glassy polymers (4-6,40-44). This model postulates that a gas dissolved in a glassy polymer consists of two distinct molecular populations:

(1) Molecules dissolved in the polymer by an ordinary dissolution process, similar to that above Tg, and

(2) Molecules dissolved in a limited number of fixed, pre-existing microcavities, or at fixed sites, in the polymer matrix.

The concentration of molecules dissolved by the ordinary dissolution process, $c_D$, is related to the penetrant equilibrium pressure, p, by a Henry's law isotherm:

$$c_D = k_D p \tag{17}$$

where $k_D$ is a solubility coefficient in the Henry's law limit. The concentration of molecules dissolved in microcavities, $c_H$, is related to p by the Langmuir isotherm:

$$c_H = c_H' bp/(1+bp) \tag{18}$$

where $c_H'$ is a "Langmuir saturation" constant, and b is a "Langmuir affinity" constant.

Local equilibrium is assumed to exist between the two molecular populations. The total concentration of the dissolved penetrant (the solubility), c, at a given p and temperature is then obtained from the sum of Equations 17 and 18 (see Figure 5):

$$c = c_D + c_H = k_D p + c_H' bp/(1+bp) \tag{19}$$

The solubility of most gases and vapors in a polymer decreases with increasing temperature. This is due to the fact that the parameters $k_D$, $c_H'$, and b decrease as the temperature is raised; $c_H'$ is zero above Tg, in which case Equation 19 reduces to Henry's law (see Figure 6).

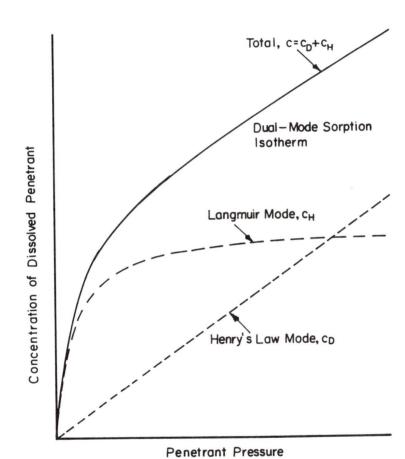

Figure 5. Typical dual-mode sorption isotherm and its components.

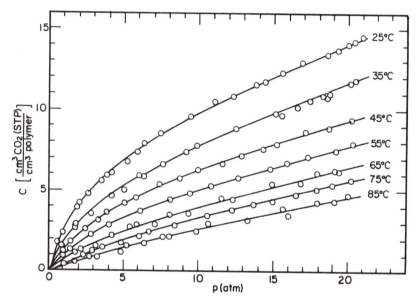

Figure 6.    Solubility isotherms for $CO_2$ in poly(ethylene
terephthalate) over a range of temperatures that
encompasses the glass transition temperature of the
polymer (Tg $\simeq$ 85°C).   (W.J. Koros and D.R. Paul,
J. Polym. Sci.: Polym. Phys. Ed. 16, 1947, 1978,
copyright © 1978 John Wiley & Sons, Inc.  Reprinted by
permission of John Wiley & Sons, Inc.)

The transport of gases in and through glassy polymers is best represented by the additional assumption that the two penetrant populations of concentrations $c_D$ and $c_H$ have different mean mobilities (the "partial immobilization" hypothesis) (6, 41-43). These mobilities are characterized by the mutual diffusion coefficients $D_D$ and $D_H$, respectively. Then, the local penetrant flux, $J$, normal to a plane at some position coordinate $x$ in a glassy polymer can be expressed by Fick's first law; cf. Equation 1 for $v=1$, $\omega=1$, $r=x$:

$$J = -D_{eff} (\partial c/\partial x) \tag{20}$$

where $D_{eff}(c)$ is an effective diffusion coefficient which is related to $D_D$ and $D_H$ by the expression (41, 42):

$$D_{eff} = D_D \left[ \frac{1 + FK/(1+\alpha c_D)^2}{1 + K/(1+\alpha c_D)^2} \right] \tag{21}$$

where $F = D_H/D_D$, $K = c'_H b/k_D$, and $\alpha = b/k_D$; also $\alpha c_D = bp$; $D_H$ and $D_D$ are assumed to be constant, cf. Figure 7. Equation 21 shows that $D_{eff}$ increases with increasing pressure (and, hence, with increasing $c_D$) from a minimum of $D_{eff} = D_D [(1+FK)/(1+K)]$ when $p \to 0$ to a maximum of $D_{eff} = D_D$ as $p \to \infty$. The permeability coefficient, $\bar{P}$, for gas transport across a glassy polymer membrane is then given by the relation (41, 42):

$$\bar{P} = k_D D_D \left( 1 + \frac{FK}{1+bp} \right) \tag{22}$$

where $p (\equiv p_h)$ is now the "upstream" pressure of the penetrant; the "downstream" pressure, $p_\ell$, is assumed to be negligibly small. According to Equation 22, $\bar{P}$ decreases with increasing $p$ from a maximum of $\bar{P} = k_D D_D (1+FK)$ when $p \to 0$ to a minimum of $\bar{P} = k_D D_D$ as $p \to \infty$, cf. Figure 8.

The dual-mode sorption model does not provide a physical description of the polymer environments in which the two penetrant

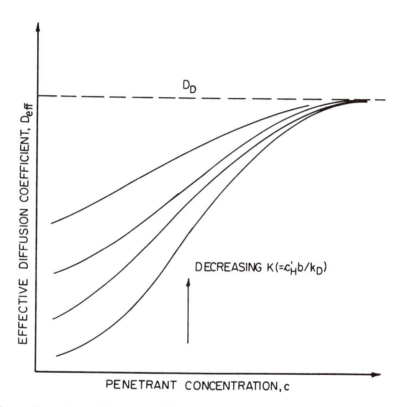

Figure 7.    The effective diffusion coefficient for small molecules
in glassy polymers as a function of penetrant concentration. $D_D$ is the mutual diffusion coefficient for the
penetrant population dissolved by the Henry's-law mode.
Other symbols are defined in the text.

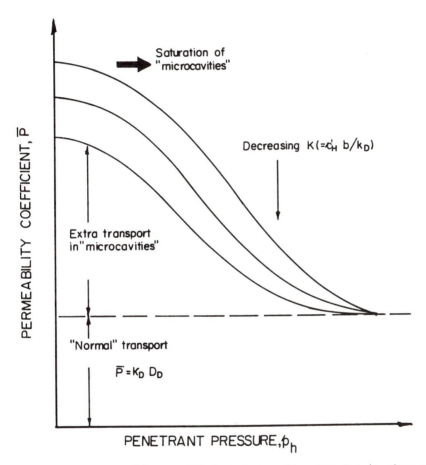

Figure 8.    Permeability coefficient for small molecules in glassy polymers as a function of penetrant pressure. Symbols are defined in the text. (Source: W.J. Koros).

populations are dispersed. Investigators seem to agree that the environment of the population dissolved by the Henry's law mode, i.e., by ordinary dissolution, has a relaxed, quasi-liquid structure, similar to that of polymers above Tg. The environment of the population dissolved in microcavities by the Langmuir mode appears to be visualized differently by different investigators. Some view these microcavities as being formed within packets of unrelaxed, disordered polymer chains. The two types of penetrant environment are designated for expediency in this report as the "Henry's law" and the "Langmuir" domains, although the latter probably are of near molecular size. The characterization of these "domains" is not necessary for the application of the dual-mode sorption model.

Another version of dual-mode sorption with partial mobility was proposed independently by Petropoulos (45), who expressed the transport equation in terms of the gradient of the chemical potential rather than the concentration gradient.

3.  Extensions of the Dual-Mode Sorption Model

The formulations of the dual-mode sorption model presented above assume that the solubility of the penetrant gas in a glassy polymer is very low, and that consequently the polymer is not plasticized to any significant extent by the penetrant. However, gases with higher critical temperatures, such as organic vapors, exhibit sufficiently high solubilities to plasticize glassy polymers to various extents. Therefore, several extensions (generalizations) of the dual-mode sorption model have been proposed in recent years.

Thus, Mauze and Stern (46, 47) have extended the model to cases where the solubility of the penetrant population dissolved by the ordinary dissolution process exceeds the Henry's law limit. Also, Stern and coworkers (48-50) have generalized the dual-mode sorption model to cases where the mutual diffusion coefficients $D_D$ and $D_H$ are not constant but increase exponentially with increasing penetrant concentration. One of these extensions (50) provides an insight into the effects of plasticization of Henry's law and Langmuir domains separately, cf. Figure 9. These extensions also describe the effects of antiplasticization on penetrant transport.

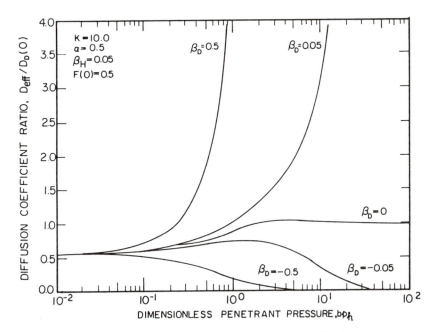

Figure 9.   Dimensionless effective diffusion coefficient, $[D_{eff}/D_D(0)]$, as a function of dimensionless penetrant pressure for different values of $\beta_D$, an empirical constant that characterizes the effects of plasticization on penetrant transport in the Henry's-law domains. Similarly, $\beta_H$ characterizes the plasticization of the Langmuir domains. Other symbols are defined in the following reference. (S. Zhou and S.A. Stern, J. Polym. Sci: Part B: Polym. Phys., 27, 205, 1989, copyright c 1989 John Wiley & Sons, Inc. Reprinted by permission of John Wiley & Sons, Inc.)

An important modification of the dual-mode sorption model has been made by Koros and coworkers (6), who have extended it to binary mixtures. These investigators have shown that the predictions of their extended model were in satisfactory agreement with sorption (solubility) and transport measurements.

Fredrickson and Helfand (51) have recently pointed out that the original "partial immobilization" version of the model does not provide a clear picture of the penetrant mobility in the Langmuir "domains" (microcavities). In fact, the use of two mutual diffusion coefficients, $D_D$ and $D_H$, could imply that the transport of penetrant molecules in a glassy polymer takes place through two different pathways. In other words, not only could it be inferred that the Langmuir domains are large enough to sustain diffusion (51), but that these domains, as well as the "Henry's law" domains, are separately continuous. This problem arises from the fact that the transport equation in the above-mentioned version of the model does not contain "coupling" terms that describe the transfer of penetrant molecules between the Henry's law and Langmuir domains (49).

Fredrickson and Helfand incorporated coupling terms in the transport equation and showed that "these terms provide mobility to molecules that are absorbed into microvoids, even if the molecules have an intrinsic diffusion coefficient in the hole phase that vanishes." These investigators proposed experiments that could provide information on the size and topological connectivity of microcavities (51). Related studies have also been made by Chern, Koros, et al. (52) and by Barrer (53). However, Petropoulos (54) has expressed the opinion that the treatment of Fredrickson and Helfand (51) as well as that of Barrer (53) "introduces more diffusion parameters than can reasonably be expected to be measurable on the basis of past experience". Petropoulos also showed that these treatments impose certain limitations on the physical meaning of the diffusion parameters. In the same study, Petropoulos has examined the modification of his dual-mode sorption model (45) which are necessary if the Langmuir domains are sufficiently extensive to constitute a macroscopically recognizable phase rather than scattered individual sites (or microcavities)(54).

Two other pertinent studies are of interest. Sangani (55) has derived a relation between the average penetrant flux and the concentration gradient in the polymer, using a model related to the classical treatment of diffusion and thermal conduction in composite media. Sangani attributes a different meaning to the dual-mode sorption parameters $D_D$ and $D_H$ (see above). In a very interesting recent study, Petropoulos (56) examined the dual-mode sorption model with "partial immobilization" from the viewpoint of random walk topology. He showed that this model can be considered as rigorous only under conditions of homogeneous random walk, and that the experimental values of the parameter $D_H$, as currently determined, "cannot be considered to be quantitatively meaningful." Petropoulos also presents a generalized dual-mode sorption model which is topologically consistent.

## 4. Search for Additional Evidence of Dual-Mobility

Evidence for the dual-mobility ("partial immobilization") hypothesis has been obtained almost entirely from studies of an indirect nature, such as steady-state permeability, "time-lag", and absorption/desorption rate measurements. These measurements yield a single effective diffusion coefficient as a function of pressure (or concentration) for the penetrant/polymer system of interest. The (constant) mutual diffusion coefficients $D_D$ and $D_H$ can then be extracted from Equation 21, using the values of parameters $k_D$, $c'_H$, and b obtained by fitting Equation 19 to equilibrium solubility data. Several investigators have attempted to use nuclear magnetic resonance to ascertain the validity of the dual-mobility hypothesis (57-59). However, the results of these studies are ambiguous or contradictory. Also, the quenching of the excited triplet state of film-incorporated bromopyrene by $O_2$ has been used to infer oxygen diffusion coefficients (60). The pressure dependence was in agreement with "dual-mode sorption" theory, but direct resolution of two characteristic diffusion coefficients was not achieved.

Recently, Stern, Ware, et al. (Stern, S.A.; Zhou, S.; Araux-Lara, J.L.; Ware, B.R., to be published in *Polym. Letters*)

have used a new technique known as "laser fluorescence photobleaching recovery" (FPR) to study the in-plane diffusion of a fluorophore
(a fluorescent dye) in glassy poly(1-trimethylsylil-1-propyne) membranes. These investigators found that the fluorophore dispersed
in these membranes consisted of two molecular populations with
widely different mobilities. The diffusion coefficients of these
populations differed by two orders of magnitude, cf. Figure 10.
However, the exchange between the two populations appeared to be
slower than the experimental FPR time-scale. This lack of rapid
exchange is not consistent with the assumptions of the dual-mode
sorption model, and was attributed to the large size of the
fluorophore molecules used.

    The transition to dual-mode sorption behavior as the tempera-
ture is lowered through Tg is often reflected in changes in the
slope of van't Hoff-type plots of log S versus 1/T and of Arrhenius-
type plots of log D versus 1/T in the glass-transition region.
These changes in slope are generally represented as "breaks" or
discontinuities in these plots. However, the dual-mode sorption
model indicates that the change in slope should be gradual, as
expected also from the fact that the glass transition takes place
over a temperature range.

    It has been observed recently that some penetrant gas/polymer
systems do not exhibit a transition to dual-mode sorption behavior
as the temperature is lowered through Tg, at least within the
experimental error of the measurements [(61), (Itoh, K; Stern,
S.A., unpublished data)]. No changes are then observed in the
slopes of van't Hoff and Arrhenius plots at or near the Tg of the
polymers. These systems involve polymers which have a small
"excess" free volume in the glassy state, i.e., for which the
parameter $c'_H$ in Equation 19 is small, at temperatures just below
Tg. Examples of such polymers are some of the poly(alkyl
methacrylates) [(61), (Itoh, K; Stern, S.A., unpublished data)].

    The parameter $c'_H$ in Equation 19 is proportional to the
product $\Delta\alpha$ (Tg-T), where $\Delta\alpha = \alpha_r - \alpha_g$, the difference between the
coefficients of thermal expansion above and below Tg, respectively
(42). The values of $\Delta\alpha$ for the above poly(alkyl methacrylates)

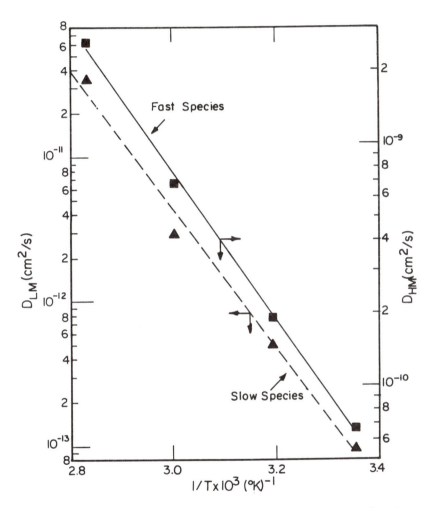

Figure 10.  Mutual diffusion coefficients, of the two molecular
populations of a fluorescent dye dispersed in
poly(1-trimethyl-1-propyne) plotted as functions of
reciprocal absolute temperature.  The technique used
was laser fluorescence photobleaching recovery.

are significantly lower than those of other glassy polymers.
Therefore, it may be expected that, for these polymers, dual-mode
sorption behavior will manifest itself at temperatures T suffi-
ciently below Tg, i.e., when $c_H^!$ [$\propto \Delta\alpha(Tg-T)$] becomes sufficiently
large. This has been confirmed recently for Ar, $CH_4$, and $C_2H_6$ in
poly(ethyl methacrylate), poly(n-propyl methacrylate), and poly(n-
butyl methacrylate). An onset of dual-mode behavior has been
observed with these penetrant/polymer systems at temperatures 30
to 65°C below the Tg of the polymers (Zhou, S.; Stern, S.A., submitted for publication in J. Polym. Sci.: Part B: Polym. Phys.).
These results require further validation.

5. Estimation of Diffusion Coefficients from
Simulation of Polymer Microstructure
The diffusion coefficient of small penetrants in glassy polymers can
also be correlated with the polymer free volume. In view of the
fact that experimental techniques for the determination of the free
volume are inherently difficult, Shah, Stern, and Ludovice (Shah,
V.M.; Stern, S.A.; Ludovice, P.J., submitted for publication in
Macromolecules) have utilized the detailed atomistic modeling of
relaxed polymer glasses developed by Theodorou and Suter (62,63) to
estimate the free volume available in polymer glasses for the diffusion of small molecules.

The framework of the model system is a cube with periodic
boundary conditions filled with segments from a single "parent"
chain. An initial-guess structure, generated by means of a modified
Markov process, is "relaxed" to mechanical equilibrium by potential
energy minimization using analytical derivatives. Model estimates
of the cohesive energy density and the Hildebrand solubility parameter are in excellent agreement with experiment.

Using a Monte Carlo technique, Shah, Stern, and Ludovice
evaluated the total free volume and the distribution of free volume
available to spherical penetrants of different sizes in model
structures of poly(propylene) and poly(vinyl chloride) at temperatures just below their glass transitions. Investigation of the

free-volume overlap or connectivity between cavities of free volume
is currently in progress.

## MODELS FOR DIFFUSION IN BOTH RUBBERY AND IN GLASSY POLYMERS

Some models are applicable to diffusion of small molecules in glassy
as well as in rubbery polymers. These too fall into the general
categories of molecular models and free-volume models. Recent molecular dynamics simulations of simple polymer/penetrant systems will
also be discussed.

### 1. Molecular Models

#### The Model of Pace and Datyner

Pace and Datyner (64-66) have proposed a more detailed molecular
model which provides an expression for the activation energy
that is free of adjustable parameters. This model was successfully
applied to polymer/penetrant systems both above and below Tg (64-
66). The glassy state is accounted for only by its lower thermal
expansivity.

The Pace-Datyner theory incorporates the main features of both
the DiBenedetto-Paul model (16) and the model of Brandt (13).
Specifically, it assumes that non-crystalline polymer regions
possess structure in the form of parallel bundles with a coordination number of 4. Two distinct penetrant motions are considered:
(1) along the axis of the tube formed by the main bundle, as in the
DiBenedetto-Paul model, and (2) perpendicular to this tube axis when
two chains bend and separate sufficiently, as in the Brandt model,
cf. Figure 1c. The first process is assumed to occur much more
rapidly than the second and is regarded as having no activation
energy. Energy of activation for the second motion is needed to
produce a sufficient chain separation. If the diffusion process is
to exhibit appreciable activation energy, processes 1 and 2 must
occur in series rather than in parallel, with the second process
being rate limiting.

In order to obtain the energy needed for the symmetric
separation of two chains required for the passage of a penetrant
molecule, the chains are assumed to interact according to the
DiBenedetto potential, which takes the chain to be a sequence of

Lennard-Jones centers. The complex expression for $E_d$, presented in Figure 11, is a consequence of the more elaborate nature of this model as compared to earlier ones.

Pace and Datyner assume that thermal expansion occurs, on average, by a uniform increase in interchain spacing for temperatures both above and below Tg. The temperature dependence of $E_d$ on T through $\rho$ is linear over temperature intervals of ~50K, and $E_{app}$ is therefore constant in this range. $E_{app}$ is related to $E_d$ by

$$E_{app} = d(E_d/T)/d(1/T) \qquad (23)$$

The dependence of $E_{app}$ on $\sigma$, the molecular diameter of the penetrant, calculated for several species of penetrants in a number of amorphous and semicrystalline polymers to agree acceptably well with experimental values (64).

Pace and Datyner resorted to the theory of stochastic processes to formulate an expression for the diffusion coefficient. Contained in this complex expression (see Figure 11) is a parameter that represents the mean-square jump length of the penetrant molecule. Jump lengths cannot be independently estimated within the limits of the theory and values implied from experimental diffusion coefficients are unrealistically large (~400 Å). In the temperature range Tg ≤ T ≤ Tg + 150 K, the expression for D contains two additional adjustable parameters.

Pace and Datyner (65, 65) have also proposed a model for the absorption (solution) of small molecules in polymers applicable at temperatures above and below Tg, which incorporates the dual-mode sorption model for the glassy region. The presence of microvoids is assumed for rubbery polymers as well as for polymer glasses. "Hole filling" is suggested as an important sorption mode above as well as below Tg, with one crucial difference between the sorption mechanism in the rubbery and glassy regions: hole saturation does not occur in the rubbery state because new microvoids are formed to replace those filled with penetrant molecules.

The microcavity population is assumed to be constant below Tg and an increasing function of temperature above Tg. The parameter

## ENERGY OF ACTIVATION FOR DIFFUSION

$$E_d = 5.23 \ \ \beta^{1/4} \left( \frac{\varepsilon^* \rho^*}{\lambda^2} \right)^{3/4} (\sigma')^{-1/4}$$

$$x \left\{ 0.077 \left[ \left( \frac{\rho^*}{\rho} \right)^{11} (\rho - 10\sigma') - \rho^* \left( \frac{\rho}{\rho^* + \sigma} \right)^{10} \right] \right.$$

$$\left. - 0.58 \left[ \left( \frac{\rho^*}{\rho} \right)^{5} (\rho - 4\sigma') - \rho^* \left( \frac{\rho^*}{\rho^* + \sigma} \right)^{4} \right] \right\}^{3/4}$$

$\beta$ = average single chain–bending modulus per unit length
$\rho$ = equilibrium chain separation
$\lambda$ = mean backbone element separation along chain axis
$\sigma$ = diameter of penetrant molecule
$\varepsilon^*$ = average Lennard–Jones energy parameter
$\rho^*$ = average Lennard–Jones distance parameter
$\sigma' = \sigma + \rho^* - \rho$

## DIFFUSION COEFFICIENT

$$D = \frac{1}{6} L^2 \upsilon$$

$L^2$ = mean – square jump displacement of penetrant molecule
$\upsilon$ = frequency of chain openings that permit passage of penetrant molecules

$$\upsilon = 0.0546 (\frac{1}{\lambda^2})(\frac{\varepsilon^*}{\rho^*})^{5/4} (\frac{\sqrt{\beta}}{m^*})^{1/2} \frac{\sigma'}{(\partial E_d / \partial \sigma)} \exp(-E_d / RT)$$

$m^*$ = mass of backbone element

Figure 11.  Expressions for the energy of activation for diffusion and the diffusion coefficient from the theory of Pace and Datyner (64).

$c_H^{'}$ is taken to be constant in this region, contrary to experimental evidence that $c_H^{'}$ decreases with decreasing temperature and becomes zero in the vicinity of Tg. Pace and Datyner argue that the significant amount of penetrant dissolved by Henry's-law mode acts to plasticize the polymer, causing a relaxation of the initial microvoids during the experiment. They further assert that, since the plasticization effect decreases with decreasing (Tg-T), the apparent $c_H^{'}$ will display the temperature dependence mentioned above.

Using a statistical-mechanical analysis, Pace and Datyner have estimated values of the model parameters. Semiquantitative agreement was obtained between solubility data generated from the model and experimental data for a number of simple gases in polyethylene, poly(ethylene terephthalate) and, to a lesser extent, for polystyrene. For polymers with flexible side groups, the agreement was only qualitative.

The Pace-Datyner theory has recently been criticized by Kloczkowski and Mark (Kloczkowski, A.; Mark, J.E., submitted for publication in J. Polym. Sci., Polym. Phys. Ed.). An analytically exact potential was obtained by these workers by translationally averaging the DiBenedetto potential. Also, the problem of an unnatural boundary condition inherent in the theory was resolved by imposition of proper integration limits. Corrected activation energies are lower by only 5 - 7%, but mean-square jump lengths are still too large and are physically questionable.

## 2. Free-Volume Models

### The Model of Vrentas and Duda

The free-volume model proposed by Vrentas and Duda (67-69) is based on the models of Cohen and Turnbull and of Fujita, while utilizing Bearman's (70) relation between the mutual diffusion coefficient and the friction coefficient as well as the entanglement theory of Bueche (71) and Flory's (72) thermodynamic theory. The formulation of Vrentas and Duda relaxes the assumptions deemed responsible for the deficiencies of Fujita's model. Among the latter is the assumption that the molecular weight of that part of the polymer chain involved in a unit "jump" of a penetrant molecule is equal to the molecular weight of the penetrant. Also, Fujita's assumption of a polymer-fixed frame of reference for $D_T$ implicitly assumes the equivalence of $D_T$ and the penetrant self-diffusion coefficient.

Vrentas and Duda's theory formulates a method of predicting the mutual diffusion coefficient D of a penetrant/polymer system. The revised version (68) of this theory describes the temperature and concentration dependence of D but requires values for a number of parameters for a binary system. The data needed for evaluation of these parameters include the Tg of both the polymer and the penetrant, the density and viscosity as a function of temperature for the pure polymer and penetrant, at least three values of the diffusivity for the penetrant/polymer system at two or more temperatures, and the solubility of the penetrant in the polymer or other thermodynamic data from which the Flory interaction parameter $\chi$ (assumed to be independent of concentration and temperature) can be determined. An extension of this model has been made to describe the effect of the glass transition on the free volume and on the diffusion process (73).

The model of Vrentas and Duda is particularly applicable to penetrants with high solubilities, e.g., organic vapors and liquids, which can plasticize the polymer. The model has been extended (69,73,74) to apply to small molecular penetrants, but its validity has not been sufficiently tested.

3. Molecular Dynamics Simulation
Estimation of Diffusion Coefficients
In recent studies [(Trohalaki, S.; Rigby, D.; Kloczkowski, A.; Mark, J.E.; Roe, R.J., submitted for publication) (Takeuchi, H.; Okazaki, K., submitted for publication)], molecular dynamics simulations of an assembly of *n*-alkane-like chains together with a low concentration of a simple penetrant gas have been performed in order to model the diffusion of the penetrant in polyethylene. The chain molecules, previously shown to exhibit realistic polymeric properties (75), were subject to potentials restricting bond lengths, bond angles, and dihedral angles.

At temperatures 250 and 300 K above the glass transitions of the systems, Takeuchi and Okazaki (Takeuchi, H.; Okazaki, K., submitted for publication) determined the self-diffusion coefficient of $O_2$, and examined the effect of chain flexibility by altering the torsional potential of the chains. The free volume was verified as a dominant factor for diffusion, but, for a fixed amount of free volume, high chain flexibility results in large diffusion coefficients and low apparent energies of activation for diffusion. Chain motions with characteristic times on the order of relaxation times of internal rotations were determined to be responsible for the diffusion of the penetrant.

Trohalaki, et al. calculated self-diffusion coefficients for $CO_2$ in polyethylene from simulations of this system. Also, in order to investigate the effect of penetrant size, additional simulations employed values of the Lennard-Jones radius appropriate for H, He, Ar, and $CH_2$ for the penetrant, holding all other parameters constant. The self-diffusion coefficients for the largest and smallest penetrant particles were evaluated at four temperatures in the range of 240 K to 420 K. The activation energies for the largest and smallest penetrants were found not to differ significantly. The diffusion coefficients and activation energies from both the above studies are in good agreement with experimental values. Molecular dynamics simulation is not restricted to temperatures above Tg.

CONCLUSIONS

A substantial number and variety of models of gas transport in polymers have been proposed during the last 20-30 years, in view of the great practical and scientific importance of this process. Molecular-type models are potentially most useful, since they relate diffusion coefficients to fundamental physicochemical properties of the polymers and penetrant molecules, in conjunction with the pertinent molecular interactions. However, the molecular models proposed up to now are overly simplified and contain one or more adjustable parameters. Phenomenological models, such as the dual-mode sorption model and some free-volume models, are very useful for the correlation and comparison of experimental data, but the model parameters are not directly related to the polymer structure. Consequently, such models cannot be used for the prediction of diffusion coefficients from fundamental properties of gas/polymer systems alone.

A significant role in future research will be played by molecular dynamics simulation as well as other new and very promising computational techniques, made possible by the advent of supercomputers. Also of importance are new and more sophisticated experimental techniques, such as NMR, laser correlation spectroscopy, neutron diffraction, and laser fluorescence bleaching recovery. These new techniques will surely yield new insights into the mechanisms of transport of small molecules in both rubbery and glassy polymers.

ACKNOWLEDGMENTS

The support of Union Carbide Industrial Gases and of the U.S. Department of Energy through its Office of Basic Energy Sciences is gratefully acknowledged.

LITERATURE CITED

1. Rogers, C.E. In Physics and Chemistry of the Organic Solid State; Fox, D.; Labes, M.M.; Weissberg, A., Eds.; Interscience: New York, 1965; Chapter 6, pp 509-635.
2. Barrer, R.M. In Permeability of Plastic Films and Coatings; Hopfenberg, H.B. Ed.; Plenum Press: New York, 1975; pp 113-124.
3. Crank, J. The Mathematics of Diffusion; Clarendon Press: Oxford, 1975; 2nd Ed.
4. Stern, S.A.; Frisch, H.L. Ann. Rev. Mater. Sci. 1981, 11, 523.
5. Frisch, H.L.; Stern, S.A. Crit. Revs. Solid State and Mat. Sci., CRC Press:, Boca Raton, FL, 1983, 11(2), 123.
6. Koros, W.J.; Chern, R.T. In Handbook of Separation Process Technology; Rousseau, R.W., Ed.; Wiley-Intersciences, New York, 1987; pp 862-953.
7. Barrer, R.M. In Diffusion in Polymers; Crank, J; Park, G.S. Eds., Academic, New York, 1968; Chapter 6, pp 165-217.
8. Frisch, H.L. Polym. Eng. Sci. 1980, 20, 241.
9. Hopfenberg, H.B.; Stannett, V. In The Physics of Glassy Polymers; Haward, R.N., Ed.; J. Wiley & Sons, New York, 1973; Chapter 9, pp 504-547.
10. Stannett, V.T.; Hopfenberg, H.B.; Petropoulos, J.H. In Macromolecular Science; Bawn, C.E.H., Ed.; MTP International Revs. of Science; Butterworths, London, 1972; Vol. 8, pp 329-369.
11. Durning, C.J. J. Polym. Sci., Polym. Phys. Ed. 1985, 23, 1831.

12.  Meares, P.  J. Am. Chem. Soc. 1954, 76, 3415.
13.  Brandt, W.W.  J. Phys. Chem. 1985, 63, 1080.
14.  Barrer, R.M.  J. Phys. Chem. 1957, 61, 178.
15.  Brandt, W.W.; Anysas, B.A.  J. Appl. Polym. Sci. 1963, 7, 1919.
16.  DiBenedetto, A.T., Paul, D.R. J. Polym. Sci. 1964, A-2, 1001.
17.  DiBenedetto, A.T, J. Polym. Sci. 1963, A-1, 3459.
18.  Cohen, M.H.; Turnbull, D.  J. Chem. Phys. 1959, 31(5), 1164.
19.  Fujita, H., Fortschr. Hochpolym. Forsch. 1961, 3, 1.
20.  Fujita, H.; Kishimoto, A.  J. Chem. Phys. 1961, 34, 343.
21.  Fujita, H.  In Diffusion in Polymers; Crank, J.; Park, G.S.,
     Eds., Academic  New York, 1968; Chapter 3, pp 75-105.
22.  Fujita, H.; Kishimoto,A.; Matsumoto, K.  Trans. Faraday Soc.
     1960, 56, 424.
23.  Williams, M.L.; Landel, R.F.; Ferry, J.D. J. Am. Chem. Soc.
     1955, 77, 3701.
24.  Kishimoto, A.; Enda, Y. J. Polym. Sci. 1963, A-1, 1799.
25.  Newns, A.C. Trans. Faraday So. 1963, 59, 2150.
26.  Kishimoto, A.  J. Polym. Sci. 1964, A-2, 1421.
27.  Suwandi, M.S.; Stern, S.A.  J. Polym. Sci., Polym. Phys. Ed.
     1973, 11, 663.
28.  Kishimoto, A.; Mackawa, E.; Fujita, H. Bull. Chem. Soc. Japan
     1960, 33, 988.
29.  Frisch, H.L.; Klempner, D.; Kwei, T.K. Macromolecules 1971, 4,
     237.
30.  Kulkarni, S.S.; Stern, S.A. J. Polym. Sci., Polym. Phys. Ed.
     1983, 21, 441.
31.  Stern, S.A.; Fang, S.-M.; Frisch, H.L.  J. Polym. Sci.: Part
     A-2 1972, 10, 201 ibid., 1972, 10, 575.
32.  Stern, S.A.; Kulkarni, S.A.; Frisch, H.L.  J. Polym. Sci.,
     Polym. Phys. Ed. 1983, 21, 467.
33.  Stern, S.A.; Sampat, S.R.; Kulkarni, S.S.  J. Polym. Sci.:
     Part B: Polym. Phys. Ed. 1986, 24, 2149.
34.  S.M. Fang, S.A. Stern, and H.L. Frisch, Chem. Eng. Sci. 1975,
     30, 773.
35.  Stern, S.A.; Mauze, G.R.; Frisch, H.L.  J. Polym. Sci., Polym.
     Phys. Ed. 1983, 21, 1275.
36.  Kreituss, A.; Frisch, H.L.  J. Polym. Sci., Polym. Phys. Ed.
     1981, 19, 889.
37.  Frisch, H.L.; Kreituss, A.  In Polymer Separation Media;
     Cooper, A.R., Ed.; Plenum Press, New York, 1982, pp 21-25.
38.  Kumins, C.A.; Kwei, T.K.  In Diffusion in Polymers; Crank, J.;
     Park, G.S., Eds.; Academic:, New York, 1968; pp 107-139.
39.  Rogers, C.E.; Machin, D.  CRC Crit. Revs. Macromolec. Sci.,
     CRC Press: Boca Raton, FL, 1972; pp 245-313.
40.  Vieth, W.R. Membrane Systems: Analysis and Design; Hanser
     Publishers: New York, 1988; pp 9-92.
41.  D.R. Paul and W.J. Koros, J. Polym. Sci., Polym. Phys. Ed.
     1976, 14, 675.
42.  Koros, W.J.; Paul,D.R.  J. Polym. Sci., Polym. Phys. Ed. 1978,
     16, 1947.
43.  Paul, D.R.  Ber. Bunsenges. Phys. Chem. 1979, 83, 294.
44.  Koros, W.J. ; Paul, D.R.; Huvard, G.S.  Polymer 1974, 20, 956.
45.  Petropoulos, J.H. J. Polym. Sci.: Part A-2 1970, 8, 1797.
46.  Mauze, G.R.; Stern, S.A.  J. Membrane Sci. 1982, 12, 51.

47. Mauze, G.R.; Stern, S.A. Polym. Eng. Sci. 1983, 23, 548.
48. Stern, S.A.; Saxena, V. J. Membrane Sci. 1980, 7, 47.
49. Saxena, V.; Stern, S.A. J. Membrane Sci. 1982, 12, 65.
50. Zhou, S; Stern, S.A. J. Polym. Sci.: Part B: Polym. Phys. 1989, 27, 205.
51. Fredrickson, G.H.; Helfand, E. Macromolecules 1985, 18, 2201.
52. Chern, R.T.; Koros, W.J.; Sanders, E.S.; Chen, S.H.; Hopfenberg, H.B. ACS Symposium Series No. 223; American Chemical Society: Washington, D.C., 1983, pp 47-73.
53. Barrer, R.M. J. Membrane Sci. 1984, 18, 25.
54. Petropoulos, J.H., J. Polym. Sci.: Part B: Polym. Phys. 1988, 26, 1009.
55. Sangani, A.B. J. Polym. Sci.: Part B: Polym. Phys. 1986, 24, 563.
56. Petropoulos, J.H. J. Polym. Sci.: Part B: Polym. Phys. 1989, 27, 603.
57. Assink, R. J. Polym. Sci., Polym. Phys. Ed. 1975, 13, 1665.
58. Sefcik, M.D.; Schaeffer, J. J. Polym. Sci., Polym. Phys. Ed. 1983, 21, 1055.
59. Wen, W.-Y.; Cain, E.; Inglefield, P.T.; Jones, A.A. Polym. Preprints 1987, 28(1), 225.
60. Chu, D.Y.; Thomas, J.K.; Kuczynski, J. Macromolecules 1988, 21, 2094.
61. Stern, S.A.; Vakil, V.M.; Mauze, G.R. J. Polym. Sci.: Part B: Polym. Phys. Ed. 1989, 27, 405.
62. Theodorou, D.N.; Suter, V.W. J. Chem. Phys. 1985, 82(2), 955.
63. Theodorou, D.N.; Suter, V.W. Macromolecules 1985, 18, 1467.
64. Pace, R.J.; Datyner, A. J. Polym. Sci., Polym. Phys. Ed. 1979, 17, 437, 453, 465, 1675.
65. Pace, R.J.; Datyner, A. J. Polym. Sci., Polym. Phys. Ed. 1980, 18, 1103.
66. Pace, R.J.; Datyner, A. Polym. Eng. Sci. 1980, 20, 51.
67. Vrentas, J.S.; Duda, J.L. J. Polym. Sci., Polym. Phys. Ed. 1977, 15, 403.
68. Vrentas, J.S.; Duda, J.L. AIChE J. 1982, 28, 279.
69. Vrentas, J.S.; Duda, J.L. Encyclopedia of Polymer Science and Engineering; J. Wiley & Sons: New York, 1986; Vol. 5, 2nd Ed., pp 36-68.
70. Bearman, R.J. J. Phys. Chem. 1977, 15, 403, 417.
71. Bueche, F. Physical Properties of Polymers; Interscience: New York, 1962.
72. Flory, P.J. Statistical Mechanics of Chain Molecules; Interscience, New York, 1969.
73. Vrentas, J.S.; Duda, J.L. J. Appl. Polym. Sci. 1978, 22, 2325.
74. Vrentas, J.S.; Duda, J.L. Macromolecules 1976, 9, 785.
75. Rigby, D; Roe, R.J. J. Chem. Phys. 1987, 87(12), 7285.

RECEIVED December 5, 1989

# Chapter 3

# Effects of Structural Order on Barrier Properties

D. H. Weinkauf and D. R. Paul

Department of Chemical Engineering and Center for Polymer Research, The University of Texas at Austin, Austin, TX 78712

The effects of molecular order on the gas transport mechanism in polymers are examined. Generally, orientation and crystallization of polymers improves the barrier properties of the material as a result of the increased packing efficiency of the polymer chains. Liquid crystal polymers (LCP) have a unique morphology with a high degree of molecular order. These relatively new materials have been found to exhibit excellent barrier properties. An overview of the solution and diffusion processes of small penetrants in oriented amorphous and semicrystalline polymers is followed by a closer examination of the transport properties of LCP's.

The barrier properties of polymeric materials are determined by the chemical structure of the chain and the system morphology. The parameters derived from chemical structure, such as degree of polarity, inter-chain forces, ability to crystallize, and chain stiffness, are essentially determined upon the selection of the particular polymer. Here, we will focus on how molecular order influences the barrier properties of polymers, including molecular orientation and degree of crystallinity; however, it must be recognized that the range in which these quantitites can be manipulated will depend on chemical structure.

For many years, molecular orientation and crystallinity have been observed to improve the barrier properties of polymers (1-3). In extreme cases, drawing of semicrystalline polymers has been shown to reduce permeability by as much as two orders of magnitude. A crude understanding of the dependence of the transport parameters on penetrant size and chain packing can be

0097–6156/90/0423–0060$09.00/0
© 1990 American Chemical Society

obtained through the concept of fractional free volume, since sorption and diffusion processes depend upon the availability of unoccupied volume in the polymer matrix. In the lamellar crystals of semicrystalline materials and the extended chain structure of oriented polymers, chain packing is usually much more efficient than in the amorphous, isotropic state. The efficiency of chain packing in the crystalline phase reduces the free volume available for transport to such an extent that, as a first approximation, the crystalline phase may be regarded as impermeable relative to the amorphous phase.

The unique molecular packing of rod-like chains in liquid crystalline polymers (LCP) closely resembles the extended chain structure of highly oriented flexible chain polymers, suggesting that these materials are good candidates for barrier applications. Thermotropic LCP's, first developed in the early 1970's, have been the object of much interest because of their excellent mechanical properties and ease of product fabrication. Preliminary observations have shown that a commercially available wholly aromatic thermotropic copolyester has gas permeability coefficients that are lower than those of polyacrylonitrile (4). These results raise some fundamental questions as to the nature of the mechanism for transport of small molecules through a matrix of ordered rigid rod-like chains.

We first give a concise review of the effects of orientation and crystallinity on the barrier properties of polymeric materials, paying particular attention to their effects on the solubility and diffusion coefficients. This will provide useful background for considering the transport properties of liquid crystal polymers which, because of their unique properties, may have some role to play in the quest for improved barrier polymers.

The Transport Mechanism

The mechanism by which small gas molecules permeate through rubbery or glassy amorphous polymers has been described by many authors (5-7). It will be useful to briefly review the underlying parameters which affect permeation.

Because a solution-diffusion process is involved, the permeability coefficient of amorphous polymers is the product of the effective diffusion, $D$, and solubility, $S$, coefficients.

$$P = D S \tag{1}$$

The solubility coefficient is the ratio of the equilibrium concentration of the dissolved penetrant to its partial pressure in the the gas phase. Investigations of sorption isotherms for gases in glassy polymers have lead to the hypothesis that the equilibrium concentration is the result of contributions from both Henry's law and Langmuir (hole filling) type sorption mechanisms, or the so-called "dual sorption" model. Thus, the solubility is dependent upon both polymer-penetrant interactions and the volume available for "hole filling." Over a reasonable temperature range, the apparent solubility coefficient, $S$, can be expressed in terms of a van't Hoff relationship, and a preexponential factor, $S_0$.

$$S = S_0 \exp\left(\frac{-\Delta H_S}{RT}\right) \qquad (2)$$

Here, the heat of sorption, $\Delta H_S$, is a composite parameter involving both sorption mechanisms. The Henry's law mode requires both the formation of a site and the dissolution of the species into that site. The formation of a site involves an endothermic contribution to this process. In the case of the Langmuir mode, the site already exists in the polymer matrix and, consequently, sorption by hole filling yields more exothermic heats of sorption.

A form of Fick's law describes the diffusion of gases through the amorphous polymer matrix. The diffusion coefficient has been observed to follow an Arrhenius relationship, characteristic of an activated process.

$$D = D_0 \exp\left(\frac{-E_D}{RT}\right) \qquad (3)$$

In this equation, $E_D$ is the activation energy required to generate an opening between polymer chains large enough to allow the penetrant molecule to pass. The activation energy is a function of the intra- and inter-chain forces that must be overcome in order to create the space for a unit diffusional jump of the penetrant. The activation energy of diffusion will be greater the larger the penetrant molecule, the stronger the polymer cohesive energy, and the more rigid the chains.

By substituting Equations 2 and 3 into Equation 1, the permeability can also be expressed in terms of an Arrhenius type relationship.

$$P = P_0 \exp\left(\frac{-E_P}{RT}\right) \qquad (4)$$

where

$$E_P = E_D + \Delta H_S \qquad (5)$$

A conceptual alternative to the formulation of transport parameters in terms of the activated-state approach, particularly when concerned with polymer morphology, is the use of free volume theory (8-10). In this approach, the transport parameters are dependent on the statistical distribution of the free volume in the polymer matrix. The transport process is thought to be directly related to the probability of a penetrant finding a pathway large enough to perform a diffusional "step" from one sorption site to another. Cohen and Turnbull (8) and Fujita (9) have derived expressions for the exponential dependence of the diffusion coefficient on the fractional free volume, $f$,

$$D = D_0 \exp\left(\frac{-B}{f}\right) \qquad (6)$$

where, $D_0$ and B are constants characteristic of particular polymer-penetrant systems.

The concept of free volume has been of more limited use in the prediction of solubility coefficients; although, Peterlin (11) has suggested that the solubility coefficient is directly proportional to the free volume available in the polymer matrix . In many respects, the free volume expressions closely resemble the relationships developed in the activated state approach. In fact for the case of diffusivity, the two models can be shown to be mathematically equivalent by incorporating thermal expansion models such as the one proposed by Fox and Flory (12). The usefulness of the free volume model; however, lies in the accessibility of the fractional free volume, through the use of group contribution methods developed by Bondi (13) and Sugden (14), for correlation of barrier properties of polymers of different structure as demonstrated by Lee (15).

## The Effects of Crystallinity

The potential advantage of crystalline polymers for barrier applications was recognized by Morgan (1); although, previous investigators had reported reduced permeabilities in semi-crystalline polymers (2,3). As will be discussed, the equilibrium concentration of sorbed penetrant and the diffusion process are affected by crystallinity in uniquely different ways.

### Sorption in Semicrystalline Polymers. The work of several investigators has concluded that the chain packing in polymer crystallites is such that it precludes the dissolution of even small gas molecules (16-23). Transport in semi-crystalline polymers can, thus, be treated on the basis of a simple two phase model, consisting of a dispersed impermeable crystalline phase in a permeable amorphous matrix. Lasoski and Cobbs (17) attempted to correlate the solubility of water vapor in unoriented poly(ethylene terephthalate) and nylon 610 with

the degree of crystallinity.   Over a crystalline range of
0 to 40 %, they found the solubility to be a linear
function of the amorphous volume fraction, $\Phi$,

$$S = S_a \Phi \tag{7}$$

where $S_a$ is the solubility coefficient for the completely
amorphous material.   Michaels (19) confirmed the validity
of the simple two phase model (for a rubbery amorphous
phase) over a wide range of crystallinity from measurements
of the solubility of small probe gases in polyethylene
(Figure 1).   Surprisingly, the heats of solution of the
amorphous phase remained independent of degree or method of
crystallization.   While this is a simple and appealing
model, one might expect that the significant amount of
stress imposed by crystallization on the non-crystalline
phase would alter its sorption properties.
        In poly(ethylene terephthalate), however, Michaels et
al(21) found the decrease in solubility accompanying
crystallization was smaller than that predicted by Equation
7.   They proposed that crystallization tended to occur in
the denser regions of the amorphous matrix, thereby
effectively concentrating the volume of microvoids in the
remaining amorphous regions.   Experimental sorption
isotherms showed that the sorption capacity of the
Langmuir or "hole filling" mechanism per unit volume of
amorphous phase increased by 40% on crystallization.
        More recent work by Puleo and Paul (24) has concluded
that certain crystal structures will permit sorption and
subsequent diffusion of small gas molecules.   In this
investigation, gas sorption and transport measurements were
made using poly(4-methyl-1-pentene) (PMP) samples having a
wide range of crystallinity.   In contrast to Equation 7,
extrapolation to 100% crystallinity suggests that the
crystalline phase has a finite solubility coefficient,
roughly 30% of that for the amorphous phase, for gases such
as $CO_2$ and $CH_4$.   Since sorption in the PMP samples strictly
obeys Henry's law, the "hole filling" argument applied by
Michaels for PET is not justifiable.   The departure from
conventional crystalline behavior is attributed to the open
structure of the PMP crystal lattice where there are gaps
of sufficient size to accommodate small gas molecules
which, of course, is not the case for polyethylene.
        Vieth and Wuerth (25) found negative deviations from
the simple two phase model for semicrystalline
polypropylene suggesting that the presence of crystallites
in some way reduces the sorptive capacity of the amorphous
phase.   However, analysis of samples using x-ray
diffraction revealed the presence of a less stable
crystalline phase having a lower density.   Since the
crystalline volume fraction is commonly determined from
density measurements, the presence of a second, less dense
(however, still impermeable) crystalline phase would seem

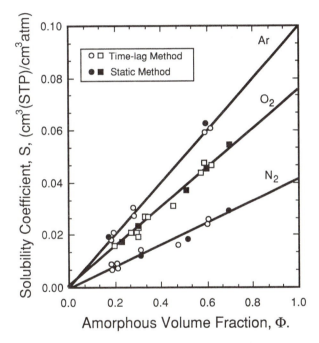

Figure 1. Effect of amorphous phase content on the solubility coefficients of various gases in semi-crystalline polyethylene. (Data taken from Ref. 19.)

to offer an explanation for these observations. Results like these suggest that we reconsider the potential of gas sorption experiments as an analytical tool for probing the microstructure of polymers.

Diffusion in Semicrystalline Polymers. The extension of the simple two phase model to include the diffusion coefficient in semicrystalline polymers requires some refinements. Given the tensorial character of the diffusion processes, one must first consider the spatial organization of the dispersed impermeable phase. Second, one might, also, expect that the "crosslinking" effect from tie molecules between crystallites and other chains merely anchored to the crystal would reduce the mobility of the amorphous phase. Both issues have been the focus of numerous investigations (19-29).

The most widely accepted view of diffusion in semicrystalline polymers stems from the extensive series of studies conducted by Michaels and coworkers (18-23). Using an electrical analog of a porous medium consisting of conducting channels, the flux of a penetrant through a semicrystalline polymer can be expressed as follows

$$J = \alpha \left(\frac{D_a}{\tau}\right)\frac{\partial C}{\partial x} \tag{8}$$

where $D_a$ is the diffusivity of the gas in the completely amorphous polymer. The term $\alpha$ is the cross-sectional area fraction of the regions available for transport. The quantity $\tau$ is the "tortuosity factor" accounting for the impedance of flow offered by the increased effective path length as well as variations in the cross sectional area of the conducting regions. If the system of impermeable crystallites is randomly dispersed, the area fraction, $\alpha$, and the amorphous volume fraction, $\Phi$, can be equated. From this model, the effective diffusion coefficient becomes

$$D = \left(\frac{D_a}{\tau}\right) \tag{9}$$

Using another electrical analog approach, Klute (26,27) arrived at a similar expression in which the impedance factor was termed the "transmission function." In both cases, however, the properties of the conducting phase are considered not to be influenced by the crystalline phase. Although this was found to be true for sorption, Michaels et al asserted that the crystallites would reduce the mobility of the chains in the amorphous phase and, thus, further reduce penetrant diffusion rates. The "chain immobilization" factor, $\beta$, was introduced to account for the change in penetrant mobility caused by the "crosslinking" or "anchoring" effect of the crystallites. The effective diffusivity can then be expressed as

$$D = \left(\frac{D_a}{\tau\,\beta}\right) \tag{10}$$

The chain immobilization term indirectly reflects the amount of increase in activation energy of diffusion that is observed in the amorphous phase upon crystallization. It is probably a function of both effective crystal surface area and penetrant size. It was noted, however, that the concept of chain immobilization loses its significance as the rigidity of the polymer backbone increases (22).

Figure 2 summarizes the results of Lowell and McCrum (28) for the apparent diffusion and solubility coefficients as a function of temperature for cyclopropane in linear polyethylene through the melting point region of the material. The activation energy for diffusion is higher in the semicrystalline material than after melting, while the derived heat of solution is nearly the same above and below the melting point. This suggests, as proposed by Michaels et al (19), that the presence of the crystallites impairs the diffusional mobility in the amorphous rubbery phase, the extent to which increases as the temperature is lowered, but does not affect the solubility characteristics of the amorphous regions.

Distinctions between the effect of tortuosity and chain immobilization can be approximated by assuming that the reduction in diffusivity for chain immobilization for a small penetrant gas such as helium is negligible, i.e., $\beta \approx 1$. Then, the effective tortuosity can be assessed by comparing the diffusivities of helium in the completely amorphous and the semicrystalline materials. Assuming that tortuosity is independent of penetrant diameter, the chain immobilization factor can then be estimated for other larger penetrants using Equation 10. At high levels of crystallinity, though, the reduced cross-sectional area of the permeable, amorphous passages tends to exclude some larger diffusing species, attributing significantly larger apparent reductions in diffusivity to the chain immobilization parameter. Understanding the limitations of such an approach, Michaels et al have discussed the correlation the chain immobilization parameter as a function of penetrant diameter and the amorphous volume fraction (20).

Attempts have also been made to relate the tortuosity factor to the amorphous volume fraction. A power law relationship has been suggested

$$\tau = \Phi^{-n} \tag{11}$$

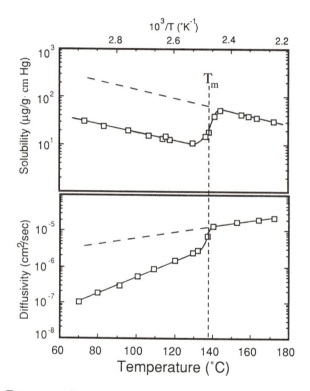

Figure 2. Temperature dependence of transport coefficients for cyclopropane in linear polyethylene above and below the melting point. (Data taken from Ref 28.)

where the exponent, n, for unoriented polymers commonly
ranges from 1 to 2. The tortuosity is dependent upon the
size, shape, and anisotropic nature of the crystalline
phase. Barring any mechanical stress, most of these
parameters are determined by the kinetics of
crystallization. For several polymer systems, 'n' has been
found to be constant over a wide range of crystallinity
(20, 25).

Annealing semicrystalline samples has been found to
result in reductions in the tortuosity parameter as much as
50% despite large increases in crystal size (23, 25). Some
workers (25) have suggested, based on the kinetics of
crystal growth, that this is the result of the formation of
permeable gaps in the crystallite.

In brief, unlike solubility, the effects of
crystallinity on the effective diffusivity intimately
involve the details of the polymer morphology. Because of
the chain immobilization effect, crystallinity may cause an
increase in the activation energy of diffusion. However,
observed decreases for the activation energy of diffusion
for helium in semicrystalline materials have been
attributed to "grain boundary" effects (29). For a first
approximation, some authors have found it sufficient to use
the following relationship for the correlation of amorphous
volume fraction and effective diffusivity,

$$D = D_a \Phi \tag{12}$$

and thus, by substituting into Eqn. 1,

$$P = S_a D_a \Phi^2 \tag{13}$$

This amounts to setting $\beta=1$ and $n=1$ in Equations 10 and 11.
Understandably, these approximations are somewhat limited
in their predictive ability.

More detailed aspects of transport in heterogeneous
media have been given in the excellent reviews of the
subject by Barrer (30) and Petropolus (31). The models
described by these authors and others include the effects
of size, shape, and anisotropy of the crystalline phase on
the tortuosity. Models of highly ordered anisotropic media
have been demonstrated to have tortuosities in the range of
30, which reflects the rather dramatic role that
orientation can have on the barrier properties of
semicrystalline polymers.

The Effects of Orientation

Stretching or drawing of polymer films is used to improve
mechanical behavior, and under certain conditions can lead
to improved barrier properties. The degree of orientation
achieved is primarily dependent upon the draw ratio and

other process conditions. Depending on the mode of
deformation, the permeability of a polymer may increase or
decrease. Much of the variance in transport data for a
given polymer that has been reported can be attributed to
the conditions that have been employed during orientation
of the material. The transport properties of amorphous and
semicrystalline polymers are affected by orientation in
different ways, and, thus will be considered separately.
The effects of uniaxial and biaxial orientation can be
quite different. Although many commercial processing
techniques impose a biaxial strain on the polymer melt, we
will deal here primarily with uniaxially deformed
materials.

    Orientation of Amorphous Polymers. Since optical
clarity is frequently a desired property of packaging
films, these materials are often essentially amorphous,
e.g. polyvinyl chloride (PVC) and polystyrene, and
orientation is one of the few approaches available for fine
tuning barrier properties. To fully understand the growing
data base of transport properties of oriented films, it
will first be necessary to discuss the mechanism by which
these polymers orient (See Figure 3). Initially, the
unoriented polymer can be envisioned as an entangled
network of randomly coiled chains. At thermal conditions
which allow some degrees of freedom, an applied stress will
cause chains to uncoil and the axes of the segments to
progressively become oriented in the direction of the
deformation. Some disentanglement of the chains may
accompany this process. If the temperature is reduced to a
point where the potential energy of the extended chains is
less than the mechanical constraints imposed by neighboring
chains, the oriented chains will be frozen into place and
will not retract upon removal of the stress. Subsequent
heating, however, will increase mobility such that the
polymer chains can contract and lose the preferred
orientation. Polymers plastically deformed in this way
commonly show decreases in permeability as a result of a
reduction in the effective diffusion coefficient. For the
case of rubbery networks externally constrained in a
deformed state, the effect of orientation appears to have
little effect on either solubility or mobility (32).
Generally, materials that are elastically deformed in the
solid state are more permeable which may be attributed to
an increase in free volume accompanying deformation (33).

    Presently, the amount of data on transport in
uniaxially oriented amorphous polymers is small in
comparison with that of semicrystalline materials. The
transport properties of oriented natural rubber (32),
polystyrene (33,34), polycarbonate (35), and polyvinyl
chloride (36,37) among others have been reported. One of
the more complete descriptions of the effects of uniaxial
orientation on gas transport properties of an amorphous
polymer is that by Wang and Porter (34) for polystyrene.

As seen in Table 1, the decrease in permeability can be directly attributed to a dramatic reduction in the effective diffusion coefficient, while there is a much smaller effect on the apparent solubility. A similar dependence of the solubility and diffusion coefficients on the draw ratio has been observed in other uniaxially oriented polymers (35-37). Because the glass transition and density of the polystyrene samples were found independent of the draw ratio, they concluded that the reduction in diffusivity was due to anisotropic redistribution of the free volume during drawing. Using an expansion coefficient related to draw ratio, the polystyrene data were successfully correlated using the Cohen-Turnbull free volume theory. However, the situation was found to be more complex for PVC (36).

Thermal and mechanical history influence microstructure of polymers and these effects should be considered when interpreting transport data. Work by El-Hibri (36) and Brady et. al. (37) on uniaxially and biaxially oriented PVC demonstrates the complex dependence of the transport parameters on the various thermal and mechanical histories employed. At constant draw ratios over a wide range of drawing temperatures, the effective diffusion coefficient increased and decreased in temperature ranges both above and below the glass transition. The effect of biaxial orientation was found to decrease the permeability, again, through a reduction in the effective diffusivity. As the biaxial stretching became more "unbalanced" (i.e., larger draw ratio along a particular axis), the permeabilities decreased further. Generally, as the mode of deformation becomes more uniaxial, the benefits in terms of improved barrier properties become larger (37).

Orientation of Semicrystalline Polymers. The anisotropic arrangement of impermeable polymer crystallites caused by deformation generally decreases the effective diffusivity by increasing the tortuosity. In addition, drawing of semicrystalline polymers has been found to improve the barrier properties through stress-induced crystallization and orientation of the remaining amorphous phase. Consequently, the reduction in permeability caused by orientation of a crystallizable polymer are often substantially larger than those observed in non-crystallizable polymers. As with amorphous polymers, the transport properties of oriented semicrystalline polymers depend on the draw ratio and the deformation mechanism. Stretching above the melting point with subsequent crystallization of the material can lead to quite different results than the stretching of already crystalline materials.

Crystallization from an oriented melt may occur during extrusion and injection molding processes of crystallizable polymers. The crystalline microstructure which results is

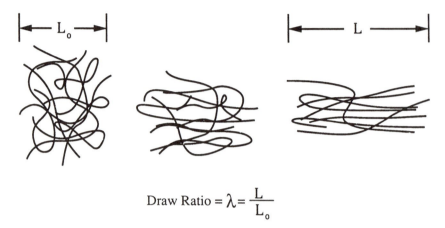

$$\text{Draw Ratio} = \lambda = \frac{L}{L_o}$$

Figure 3. Orientation of randomly coiled chains in an amorphous polymer.

Table 1. Effect of Molecular Orientation on $CO_2$ Transport Behavior in Polystyrene at 25 °C †

| Draw Ratio | Orientation Factor, f | Permeability (Barrers) | Diffusivity, D $(cm^2/sec) \times 10^{10}$ | Solubility, S $(cm^3 \ (STP)/cm^3 \ atm)$ |
|---|---|---|---|---|
| 1.0 | 0.0 | 8.0 | 24.0 | 25.1 |
| 1.8 | 0.016 | 5.8 | 19.0 | 23.4 |
| 3.1 | 0.048 | 2.9 | 7.5 | 29.6 |
| 4.4 | 0.122 | 1.5 | - | - |
| 5.0 | 0.170 | 1.0 | 4.6 | 16.0 |

† (Data taken from Ref. 34.)

highly dependent upon the degree of strain imposed (38-40).
In the region of low or zero stress, a spherulitic
morphology is formed (Figure 4a), where the lamellar
crystal structures propagate radially from the nucleation
site. At higher levels of stress in the melt, the
nucleation sites orient along thin fibrillar axes promoting
the growth of lamellar crystals perpendicular to the
direction of strain (Figure 4b). With exceedingly high
levels of strain imposed by the processing conditions, the
crystallization process results in complete replacement of
the spherulitc structure by a unidirectional crystalline
morphology, often referred to as a "shish kebab" structure
(Figure 4c). In this morphology, the lamellar crystallites
are separated by amorphous regions of loose loops, tie
molecules, and free chain ends (Figure 4d).
    Orientation of semicrystalline polymers below the
melting point is often referred to as "cold drawing."
Although some stress crystallization does occur, the
process primarily involves the transformation of existing
crystalline structures. A widely accepted model of the
deformation mechanism is that provided by Peterlin (Figure
5) (41). Prior to necking, the crystal lamellae which
constitute the spherulitic structure become arranged in a
tilted stack formation (5b). Necking marks the
transformation from spherulitic to a microfibrillar
structure. In this process, small blocks of lamellar
crystals are torn away from the original lamellae to form a
fibrillar structure of crystallites and taut tie molecules
(5c). Often, bundles of fibrils are formed from inter-
fibril tie molecules (5d). Under moderate draw ratios, the
spherulitic structure and the bundles of fibrils coexist
(5e).
    Information on how orientation during melt
crystallization affects the transport properties of
polymers is sparse; however, increases in the permeability
have been attributed to the "shish kebab" morphology (51).
Most of the work involving barrier properties of oriented
semicrystalline polymers has dealt with materials drawn at
temperatures well below the melting point. The transport
properties of cold-drawn polyethylene (34,42-46),
polypropylene (42,43), poly(ethylene terephthalate) (17,47-
49), and nylon 66 (50) among others have been reported.
    Peterlin et al (44,45) have described dramatic changes
during the cold drawing of polyethylene. Evidently, the
physical rearrangement of the crystal lamellae increases
the free volume of the amorphous regions during the initial
stages of drawing since the solubility coefficients
increase. The transformation of the spherulitic structure
to a microfibrillar structure brings about marked decreases
in the transport coefficients with the effective diffusion
coefficient being affected most dramatically. For
polyethylene at high draw ratios, the diffusion coefficient
has been observed to decrease by over a hundred fold

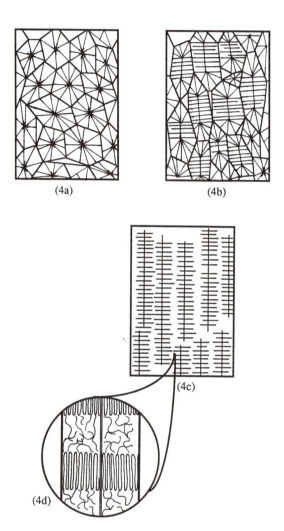

Figure 4. Crystalline morphologies of semicrystalline
polymers for different processing conditions: (a)
crystallized from a quiescent melt (spherulitic
structures); (b) crystallized under moderate stress
(oriented spherulites); (c) crystallized under high
stress ("shish-kebab" structure); (d) detailed view of
(c).

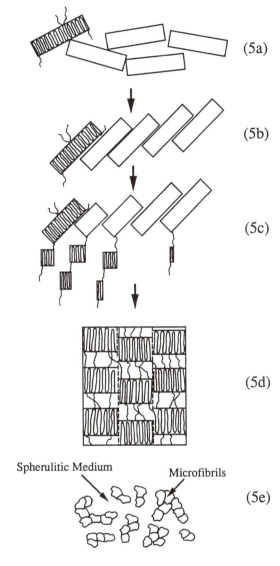

(5a)

(5b)

(5c)

(5d)

Spherulitic Medium          Microfibrils

(5e)

Figure 5. Cold drawing process of a semicrystalline
material. The dispersed crystal lamellae (a) under
stress are arranged in a tilted stack formation (b).
In the necking process, blocks of lamellae are torn
away (c) to form bundles of microbrils with highly
stressed amorphous regions (d). Part (e) shows a
cross-section of a drawn semicrystalline material not
completely transformed to the microfibrillar structure.

accompanied by a six fold decrease in the solubility
coefficient. In the moderate draw ratio ranges (Figure
5e), transport is primarily limited to the untransformed
spherulitic medium, while the bundles of relatively
impermeable microfibrils serve to increase the effective
path length of diffusion. At some point in the drawing
process, all the spherulitic medium is replaced by the
microfibrillar structure and further increases in the draw
ratio have little effect on the overall transport
properties of the material (45). The cold drawing of
semicrystalline polymers is almost always accompanied by an
increase in the activation energy for diffusion. The large
number of strained tie molecules which traverse the
amorphous regions of the microfibrils greatly diminish the
probability of diffusional jumps. According to the work
of Peterlin et al (44), the heat of sorption in the
amorphous regions of cold drawn materials are similar to
the undrawn state; however, the sorption capacity is
significantly reduced. The concentration of sorbed
penetrant has an enormous effect on the diffusion
coefficients in the microfibrillar structure. Apparently,
the strained state of the amorphous phase is subject to a
large degree of plasticization even at low concentrations.
    The effect of drawing temperature on the gas transport
properties of semicrystalline polymers has been extensively
examined for poly(vinylidene flouride) (PVF$_2$) (51). The
permeabilities and diffusivities of the drawn material were
found to steadily increase with increasing draw
temperature, while the solubility has a somewhat more
complex dependence. The draw ratio at which the system
begins to form a microfibrillar morphology decreases as the
drawing temperature is reduced. As a consequence, a larger
fraction of the spherulitic medium is converted to the
fibrillar structure for the same draw ratio employed at a
lower drawing temperature. Annealing of the drawn films at
temperatures above the drawing temperature used relaxes the
strained amorphous regions and results in increased
transport coefficients (44).
    The effect of biaxial strain on the transport
properties of semicrystalline polymers has been
demonstrated for several systems (43). However, particular
attention has been paid to poly(ethylene terephthalate) (47-
49) because of its wide use in blow molding applications.
In Figure 6, one can see, as with amorphous materials, that
the more unbalanced the biaxial orientation is the greater
are the reductions in permeability. Interpretations of
the transport data for biaxially and uniaxially drawn PET
samples can be explained by observing conformational
changes in the polymer backbone itself. Apparently, the
chain packing efficiency of the amorphous phase improves as
the number of trans isomers in the ethylene glycol unit
increases. Polarized infrared analysis of uniaxially and
biaxially oriented systems indicates that the fraction of

trans isomers levels off at 51% in balanced films but
increases linearly to 89% at a draw ratio of 5.5 in
uniaxial stretching (47).

## Polymers with Liquid Crystalline Order

Over the past two decades, liquid crystal polymers (LCP's)
have received a considerable amount of attention in both
academic and industrial laboratories. Often termed
mesomorphic (meaning having "middle form"), liquid
crystalline phases have a degree of order between that of
the zero ordered liquid and that of the three dimensional
crystal lattice. Recent reviews of liquid crystal polymers
have provided a fundamental understanding of the synthesis,
classification, morphology, and rheology of this unique
class of materials (52-54).
    Main chain thermotropic LCP's have been commercialized
under the tradenames Vectra, Xydar, and X7G. Most of the
interest in these materials centers on the excellent
mechanical properties which can be achieved through
conventional melt processing techniques. Typically, these
systems are copolymers of rigid mesogenic monomer units
(which alone would yield an intractable homopolymer) and
"flexible spacers" or "kink structures" which serve to
lower the melting point via disruption of chain regularity.
The ultra-high strength and moduli of these liquid crystal
polymers are attributed to the highly oriented morphology
that the rigid rod-like chains assume in the melt and
retain in the solid state. The unique packing arrangement
of these polymer systems has raised some fundamental
questions about the mechanism by which small molecules
permeate through them. Do solid state liquid crystalline
phases behave more like glasses or conventional crystals?
For the remainder of this discussion we will be concerned
with the gas transport properties of thermotropic LCP's.
By first taking a detailed look at the morphology of these
systems, we hope to provide some relevant background for
understanding transport in this relatively new class of
polymers.

    *Liquid Crystal Polymer Morphology.* As with semi-
crystalline polymers, the nature of the solid state of
liquid crystal polymers depends on both thermal and
mechanical histories. For the moment we will examine the
morphology of LCP's through a discussion of the their
thermal transitions free from mechanical stress. A typical
DSC thermogram for a thermotropic LCP is given in Figure 7
showing the various transitions that might occur upon
cooling. From the isotropic melt, there is a first order
transition to the liquid crystal phase. Commonly, the
entropy of transition is quite small in comparison with the
total entropy change in going from an isotropic melt to
fully ordered crystals. The small entropy change and a
small reduction in volume observed during the transition

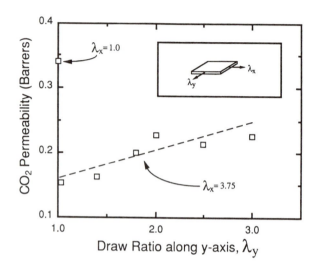

Figure 6. Effect of biaxial orientation in poly(ethylene terephthalate). (Data taken from Ref. 47.)

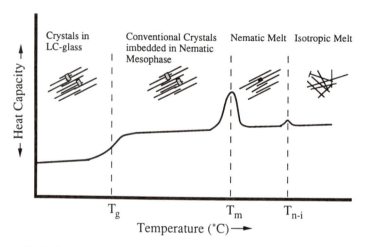

Figure 7. Typical DSC thermogram of a main chain thermotropic liquid crystalline polymer.

suggests that the order of the liquid crystals is rather imperfect. As the number of units in the flexible spacer varies, a dramatic odd-even effect is observed in temperature, enthalpies, and entropies of the nematic-isotropic transition (55,56). Corresponding to the maintenance of the linear propagation of the zig-zag conformation, the odd-even effect indicates that the units designed to disrupt the crystal structure take part in the nematic phase ordering process.

Only limited supercooling is possible upon traversing the isotropic-liquid crystal transition. As a result, it is extremely difficult to quench from an isotropic melt to an isotropic glassy state. The formation of a nematic mesophase is virtually complete. While cooling the liquid crystal phase, the polymer may undergo a first order nematic to smectic mesophase transition. However, cases of the smectic mesophase are not common in main-chain LCP systems. Figure 8 shows the molecular structure of the nematic mesophase with both the "flexible spacer" and the "mesogenic" group participating in the chain packing. While the overall order of the phase is preserved, individual units in the chain may deviate from the orientation director of the domain.

The transition from the liquid crystalline state to a fully three dimensional crystalline form is also first order. The kinetics of this process have been examined by several investigators (57,58). The degree of three dimensional crystallinity is generally quite low, and the rigid nature of the polymer chains prohibits the chain folding often observed in flexible polymers. Low heats of transition are caused by small crystal size and defect-ridden crystal structures. The defects are often the result of the flexible spacer or the mesogen having to incorporate other components into its crystal lattice. The liquid crystal to crystal transition is subject to extensive supercooling and conversion to a fully three dimensional order is usually quite incomplete. Subsequently, a large amount of the mesophasic structure is preserved in the solid-state, termed liquid crystal glass (LC-glass). The LC-glass has a glass transition with a step change in heat capacity, $\Delta C_p$, comparable to glass formation in flexible chain polymers; however, the temperature of the transition is normally slightly higher than in such systems (59-61). As evidenced by both thermal analysis (62,63) and x-ray diffraction (64), annealing of the quenched liquid crystal samples above the LC-glass transition results in crystallization as described by the above. Long annealing times at elevated temperatures have produced liquid crystal copolymer samples with up to 70 wt.% crystallinity. These high degrees of crystallinity suggest the presence of a copolymer inclusion mechanism in the formation of the crystal lattice (64). However, under slow cooling conditions, liquid crystal polymers have been

observed to exhibit copolymer exclusion of the non-
crystallizable units (57).

In many commercial LCP's, annealing at conditions 10 -
30°C below the melting point produces remarkable increases
in mechanical properties. Along with the increased
ordering of the molecular structure, many polymers exhibit
a dramatic rise in intrinsic viscosity, suggesting that
solid-state polymerization is occurring.

From the preceding discussion, a schematic diagram of
liquid crystal polymer morphology can be drawn. Free from
any mechanical stress, Figure 9 represents a typical
thermotropic liquid crystal polymer cooled from an
isotropic melt. The sample has a polydomain morphology
consisting of small crystallites imbedded within relatively
large nematic domains. The structure of the "grain
boundary" which separates the nematic domains having
different orientation direction has not been well studied,
but isotropic and splay structures have been suggested. In
any case, one can imagine that these grain boundaries could
play a significant role in the overall gas transport
mechanism.

Possibly the most important aspect of liquid crystal
polymers is their ability to orient during melt processing
and to maintain this orientation upon solidification. The
anisotropic characteristics of liquid crystal materials
increase dramatically when subjected to elongational flow
while shear flows have relatively less effect. The effect
on the polydomain morphology is seen in Figure 10. During
elongational flow, the molecules in each of the ordered
domains become aligned in the direction of flow. With
extensive draw-down, the system becomes increasingly
homogeneous while the extent of "grain boundary" regions
presumably diminishes.

Molded samples of LCP's often form a "skin-core"
structure. The phenomenon is depicted by a significant
dependence of the anisotropic properties of molded parts on
part thickness (65). Scanning electron photomicrographs of
the cross-section of a molded part reveal a highly oriented
"skin" layer surrounding a less ordered inner "core."
Apparently, the fraction of disordered core material
diminishes as the sample thickness decreases.

Gas Transport Properties of Liquid Crystal Polymers.
To date, reports of investigations on the gas transport
properties of main chain liquid crystalline polymers appear
to have been limited to the work conducted in our
laboratory. Chiou and Paul (4) have briefly described the
transport parameters of an extruded film of an LCP having a
similar structure to the commercial product Vectra. This
copolyester belongs to the family of napthylene
thermotropic polymers (NTP's) commercialized by Hoechst-
Celanese Corp. whose synthesis and properties have been
described previously (66). Transient permeation experiments
were conducted with a series of gases. The effective

Rigid Mesogenic Units

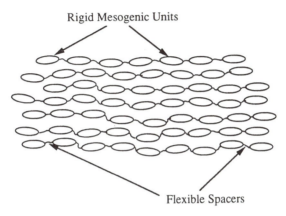

Flexible Spacers

Figure 8. Nematic mesophase of a copolymer consisting of rigid mesogenic and flexible spacers.

Nematic Mesophases                                          Imbedded Crystallites

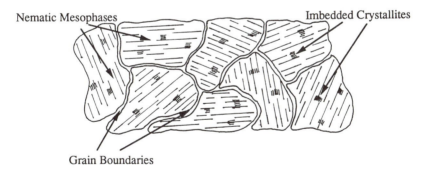

Grain Boundaries

Figure 9. Schematic drawing of the polydomain morphology of a thermotropic LCP cooled from a quiescent melt.

diffusion coefficients, D, were obtained from experimental
time lag data. The apparent solubilities were then
calculated using Eqn. 1. The limitations of this type of
analysis have been discussed previously (5). The transport
coefficients are compared with those of polyacrylonitrile
(PAN) in Table 2 (67). The permeabilites of the LCP are
generally lower than those for PAN, which is noted for its
impermeability to gases.   More surprising, though, are the
relative magnitudes of the diffusion and solubility
coefficients. While the diffusion coefficients of the LCP
are on average 40 times larger than those for PAN, the
apparent solubilities are 100-fold smaller.   Thus, it could
be  concluded that the unusual barrier properties of the
LCP stem from exceptionally low solubility rather than
intrinsically lower penetrant mobility within the matrix of
these rigid polymer chains.
        To check the validity of the time-lag method for
estimating solubility coefficients in these systems, static
sorption measurements were conducted with $CO_2$.   Such
measurements are difficult owing to the very low level of
sorption.   The solubility coefficient obtained is given in
parenthesis in Table 2.  Clearly, the agreement with those
obtained from transient permeation experiments is not
perfect; however, the values obtained by the two approaches
are close enough to confirm the general conclusions
discussed above.
        Heats of sorption and activation energies for
diffusion derived from conducting transient permeation
experiments as a function of temperature provide additional
information about the transport process. As seen in Table
3, the heats of sorption  are in accord with other gas-
polymer values, while the values of $E_D$ range from 30 to 50%
larger than values for semicrystalline PET(22).   Large
activation energies of diffusion would be expected given
the rigid nature of the LCP backbone; however, these values
are somewhat inconsistent with the high apparent diffusion
coefficients that were obtained from time-lag data.
        To expand upon the novel findings mentioned above, our
investigation of the tranport properties of liquid
crystalline polymers has been extended to poly(ethylene
terephthalate-co-p-oxybenzoate) copolymers, PET/PHB,
obtained from Eastman Kodak containing 60 and 80 mol% PHB.
Patented in 1973 (68), these semi-rigid polymers were among
the first  described as having thermotropic liquid
crystalline character (65) and have been the subject of an
extensive  number of investigations focused on their
morphology (69,70) and mechanical properties (65).   There
is general agreement that both compositions of the
copolyesters used here have some degree of chemical
heterogeneity that leads to phase separation which is more
pronounced in the 60 mol% PHB material.
        Ideally, the effect of liquid crystalline order could
be best examined by comparing gas solubility and diffusion

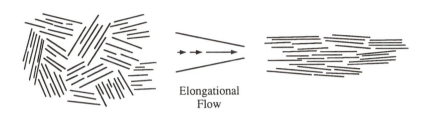

Elongational
Flow

Quiescent Melt                                    Highly Anisotropic Melt

Figure 10. Effect of elongational flow on the
morphology of thermotropic LCP's.

Table   2.   Comparison of Gas Transport Properties of a
            Wholly Aromatic Liquid Crystalline Polymer
            (Vectra   type) with Polyacrylonitrile at 35°C

| Gas | $P$ $\left(\dfrac{cm^3(STP)\cdot cm}{cm^2\cdot sec\cdot cmHg}\right)$ x $10^{15}$ | | $D$ $(cm^2/sec)$ x $10^{10}$ | | $S$ $(cm^3\ (STP)/cm^3\ atm)$ | |
|---|---|---|---|---|---|---|
|  | LCP | PAN | LCP | PAN[†] | LCP | PAN |
| He | 17,700 | 71,000 | 6,600 | 270 | 0.002 | 0.20 |
| $N_2$ | 3.0 | 2.9 | 1.4 | 0.042 | 0.0016 | 0.052 |
| $O_2$ | 47 | 54 | 7.1 | 0.14 | 0.0050 | 0.29 |
| Ar | 10 | 18 | 1.5 | 0.042 | 0.0054 | 0.33 |
| $CO_2$ | 70 | 280 | 0.96 | 0.023 | 0.051 (0.078) | 9.2 |

[†]PAN data from Ref.67.

behaviour in both isotropic and anisotropic arrangements prepared from a single polymer molecular structure. The four films used in this study are listed in Table 4 along with an outline of the methods by which the films were prepared. It was found that solution casting led to a transparent and largely isotropic film of the material containing 60% PHB. The 80% material was not soluble in the solvent used, triflouroacetic acid. Three other films were prepared from the two copolymer compositions by both melt processing and annealing of a solution cast film, as listed in Table 4.

The birefringence of thin portions of the four copolyester films was observed by placing them between crossed polarizers. The annealed and melt processed films were opaque and showed a completely anisotropic morphology. The solution cast film, EK60ca, was transparent and appeared to be isotropic to the naked eye. Although, examination of the cast film under cross polarized light reveals the presence of a large fraction of an anisotropic phase. X-ray diffraction patterns of the four films exhibit a diffuse nematic peak with an average d-spacing near 4.3 Å. The diffuse nematic peak of the EK60ca film is of a much lower intensity and is not as well-defined. The EK60an and EK80ex films have similar crystal diffraction patterns, suggesting that the EK60an casting/annealing procedure allowed for more complete phase separation which permitted PHB crystals to be formed with fewer ethylene terephthalate defects. The diffraction patterns of the melt processed films indicate that the subsequent compression molding step destroyed any macroscopic orientation. SEM photographs of fractured cross sections of the copolyester films showed no evidence of a skin-core morphology.

The anisotropic films exhibited two glass transitions whose temperature location and $\Delta C_p$ agree well with values reported by Menzcel and Wunderlich (59, 60). The first heat DSC scan of the EK60ca film shows a broad endotherm above the first glass transition roughly corresponding to the vaporization of about 3 wt% triflouroacetic acid. The presence of 3 wt% residual solvent was confirmed by thermogravimetric analysis (TGA).

Transient permeation experiments were conducted on the copolyester films using a series of gases in the same manner described previously. In Table 5, the transport coefficients of the liquid crystalline films can be compared with amorphous and semi-crystalline PET(22). Apart from the partially isotropic cast film, the permeabilities of the copolyester films are an order of magnitude lower than those for the PET materials. The low permeabilites of the copolyester films seem to stem from the apparent reduction in the solubility coefficient rather than intrinsically lower penetrant mobility. This is similar to what was obseved with the wholly aromatic Vectra material mentioned earlier (4). The permeabilities of the

Table 3. Comparison of Temperature Dependence, or Energetics, of Gas Transport in Vectra type LCP with semicrystalline Poly(ethylene terephthalate)

| Gas | $E_P$ (kcal/g mole) | | $E_D$ (kcal/g mole) | | $\Delta H_S$ (kcal/g mole) | |
|-----|-----|-----|-----|-----|-----|-----|
| | LCP | PET[†] ($\Phi = 0.58$) | LCP | PET[†] ($\Phi = 0.58$) | LCP | PET ($\Phi = 0.58$) |
| $O_2$ | 9.7 | 7.7 | 14.5 | 11.0 | -4.8 | -3.3 |
| $N_2$ | 13.3 | 7.8 | 17.5 | 10.5 | -4.2 | -2.7 |
| Ar | 12.9 | - | 16.8 | - | -3.9 | - |
| $CO_2$ | 10.5 | 4.4 | 15.6 | 12.0 | -5.1 | -7.6 |

[†] PET data from Ref. 22.
Note: $\Phi$ = amorphous volume fraction

Table 4. Film Preparation Procedure for Eastman-Kodak PET-PHB Copolyesters

| Film | mole% PHB | Film Preparation |
|------|-----------|------------------|
| EK60ca | 60 | Solution cast from 10 wt.% triflouroacetic acid solution. Dried under vacuum at 50 °C for 7 days. Dried at room temperature for 6 months. |
| EK60an | 60 | Solution cast from 10 wt.% triflouroacetic acid solution. Dried under vacuum at 50 °C for 7 days. Annealed under vacuum at 220 °C for 1 hour. |
| EK60ex | 60 | Extruded through a slit die at 290 °C. Compression molded at 260 °C for uniform thickness. |
| EK80ex | 80 | Extruded through a slit die at 260 °C. Compression molded at 300 °C for uniform thickness. |

partially isotropic EK60ca film are 5 to 10 times larger
than those of the the two anisotropic films of the same
composition. By comparing the EK60ex and EK80ex films
processed in a similar manner, it is clear that the
increased aromatic character of the polymer chain results
in a decrease in permeability caused primarily by a
reduction in the effective diffusivity. Because of the
differences in chemical and physical homogeneity of these
materials, it is difficult to interpret these observations
strictly in terms of a difference in molecular structure;
however, the trend is consistent with what one would expect
from a variation in rigid aromatic content.

To gain insight into the transport mechanism, the
energetics of oxygen transport were examined using
transient permeation experiments over a range of
temperatures. A comparison of the activation energies and
the derived heats of sorption is given in Table 6. The
heats of sorption of the highly anisotroptic copolyester
films are somewhat higher than those having essentially
isotropic character. The more positive heats of solution
may suggest that the sorption mechanism in the LCP films is
dominated by the Henry's law dissolution process. Since
the $\Delta H_S$ values are calculated from $E_P$-$E_D$, an alternative
explanation for the high heats of sorption could be a low
apparent activation energy of diffusion caused by transport
along the grain boundaries of the polydomain morphology.
Given the rigid nature of the polymer chain, the high
degree of phase separation, and the lack of macroscopic
directional order in the material, this hypothesis seems
justifiable. Thus, the network of the grain boundary
pathways and the nature of the polymer chains which may be
found along these pathways would play a significant role
in the energetics of transport and effective transport
coefficients.

Clearly, there is much to be learned about gas
transport in liquid crystalline polymers. The biphasic
structure of the PET/PHB copolymer system complicates
interpretation of the nature of the transport process.
However, it is evident from the results available to date
that conventional analysis of the transient permeation data
for materials which exhibit liquid crystalline order
consistently points to lower permeabilities resulting from
very low gas solubility. Two extreme and simplistic
possibilities can be mentioned but in the end both seem
inconsistent with the facts. First, we might attribute the
low gas solubility to the proposition that regions with
liquid crystalline order exclude gas sorption just as dense
three dimensional crystals do and that essentially the
entire volume of these materials consists of regions with
one or the other of these kinds of order. The sorption
observed then would be limited to the small volume fraction
of regions of the material having a more open structure.
Second, we might consider that the liquid crystalline phase
is the only viable region for gas sorption, but because of

Table 5. Gas Transport Properties at 35°C for PET/PHB
Copolyester Films

| Gas | Film | P (Barrers) x $10^3$ | D (cm²/sec) x $10^{10}$ | S $\left(\dfrac{cm^3 \, (STP)}{cm^3 \, atm}\right)$ x $10^3$ |
|---|---|---|---|---|
| $O_2$ | PET ($\Phi$=1.00)[†] | 110 | 95 | 88 |
| | PET ($\Phi$=0.58) | 53 | 65 | 61 |
| | EK60ca | 51 | 85 | 45 |
| | EK60an | 9 | 60 | 12 |
| | EK60ex | 11 | 83 | 10 |
| | EK80ex | 5 | 41 | 8.8 |
| $N_2$ | PET ($\Phi$=1.00) | 17 | 32 | 42 |
| | PET ($\Phi$=0.58) | 9.7 | 23 | 32 |
| | EK60ca | 7.3 | 20 | 28 |
| | EK60an | 0.93 | 17 | 4.2 |
| | EK60ex | 1.2 | 32 | 2.9 |
| | EK80ex | 0.5 | 11 | 3.9 |
| $CO_2$ | PET ($\Phi$=1.00) | 430 | 16 | 2000 |
| | PET ($\Phi$=0.58) | 180 | 11 | 1300 |
| | EK60ca | 190 | 22 | 630 |
| | EK60an | 21 | 16 | 100 |
| | EK60ex | 33 | 28 | 89 |
| | EK80ex | 12 | 12 | 76 |

† PET data from Ref.22.
Note: $\Phi$ = amorphous volume fraction

Table 6. Energetics of Oxygen Transport in PET/PHB
Copolyesters

| Film | $E_P$ (kcal/g mole) | $E_D$ (kcal/g mole) | $\Delta H_S$ (kcal/g mole) |
|---|---|---|---|
| PET ($\Phi$=1.0)[†] | 9.0 | 11.6 | - 2.6 |
| PET ($\Phi$=0.58) | 7.7 | 11.0 | - 3.3 |
| EK60ca | 8.0 | 11.2 | - 3.2 |
| EK60an | 11.2 | 9.7 | 1.5 |
| EK60ex | 10.3 | 11.5 | - 1.2 |

†PET data from Ref. 22.

its efficient chain packing the level of solubility is
orders of magnitude less than that seen in amorphous
glasses.  However, both scenarios would lead one to expect
very low diffusion coefficients.  The first because of the
small fraction of material in which diffusion occurs and
the attendent high toruousity, and the second because of
the low level of molecular free volume needed to explain
the solubility data.  However, the diffusion coefficients
calculated from the time lag data are too large to be
consistent with either of these possibilties.  Thus, it
would be premature, to assign a specific mechanism to
account for the observations in the copolyester systems
discussed here.  Currently, research efforts are
concentrating on a series of more chemically homogeneous
thermotropic materials in order to elucidate the gas
transport mechanism in liquid crystal polymers.

Conclusions

In this report the effects of structural order in the form
of conventional crystallinity and orientation have been
reviewed.  Gas transport properties have been examined
through the use of both free volume theory and an
activated-state approach.  The most dramatic improvements
in the barrier properties are observed in the orientation
of semicrystalline polymers, where both geometric impedance
and chain immobilization contribute to improvements in
barrier characteristics.  Liquid crystalline polymers have
a highly anisotropic morphology with a degree of order
which lies in between the zero ordered glass and the three
dimensional crystal lattice.  Contrary to what is observed
in semicrystalline systems, liquid crystalline order
appears to have the largest influence upon the effective
solubility.  The mechanism by which small gas molecules
permeate through the liquid crystalline matrix is under
current scrutiny.  Despite the uncertainties in the
specific mode of transport, it is evident that liquid
crystalline polymers are a class of polymers with rather
remarkable barrier properties.  The potential benefits from
using such polymers in barrier applications are compounded
by the ease of product fabrication and excellent thermal-
mechanical properties.  Future efforts in the field may
find that the effect of combining LCP's with commercial
grade polymers as blends or composites could lead to cost
effective barrier materials.

Acknowledgments

The authors wish to express their appreciation to Eastman
Kodak Co. for providing the materials used in this work and
to the U.S. Army Research Office and the Separations
Research Program at the University of Texas at Austin for
financial support.

Literature Cited

1.  Morgan, P.W. Ind. Eng. Chem. 1953, 45, 2296.
2.  Van Amerongen,G. J. Polym. Sci. 1947, 2, 381.
3.  Doty, P. M.; Aiken,W. H.; Mark, H. Ind. Eng.Chem. 1946, 38, 788.
4.  Chiou, J. S.; Paul, D.R. J. Polym. Sci. Polym. Phys Ed. 1987, 25,1699.
5.  Paul, D. R. Ber. Bunsenges Phys. Chem. 1979, 83, 294.
6.  Stannett, V.; Koros, W. J.; Paul, D. R.; Lonsdale, H. K.; Baker, R. W. Adv. Polym. Sci. 1979, 32, 69.
7.  Crank, J.; Park, G. S. Eds. Diffusion in Polymers; Academic Press:New York; 1968.
8.  Cohen, M. H. ; Turnbull, D. J. Chem. Phys. 1959, 31, 1164.
9.  Fujita, H. Fortschr. Hochpolym. Forsch. 1961, 3, 1.
10. Kumins, C. A. ; Kwei,T. K. in Diffusion in Polymers; Crank, J.; Park, G. S. Eds.; Academic Press: New York, 1968; Chapter 4.
11. Peterlin, A. J. Macromol. Sci. Phys. 1975, B11, 57.
12. Fox, T. G.; Flory, P. J. J. Appl. Phys. 1950, 21, 581.
13. Bondi, A. in Physical Properties of Molecular Crystals, Liquids, and Glasses; Wiley: New York; 1968, Chapters 3 and 4.
14. Sugden, S. J. Chem. Soc. 1927, 1786.
15. Lee, W. M. Polym. Eng. Sci. 1980, 20, 1.
16. Meyers, A. W.; Rogers, C. E.; Stannett, V.; Szwarc, M. Tappi 1958, 41, 716.
17. Lasoski, S. W., Jr. ; Cobbs, W. H., Jr. J. Polym. Sci 1959, 36, 21.
18. Michaels, A. S.; Parker, R. B., Jr. J. Polym. Sci. 1959, 41, 53.
19. Michaels, A. S.; Bixler, H. J. J. Polym. Sci. 1961, 50, 393.
20. Michaels, A. S.; Bixler, H. J. J. Polym. Sci. 1961, 50, 413.
21. Michaels, A. S.; Vieth, W. R.; Barrie, J. A. J. Appl. Phys. 1963, 34, 1.
22. Michaels, A. S.; Vieth, W. R.; Barrie, J. A. J. Appl. Phys. 1963, 34, 13.
23. Michaels, A. S.; Bixler, H. J.; Fein, H. L. J. Appl. Phys. 1964, 35, 3165.
24. Puleo, A. C.; Paul, D. R.; Wong, K. P. in Press.
25. Vieth, W.; Wuerth, W. F. J. Appl. Polym. Sci. 1969, 13, 685.
26. Klute, C. H. J. Polym. Sci. 1959, 41, 307.
27. Klute, C. H. J. Appl. Polym. Sci. 1959, 1, 340.

28.  Lowell, P. N.; McCrum, N. G. J. Polym. Sci. A-2
     1971, 9, 1935.
29.  Jeschke, D.; Stuart, H. Z. Naturforsch. 1961, 16,
     37.
30.  Barrer, R. M. in Diffusion in Polymers; Crank,
     J.; Park, G. S. Eds.; Academic Press: New York,
     1968; Chapter 6.
31.  Petropoulus, J. H. J. Polym. Sci. Polym. Phys. Ed.
     1985, 23, 1309.
32.  Barrie, J. A.; Platt, B. J. Polym. Sci. 1961, 54,
     261.
33.  Levita, G.; Smith, T. L. Polym. Eng. Sci. 1981, 21,
     936.
34.  Wang, L. H.; Porter, R. S. J. Polym. Sci. Polym.
     Phys. Ed. 1984, 22, 1645.
35.  Ito, Y. Kobunshi Kagaku 1962, 19, 412.
36.  El-Hibri, M. J.; Paul, D. R. J. Appl. Polym. Sci.
     1985, 30, 3649.
37.  Brady, T. E.; Jabarin, S. A.; Miller, G. W. in
     Permeability of Plastic Films and Coatings to
     Gases, Vapors and Liquids; Hopfenberg, H.B.,
     Ed.; Plenum Press: New York; 1968; pp 301-320.
38.  Peterlin, A. in Flow-Induced Crystallization in
     Polymer Systems; Miller, R. L., Ed.; Gordon and
     Breach Science Publishers: New York; 1979; pp 1-29.
39.  Hay, J. A.  in Flow-Induced Crystallization in
     Polymer Systems; Miller, R. L., Ed.; Gordon and
     Breach Science Publishers: New York; 1979; pp 69-
     98.
40.  Peterlin, A. Polym. Eng. Sci. 1976, 16, 126.
41.  Peterlin, A. J. Mater. Sci. 1971, 6, 490.
42.  Brandt, W. W. J. Polym. Sci. 1959, 41, 415.
43.  Yasuda, H.; Stannett, V.; Frisch, H. L. ; Peterlin,
     A. Die Macromol.Chem. 1964, 73, 188.
44.  Peterlin, A.; Williams, J. L.; Stannett, V. J.
     Polym. Sci. A-2 1967, 5, 957.
45.  Williams, J. L.; Peterlin, A. J. Polym. Sci. A-2
     1971, 9, 1483.
46.  Yasuda, H.; Peterlin, A. J. Appl. Polym. Sci.
     1974, 18, 531.
47.  Ostapchenko, G. J. 6th International Conference on
     Oriented Plastic Containers; Cherry Hill, New
     Jersey; March, 1982.
48.  Vieth, W. R.  5th International Conference on
     Oriented Plastic Containers; Cherry Hill, New
     Jersey; March, 1981.
49.  Slee, J. A.; Orchard, G. A. J.; Bower, D. I.; Ward,
     I. M. J. Polym. Sci.Polym Phys. Ed. 1989, 27, 71.
50.  Takagi, Y.; Hattori, H. J. Appl. Polym. Sci.
     1965, 9, 2167.

51. El-Hibri, M. J.; Paul, D. R. <u>J. Appl. Polym. Sci.</u> 1986, <u>31</u>, 2533.
52. Cifferi, A.; Krigbaum, W. R.; Meyer, R. B. Eds. <u>Polymer Liquid Crystals</u>; Academic Press: New York; 1982.
53. Blumstein, A. Ed. <u>Polymeric Liquid Crystals</u>; Plenum Press: New York;1985.
54. Ober, C. K.; Jin, J. I.; Lenz, R. W. <u>Adv. Polym. Sci.</u> 1984, <u>59</u>, 103.
55. Griffin, A. C.; Havens, S. J. <u>J. Polym. Sci. Polym. Phys. Ed.</u> 1981, <u>19</u>,951.
56. Blumstein, A.; Thomas, O. <u>Macromolecules</u> 1982, <u>15</u>, 1264.
57. Cheng, S. Z. D. <u>Macromolecules</u> 1988, <u>21</u>, 2475.
58. Hanna, S.; Windle, A. H. <u>Polymer</u> 1988, <u>29</u>, 207.
59. Menzcel, J.; Wunderlich, B. <u>J. Polym. Sci. Polym. Phys. Ed.</u> 1982, <u>18</u>,1433.
60. Meesiri, W.; Menzcel, J.; Guar, U.; Wunderlich, B. <u>J. Polym. Sci. Polym.Phys Ed.</u> 1982, <u>20</u>, 719.
61. Noel, C; Friedrich, C.; Laupretre, F.; Billard, J.; Bosio, L.; Strazielle, C.<u>Polymer</u> 1984, <u>25</u>, 263.
62. Krigbaum, W. R.; Salaris, F. <u>J. Polym. Sci. Polym. Phys. Ed.</u> 1978, <u>16</u>,883.
63. Sauer, T. H.; Zimmerman, H. J.; Wendorff, J. H. <u>Colloid Polym. Sci.</u> 1987, <u>265</u>, 210.
64. Sauer, T. H.; Zimmerman, H. J.; Wendorff, J. H. <u>J. Polym. Sci. Polym.Phys. Ed.</u> 1987, <u>25</u>, 2471.
65. Jackson, W. J.; Kuhfuss, H. F. <u>J. Polym. Sci. Polym. Chem Ed.</u> 1976, **14**,2043.
66. Calundann, G. W. in <u>High Performance Polymers: Their Origin and Development</u>; Seymour, R. B.; Kirshenbaum, G. S. Eds.; Elsevier Science Publishing Co., Inc.: Amsterdam; 1986; pp. 235-249.
67. Allen, A. H.; Fujii, M.; Stannett, V.; Hopfenberg, H. B.; Williams, J. L. <u>J. Membr. Sci.</u> 1977, <u>2</u>, 153.
68. Jackson, W. J.; Kuhfuss, H. F. U.S. Patent 3,778,410 (1973).
69. Nicely,V. A.; Dougherty, J. J.; Renfro, L. W. <u>Macromolecules</u> 1987, <u>20</u>, 573.
70. Sawyer, L. C. <u>J. Polym. Sci. Lett. Ed.</u> 1984, <u>22</u>, 347.

RECEIVED December 5, 1989

# Chapter 4

# Transport of Plasticizing Penetrants in Glassy Polymers

## Alan R. Berens

### RD #2, Box 3510, Middlebury, VT 05753

This survey of recent studies correlates the transport kinetics and equilibria of organic vapors and liquids in glassy polymers with the molecular size, interaction parameter, and activity of the penetrant. Sorption isotherms follow a generalized sigmoidal form combining dual-mode form for the glassy state and Flory-Huggins form for the rubbery state, with an inflection at the glass transition. The position and shape of specific isotherms are determined by the polymer-penetrant interaction parameter and the glass composition of the system. Sorption kinetics are Fickian at very low penetrant concentration, with diffusivity strongly dependent on size and shape of the penetrant molecule. Kinetics become anomalous and then Case II when the glass composition is approached and exceeded. Recent results on transport of carbon dioxide at high pressure are consistent with the correlations developed for organic liquids and vapors.

The barrier properties of glassy polymers may be seriously impaired by the presence of organic vapors or liquids which penetrate and plasticize the polymer. It therefore would be useful to foretell conditions and penetrants which might cause such plasticization and to predict the consequences on transport behavior. The complex and anomalous equilibrium and kinetic behavior of small organic molecules in glassy polymers has long been recognized ($\underline{1}$), but there are few correlations allowing predictions of transport behavior from material properties. In this paper we survey a number of recent studies, principally in the author's laboratory at The BFGoodrich Company, to relate transport properties to the molecular dimensions and thermodynamic activity of the penetrant and its interaction parameter with the polymer. These correlations seem to provide a basis for at least qualitatively useful predictions of sorption equilibria and diffusion kinetics. Some recent results on carbon dioxide seem

0097–6156/90/0423–0092$06.00/0

consistent with these correlations and reflect this penetrant's unique combination of the high diffusivity of a small gas molecule, the high solubility of an organic swelling agent, and an unusually high plasticizing efficiency.

## Experimental

Most of the results discussed here have been obtained from gravimetric sorption/desorption experiments, i.e., by following the weight of appropriate polymer samples as they absorb penetrants from a controlled vapor or liquid environment or release them into a vacuum or low concentration atmosphere. Experiments with vapors at sub-atmospheric pressure were performed with a Cahn recording vacuum microbalance using polymer powders of known geometry (2,3). Liquid sorption experiments were carried out by periodic blotting and weighing of polymer film or sheet samples during prolonged immersions in pure or mixed organic liquids (4). Kinetic and equilibrium data for polymer/$CO_2$ systems were derived from rapid periodic weighings of polymer film samples during desorption at atmospheric pressure following exposure to high-pressure $CO_2$ (5).

## Sorption Isotherms

The system poly(vinyl chloride) (PVC)/vinyl chloride monomer (VCM) seems to have been the first glassy polymer/organic vapor system whose sorption equilibrium and kinetic behavior was studied over the entire range of activities and at temperatures both below and above the glass transition (6). Figure 1 shows several isotherms for this system, plotted *vs* VCM activity, i.e. pressure relative to the saturated vapor pressure, $P/P_O$, at each temperature (6). At 90°C, above the 85° $T_g$ of PVC, the complete isotherm is well described by the Flory-Huggins equation (7) with the interaction parameter $\chi = 0.98$. The superposition of the isotherms at high activities indicates that $\chi$ is nearly independent of temperature, and thus that the solution of VCM in PVC involves a near-zero heat of mixing. Below $T_g$ the isotherms at low activities show the downward curvature characteristic of the dual-mode sorption model (8); the contribution of the Langmuirian sorption mode increases with decreasing temperature. The disappearance of the Langmuirian portion has been identified with the penetrant concentration, $C_g(T)$, which produces the glass-to-rubber transition at the isotherm temperature $T(9)$. The overall form of the PVC/VCM isotherms, then, is sigmoidal, with dual-mode behavior in the glassy state, Flory-Huggins form in the rubbery state, and an inflection at the composition-dependent glass transition.

Solubility data for methanol at low activities in glassy PVC, illustrated in Figure 2, again show the downward curvature of the dual-mode model. These data also exhibit a pronounced dependence on the previous thermal history of the polymer; it appears that the Langmuirian sorption capacity of the polymer parallels the history dependence of free volume in the glassy state (6,9).

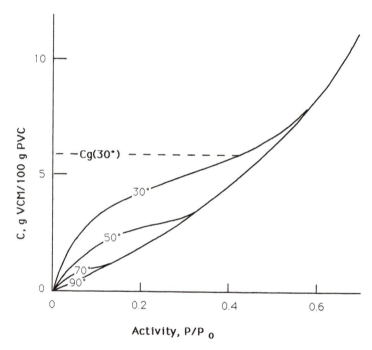

Figure 1.    Sorption isotherms for VCM in PVC. (Reproduced from
             Ref. 6. Copyright 1974 ACS Division of Polymer
             Chemistry.)

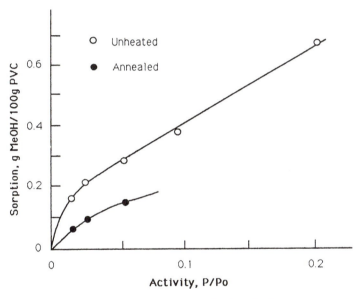

Figure 2.    Solubility of methanol in PVC at 30°C vs. methanol activity, for PVC powders of differing thermal history.

Sorption isotherm data for the toluene/PVC system at 30°C are shown in Figure 3 (4). Here the Flory-Huggins equation with $\chi = 0.75$ describes the data fairly well over the whole range of activity. There is a suggestion of dual-mode behavior at low activity, but experiments in this range are limited by the very low diffusivity of toluene in glassy PVC.

Sigmoidal sorption isotherms, combining dual-mode form at low activities with Flory-Huggins form at high activity, have recently been reported for the lower alcohols in poly(methyl methacrylate) (PMMA) (10), and for hydrocarbon vapors in polystyrene (11). It now appears that such sigmoidal isotherms, with an inflection marking the transition from glassy to rubbery behavior, are a general characteristic of polymer/penetrant systems when the range of penetrant concentrations traverses $C_g(T)$.

A generalized form of the sorption isotherms for plasticizing penetrants in glassy polymers, and the effects of the principal governing factors, are suggested schematically in Figure 4. The position of the high-activity, Flory-Huggins portion of the isotherm is determined by the value of the interaction parameter $\chi$: The curve, shown for $\chi = 1$, is shifted upward with increasing solvent strength (lower $\chi$ values), and downward for poorer solvents (higher $\chi$). The position of the inflection is determined by $T$ and also by $C_g(T)$, i.e., by the glass temperature of the polymer and the plasticizing effectiveness of the penetrant. Thus the position and form of the isotherm are governed by the interaction of $\chi$ and $C_g(T)$, and knowledge of these two parameters allows at least qualitative predictions of the isotherms. For example, a combination of a high $\chi$ and high $C_g(T)$ (i.e., a poor solvent with low plasticizing efficiency) might produce an isotherm of dual-mode form over the whole activity range. Lower $\chi$ and $C_g(T)$ values (stronger swelling agents or solvents, more efficient plasticizers) would tend to give isotherms following Flory-Huggins form over a greater portion of the activity scale.

The parameters needed for these predictions of isotherms, $\chi$ and $C_g(T)$, are readily accessible: The value of $\chi$ can be estimated from a single measurement of the equilibrium penetrant solubility at a known high activity, assuming applicability of the Flory-Huggins equation. The plasticizing effectiveness, $C_g(T)$, may be estimated either experimentally or theoretically, as discussed in the next section.

Plasticization

The degree of plasticization of a glassy polymer may be expressed as the depression of the glass transition temperature , which can be directly measured by DSC determinations of $T_g$ on polymer samples containing known concentrations of penetrant. As an example, Figure 5 shows $T_g$ as a function of volume fraction toluene in PVC, measured by DSC on polymer film samples equilibrated with toluene at varied activity (4). These data, and similar results for acetone and benzene, show

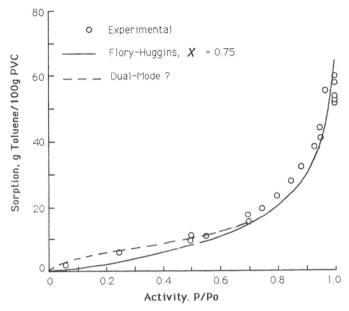

Figure 3.    Solubility of toluene in PVC at 30°C vs. toluene activity; vapor and liquid sorption data for PVC films and powders.

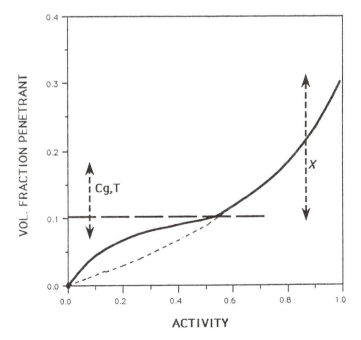

Figure 4.    Generalized sorption isotherm for plasticizing penetrants in glassy polymers.  (Schematic.)

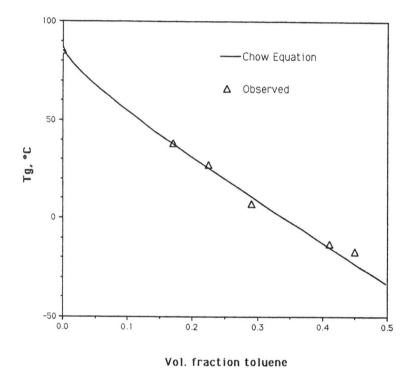

**Vol. fraction toluene**

Figure 5.        Glass transition temperature vs. composition for the system PVC toluene.

that values of $C_g(30°)$ for these typical swelling agents in PVC fall in the range 0.2 - 0.3 volume fraction solvent.

The depression of $T_g$ can also be calculated, as a function of penetrant concentration, through a theoretical expression derived by Chow on the basis of both classical and statistical thermodynamics (12). Calculated results for the toluene/PVC system, as shown in Figure 5, are in excellent agreement with experimental data. Chow's treatment shows that the plasticizing efficiency increases with decreasing molecular weight and size of the penetrant. For the VCM/PVC system, $C_g(30°)$ calculated from the Chow equation is 0.15 volume fraction VCM, comparable to an experimental estimate of about 0.11 (9). The greater plasticizing efficency of the smaller VCM molecule, compared to toluene, is consistent with the theoretical prediction.

Sorption Kinetics

It has been recognized for some time that the kinetics of sorption of solvents into glassy polymers vary in form with the penetrant activity; Fickian, anomalous, and Case II kinetics may be observed as activity is increased in a given solvent/polymer system (13). The recent study of several organic liquid and vapor penetrants in rigid PVC indicates that the the interaction parameter and plasticizing efficiency of the penetrant, as well as its activity, are factors affecting the form of the sorption kinetics (4).

As examples of the changes of sorption kinetics with penetrant activity, Figure 6 shows data for the sorption of toluene vapor at varied pressures into PVC powder, and Figure 7 shows data for the sorption of trichloroethylene (TCE) from liquid polyethylene glycol 400 mixtures of varied TCE activity into PVC films (4). Both sets of data show Fickian form (sorption proportional to $t^{1/2}$) at the lower activities. The onset of anomalous, non-Fickian kinetics occurs at about 0.4 activity for toluene, and about 0.7 for TCE. Sorption data for acetone, a stronger swelling agent for PVC, show anomalous kinetics beginning at about 0.3 activity. While the activity range for Fickian behavior is quite different for these three penetrants, the onset of anomalous kinetics in each case occurs when the final penetrant uptake is approximately one-half the $C_g$ for that penetrant, i.e., at a similar degree of plasticization.

In a like fashion, the activities and final concentrations at which sorption kinetics change from anomalous to Case II (uptake linear with time) were estimated for several penetrants in PVC (4). Figure 8, for example, shows that this kinetic transition occurs at a penetrant activity of about 0.9 in the TCE/PVC system. It appears that Case II kinetics are observed only when the final uptake is at least equal to $C_g$, i.e., when the polymer/penetrant system undergoes the glass-rubber transition during the sorption process.

These observations on PVC/penetrant systems suggest a correlation of sorption kinetics with the generalized isotherm for swelling penetrants in glassy polymers, as shown schematically in Figure 9. For sorption into an initially penetrant-free glassy

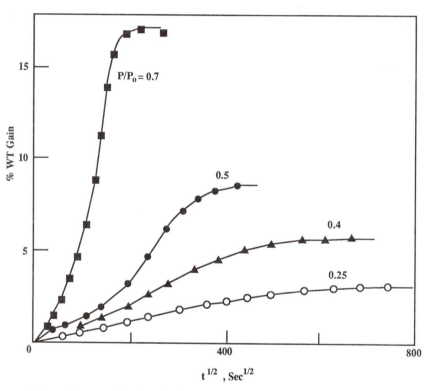

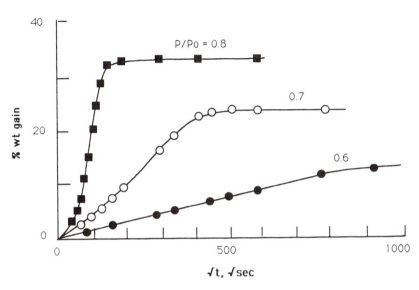

Figure 7.    Kinetics of TCE sorption from liquid mixtures into PVC films at 30°C; TCE activity 0.6 to 0.8, plotted vs. square root of time. (Reproduced with permission from Ref. 4. Copyright 1989 John Wiley & Sons.)

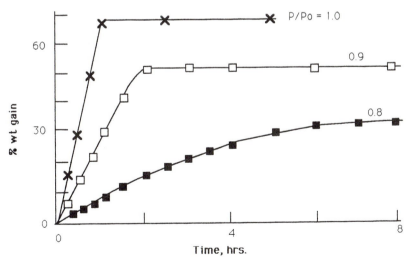

Figure 8    Kinetics of TCE sorption from liquid mixtures into PVC films at 30°C; TCE activity 0.8 to 1.0, plotted vs. time. (Reproduced with permission from Ref. 4. Copyright 1989 John Wiley & Sons.)

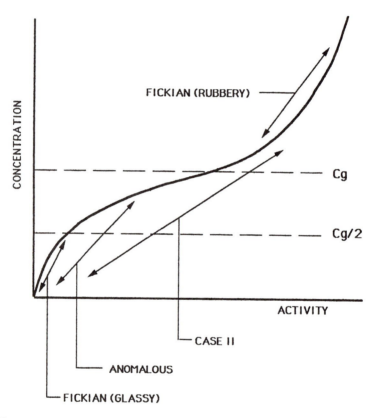

Figure 9.    Relation of sorption kinetics to generalized sorption
             isotherm. (Schematic.)

polymer, Fickian kinetics may be anticipated when the concentration range stays below about $C_g/2$; anomalous behavior is expected for final concentrations between $C_g/2$ and $C_g$; and Case II kinetics are probable when the concentration traverses $C_g$. The activities at which the kinetic transitions occur will depend on the position of the isotherm, and thus upon both the interaction parameter and plasticizing efficiency for the polymer/penetrant pair. Data for additional polymer/solvent systems at varied activity would be desirable to confirm the validity of these correlations for glassy polymers other than PVC.

Diffusion Coefficients

In the low concentration range, where plasticization is negligible and sorption kinetics are Fickian, the diffusion coefficients, $D$, of gases and vapors in glassy polymers are extremely strong functions of the size and shape of the penetrant molecules. In Figure 10, diffusivities of a number of gases and organic vapors at very low concentrations in PVC are plotted *vs* the mean diameter of the penetrant molecule; similar correlations have also been established for polystyrene and PMMA (3). It appears that there is an approximate linear relation between log $D$ and diameter of isotropic, roughly spherical molecules. More elongated molecules have $D$ values greater by factors of up to $10^3$ than spherical molecules of similar molar volume. The correlation between $D$ and molecular size and shape seems capable of predicting $D$, within perhaps an order of magnitude, for other penetrant/polymer systems, once values for a few representative penetrants in a given polymer are determined.

Fickian kinetics are also observed for diffusion of gases and organic vapors in rubbery polymers (14), but diffusivities are much higher and much less steeply dependent upon molecular size than in glassy polymers. The same is true for glassy polymers already plasticized into the rubbery state, i.e., for experiments carried out entirely above $T_g$ or $C_g$, as indicated in Figure 9. Diffusivities of several gases, vapors and liquids in plasticized PVC are compared to values in glassy PVC on a plot of log $D$ *vs* molecular diameter in Figure 11 (15). The difference in diffusivity between the glassy and rubbery states increases dramatically with increasing size of the penetrant: For the small gas molecules, diffusivity is increased by about one order of magnitude upon plasticization. For common solvent molecules of 5 to 6 Å diameter, the ratio of rubbery- to glassy-state diffusivity may be $10^6$ to $10^8$, and rough extrapolation suggests this ratio might be as great as $10^{12}$ for plasticizers or other additives of 8 to 10 Å diameter. These trends have interesting consequences for transport in systems involving carbon dioxide.

Carbon Dioxide

The transport of carbon dioxide in polymers has historically been analyzed in the same manner as other simple gases (1). A number of recent studies have shown, however,

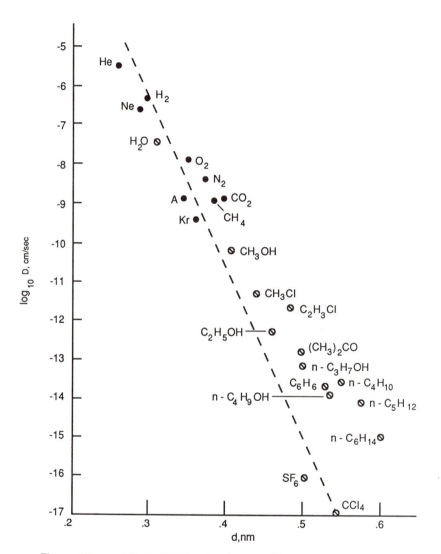

Figure 10.    Effect of molecular size on diffusivity at low penetrant
concentration in PVC at 30°C: log D vs. mean molecular
diameter of penetrant.   (Reproduced with permission
from Ref. 3. Copyright 1982 Elsevier.)

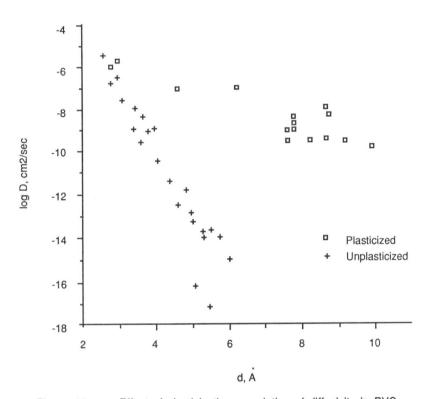

Figure 11.   Effect of plasticization on relation of diffusivity in PVC to molecular size of penetrant. (Reproduced with permission from Ref. 15. Copyright 1989 IUPAC.)

that $CO_2$ at high pressure resembles common organic solvents in its ability to swell and plasticize polymers (16-22). It therefore is appropriate to include studies of $CO_2$ transport in a discussion of plasticizing penetrants in glassy polymers.

We have recently applied a simple gravimetric technique to study sorption equilibria and kinetics for near-critical $CO_2$ in a number of glassy polymers at pressures up to the saturated vapor pressure of liquid $CO_2$ (5). Representative sorption isotherms for four glassy polymers at 25°C are shown in Figure 12, plotted against $CO_2$ pressure. All of these isotherms may be regarded as examples of the generalized isotherm of Figure 4, with appropriate values of the interaction parameter and glass composition. For poly(vinyl acetate) (PVA), a relatively low $\chi$ (strong interaction with $CO_2$) and low $Cg$ (since the $T_g$ of PVA itself is only 30°) produce an isotherm of Flory-Huggins, rubbery form over most of the pressure range. For PVC and polycarbonate (PC), the lower solubility of $CO_2$ and the higher $T_g$ of the polymers result in isotherms of dual-mode form, indicating that these polymers remain in the glassy state at all $CO_2$ pressures. For PMMA, the data suggest a sigmoid isotherm, with an inflection indicative of a glass transition at an intermediate $CO_2$ pressure.

Further examples of the applicability of the generalized sigmoidal isotherm to $CO_2$/glassy polymer systems have been obtained from published isotherms for several other polymers by converting the original pressure axis to an activity scale (5). Activity was taken as the fugacity ratio $f/f_o$, defining the reference state above the critical temperature by extrapolating the saturated vapor pressure from sub-critical temperatures. Figure 13, for example, shows such isotherms derived from data of Kamiya, et al. for poly(vinyl benzoate) (PVBz) (23). Like the VCM/PVC system (Figure 1), the $CO_2$/PVBz isotherms show sigmoidal form, and the dual-mode portion diminishes with increasing temperature. Superposition of the Flory-Huggins portions for different temperatures suggests a near-zero heat of mixing for $CO_2$ with PVBz; similar results were found for PMMA and PC (5).

Glass transitions of $CO_2$/polymer systems, determined from the isotherm inflections, occur at significantly lower weight concentrations of penetrant than is the case for organic solvents. This greater plasticizing efficiency of the smaller $CO_2$ molecule is consistent with the predictions of the Chow equation (12): Figure 14 shows $T_g$ vs weight composition calculated for $CO_2$, VCM and toluene in PVC. The calculated $T_g$ of PVC containing 8 weight % . $CO_2$, the limiting solubility at 25° and unit activity, is about 27°C. Thus, the solubility and $Cg$ of $CO_2$ in PVC at room temperature are nearly equal; i.e., liquid $CO_2$ can plasticize PVC virtually into the rubbery state at room temperature. For polymers in which $CO_2$ is more soluble, such as PMMA or poly(vinyl benzoate), depression of the glass transition to below room temperature, as evidenced by the isotherm inflections, is quite in accord with the predicted high plasticizing efficiency of $CO_2$.

The sorption kinetics for $CO_2$ in the glassy polymers studied (5) appear to be Fickian over the entire activity range; diffusivities have the high values anticipated for a

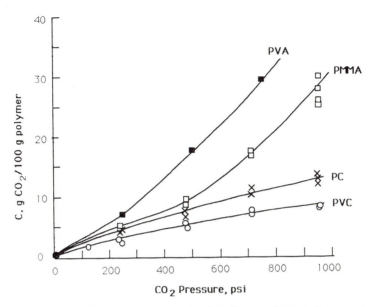

Figure 12.    Sorption isotherms for $CO_2$ in four glassy polymers at
25°C. (Reproduced from Ref. 5. Copyright 1989 ACS.)

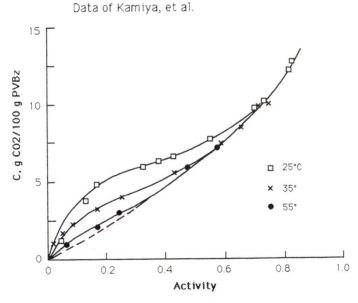

Figure 13.    Sorption isotherms for $CO_2$ in poly(vinyl benzoate)
(23)plotted vs. $CO_2$ activity.  (Reproduced from Ref.
5. Copyright 1989 ACS.)

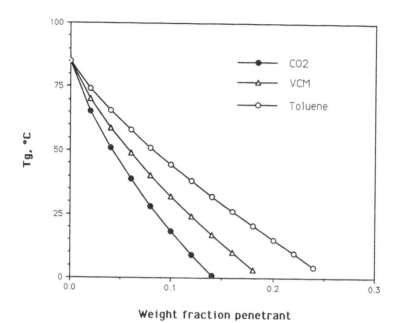

Figure 14.    Glass transition temperature vs. composition for
              PVC/penetrant systems, calculated from the equation of
              Chow (12).

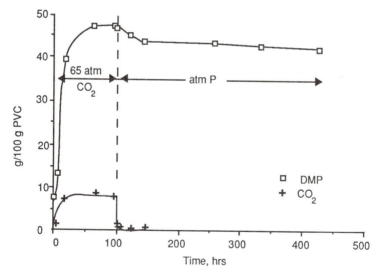

Figure 15.    Gravimetric sorption/desorption vs. time date for the
              system PVC/DMP/$CO_2$. (Reproduced with permission
              from Ref. 15.   Copyright 1989 IUPAC.)

small gas molecule and increase with increasing $CO_2$ concentration. By analogy with solvent/polymer systems, Case II sorption kinetics might be expected when the sorption of $CO_2$ produces the glass-rubber transition, as in PMMA at high $CO_2$ activity. The absence of Case II kinetics may be due to the relatively small change of $CO_2$ diffusivity between the glassy and rubbery states. The observed increase of $D$ across the glass transition is from 10 to 100-fold for $CO_2$, far less than the change for organic solvents (cf. Figure 11). According to one recent theory of Case II transport (24), this mechanism is associated with a very sharp increase of $D$ in the vicinity of $T_g$. Perhaps the change of $D$ at $T_g$ for $CO_2$ is too small to produce the sharp concentration step involved in the Case II mechanism.

The high diffusivity, polymer-solubility, and plasticizing efficiency of $CO_2$ lead to some interesting effects on the transport of other low-molecular weight penetrants in glassy polymers (25, 26). When polymer film samples are exposed simultaneously to an additive substance and to $CO_2$ under high pressure, the $CO_2$ is rapidly absorbed, plasticizing the polymer and thereby sharply increasing the diffusivity of the additive. Upon the release of pressure, the $CO_2$ is quickly desorbed, the degree of plasticization and additive diffusivity are sharply reduced, and the additive is effectively "trapped" in the polymer. It has been found that high-pressure $CO_2$ remarkably accelerates the absorption of many polymer-soluble compounds whose diffusion into the polymer alone is kinetically limited.

This "$CO_2$-assisted impregnation" process is illustrated in Figure 15 with data for the model system PVC/$CO_2$/dimethyl phthalate (DMP). In the absence of $CO_2$, the absorption of DMP by PVC at room temperature is extremely slow; no more than 1 wt % DMP was absorbed by PVC films immersed in excess liquid DMP for 64 hours. In the presence of liquid $CO_2$, in contrast, the DMP content of the PVC reached 40 wt % in 16 hours. After the pressure was released, over 95% of the absorbed $CO_2$ had escaped in 24 hours, but more than 80% of the absorbed DMP remained even after over 1000 hours.

In addition to its practical potential in incorporating additives into polymers, the effect of compressed $CO_2$ on third-component diffusivities may also be relevant to the supercritical fluid extraction of low-molecular compounds from polymers, and to the attack of barrier polymers by other swelling agents during service in high-pressure $CO_2$ environments.

Literature Cited

1. Crank, J.; Park, G. S. Diffusion in Polymers, Academic Press, London, 1968.
2. Berens, A. R. Polymer 1977, 18, 697.
3. Berens, A. R.; Hopfenberg, H. B. J. Membrane Sci. 1982, 10, 283.
4. Berens, A. R. J. Appl. Polym. Sci. 1989, 37, 901.

5. Berens, A. R.; Huvard, G. S. In <u>Supercritical Fluid Science and Technology;</u> Johnston, K. P., and Penninger, J. M. L., Eds.; ACS Symposium Series No. 406; American Chemical Society, Washington, DC, 1989; pp. 207-223.

6. Berens, A. R. <u>Polym. Prepr.</u> 1974, <u>15</u>, 197, 203.

7. Flory, P. J. <u>Principles of Polymer Chemistry</u>, Cornell University Press, Ithaca, NY, 1953; p. 514.

8. Michaels, A. S.; Veith, W. R.; Barrie, J. A. <u>J. Appl. Phys.</u> 1963, <u>34</u>, 1.

9. Berens, A. R. <u>Polym. Eng. Sci.</u> 1980, <u>20</u>, 95.

10. Connelly, R. W.; McCoy, N. R.; Koros, W. J.; Hopfenberg, H. B.; Stewart, M. E. <u>J. Appl. Polym. Sci.</u> 1987, <u>34</u>, 703.

11. Stewart, M. E.; Hopfenberg, H. B.; Koros, W. J.; McCoy, N. R. <u>J. Appl. Polym. Sci.</u> 1987, <u>34</u>, 721.

12. Chow, T. S. <u>Macromolecules</u> 1980, <u>13</u>. 362.

13. Hopfenberg, H. B.; Frisch, H. L. <u>Polym. Letters</u> 1969, <u>7</u>, 405.

14. van Amerongen, G. J. <u>Rubber Chem. & Tech.</u> 1964, <u>27</u>, 1065.

15. Berens, A. R. <u>Makromol. Chem., Macromol. Symp.</u> 1989, <u>29</u>, 95.

16. Fleming, G. K.; Koros, W. J. <u>Macromolecules</u> 1986, <u>19</u>, 2285.

17. Sefcik, M. D. <u>J. Polym. Sci, Polym. Phys.</u> 1986, <u>24</u>, 935.

18. Hirose, T.; Mizoguchi, K.; Kamiya, Y. <u>J. Polym. Sci, Polym. Phys.</u> 1986, <u>24</u>, 2107.

19. Wissinger, R. G.; Paulitis, M.E. <u>J. Polym. Sci., Polym. Phys.</u> 1987, <u>25</u>, 2497.

20. Wang, W. V.; Kramer, E. J.; Sachse, W. H. <u>J. Polym. Sci, Polym. Phys.</u> 1982, <u>20</u>, 1371.

21. Chiou, J. S.; Barlow, J. W.; Paul, D. R. <u>J. Appl. Polym Sci.</u> 1985, <u>30</u>, 2633.

22. Sefcik, M. D. <u>J. Polym. Sci, Polym. Phys.</u> 1986, <u>24</u>, 957.

23. Kamiya, Y.; Mizoguchi, K.; Naito, Y.; Hirose, T. <u>J. Polym. Sci, Polym. Phys.</u>, 1986, <u>24</u>, 535.

24. Hui, C.-Y.; Wu, K.-C.;. Lasky, R. C.; Kramer, E. J. <u>J. Appl. Phys.</u> 1987, <u>61</u>, 5137.

25. Berens, A. R.; Huvard, G. S.; Korsmeyer, R. W. <u>AIChE National Meeting</u>, Washington, DC, November 28 - December 2, 1988 (to be published).

26. Berens, A. R.; Huvard, G. S.; Korsmeyer, R. W., U. S. Patent 4 820 752, April 11, 1989.

RECEIVED December 5, 1989

# Chapter 5

# Structure of Amorphous Polyamides

## Effect on Oxygen Permeation Properties

Timothy D. Krizan, John C. Coburn, and Philip S. Blatz

Polymer Products Department, Experimental Station, E. I. du Pont de Nemours and Company, Wilmington, DE 19880

The structure of an amorphous polyamide prepared from hexamethylenediamine and isophthalic/terephthalic acids was modified in order to determine the effect of chemical structure on the oxygen permeation properties. The greatest increase in permeation was obtained by lengthening the aliphatic chain. Placement of substituents on the polymer chain also led to increased permeation. Reversal of the amide linkage direction had no effect on the permeation properties. Free volume calculations and dielectric relaxation studies indicate that free volume is probably the dominant factor in determining the permeation properties of these polymers.

Barrier resins, polymers which have relatively low rates of small molecule permeation, have revolutionized the packaging industry in recent years. For food packaging applications, it is specifically desirable to impede oxygen permeation. Each food type has its own particular packaging requirements, which leads to the use of many polymer classes at a variety of temperatures and relative humidities in these applications.

Figure 1 shows the effect of relative humidity (RH) upon the oxygen permeation values (OPV) of a few representative polymers. This data is reported in the units of cc-mil/(100 sq.in.-day-atm). For many polymers such as polyethylene, OPV is essentially unaffected by changes in RH.

For polymers such as nylon 6 or poly(vinyl alcohol) which contain hydrogen bonds, OPV increases dramatically with increasing RH. The increase in permeation is attributed to plasticization of the polymer structure by the water (1), which disrupt the polymer hydrogen bonds.

Selar PA, poly(hexamethylene isophthalamide/terephthalamide) or 6-I/T (the diamine components are listed first, then the diacid components), is an amorphous polyamide which is marketed by Du Pont. As shown in Figure 1, it has unique properties for a barrier resin in that the oxygen barrier properties actually

0097–6156/90/0423–0111$06.00/0

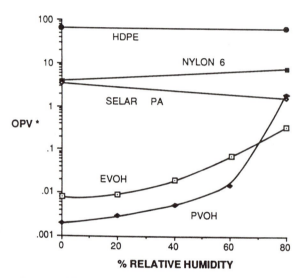

* cc-mil/100 sq. in./day/atm

Figure 1.   Effect of Relative Humidity on OPV of Selected Resins.

improve (OPV decreases) as RH increases.  This improvement is
opposite from what would be expected for a polymer which contains
a significant amount of hydrogen bonding.
    It was of interest to examine the permeation properties of
this class of aliphatic-aromatic polyamides.  More specifically,
it was desired to determine the effect of changes in chemical
structure upon OPV and upon the RH dependence of OPV.  It was also
desired to determine the factors which lead to these observed
structural effects.

Experimental

Figure 2 depicts the monomers used in this study with the abbrevi-
ations used for each monomer.  All polyamides made from aliphatic
diamines and aromatic diacid chlorides were prepared interfacially
(2).  Those made from aromatic diamines and aliphatic diacids were
prepared by a solution method using triphenylphosphite and
pyridine in N-methylpyrrolidinone (3).  All polymer samples tested
for oxygen permeation had a minimum inherent viscosity of 0.6 dL/g
in sulfuric acid.  Films of these polyamides were prepared by
pressing from the melt.  OPV data of these films were measured on
a Modern Controls Ox-Tran 10/50 at 30°C.  Densities were measured
in a carbon  tetrachloride/toluene density gradient tube.
Differential scanning calorimetry data (DSC) were obtained on a
Du Pont Instruments DSC at a heating rate of 20°C/minute.
Dielectric measurements were made on a Polymer Labs Dielectric
Thermal Analyzer.  Tests performed on wet samples were conducted
after immersing the films in water at 25°C for a minimum of 72
hours.  The samples were blotted dry prior to testing.

Results and Discussion

The effects of the following structural changes on the OPV of
aliphatic-aromatic polyamides were determined:  alteration of
the aliphatic chain length; reversal of the amide linkage;
substitution of groups upon either the amide nitrogen, the
aliphatic chain, or aromatic ring; replacement of the linear
aliphatic chain with a cycloaliphatic group; and use of other
aromatic ring systems.  The effect of placing other functional
groups in the chain was also studied, but those results will not
be discussed in this paper (Krizan, T. D., Du Pont, unpublished
data).
    In order to determine the effects of a given monomer on
polyamide permeation properties, data obtained from copolymers
where the monomer of interest was diluted by another diamine or
diacid were often used.  It is assumed that the OPV data for
copolymers are weighted averages of the OPV data for the
constituent homopolymers.  The use of copolymers was necessitated
by several reasons.  It was often too difficult to form the
homopolymer of interest with high enough molecular weight to allow
formation of cohesive films.  In other cases, the homopolymer was
semi-crystalline, which, as will be described in the next
paragraph, is undesirable for this study.
    In order to make meaningful comparisons of permeation proper-
ties, it was necessary to insure that no complicating factors were
present in the polymers under study.  The major precaution was to

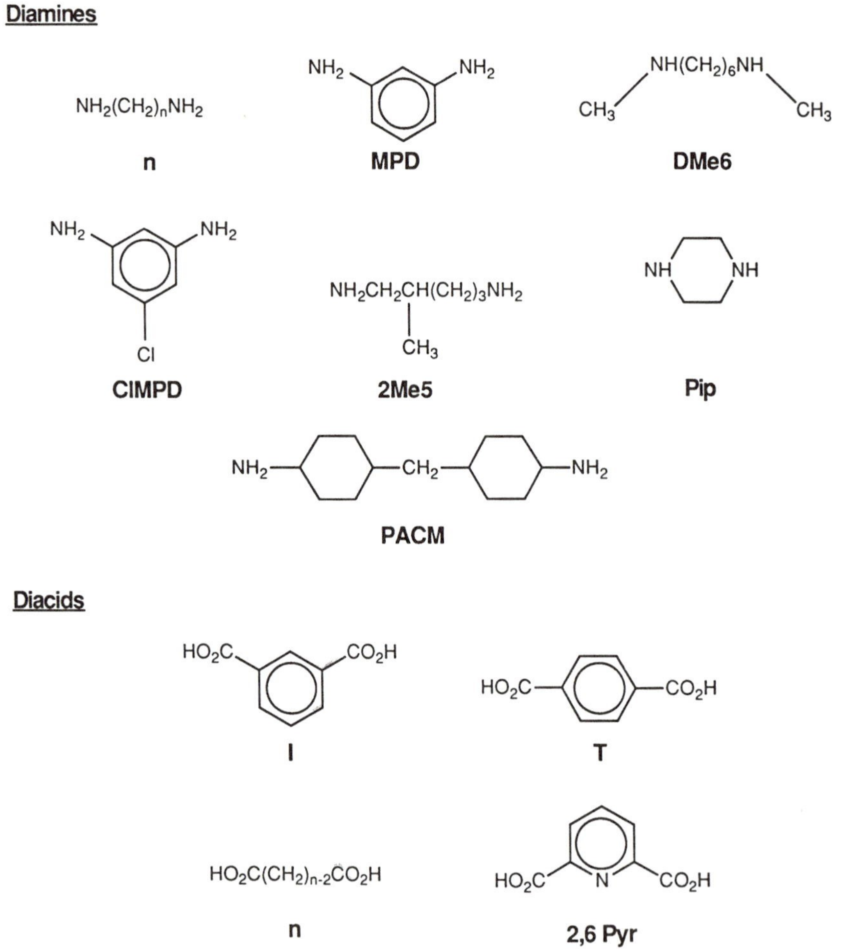

## Diamines

$NH_2(CH_2)_nNH_2$

**n**

$NH_2$ — — $NH_2$ (meta on benzene ring)

**MPD**

$NH(CH_2)_6NH$ with $CH_3$ and $CH_3$

**DMe6**

$NH_2$ — — $NH_2$ benzene ring with $Cl$

**ClMPD**

$NH_2CH_2CH(CH_2)_3NH_2$ with $CH_3$

**2Me5**

$NH$ — $NH$ (piperazine ring)

**Pip**

$NH_2$ — cyclohexane — $CH_2$ — cyclohexane — $NH_2$

**PACM**

## Diacids

$HO_2C$ — benzene (meta) — $CO_2H$

**I**

$HO_2C$ — benzene (para) — $CO_2H$

**T**

$HO_2C(CH_2)_{n-2}CO_2H$

**n**

$HO_2C$ — pyridine(N) — $CO_2H$

**2,6 Pyr**

Figure 2.  Monomers Used In This Study.

insure that the polymers had no observable crystallinity (by DSC).
In semi-crystalline polymers, it is generally assumed that
permeation occurs only through the amorphous regions while the
crystalline regions are essentially impervious (4). For this
study, the simplest way to prepare completely amorphous polymers
was to use meta-substituted benzenes as the sole aromatic
component in the aliphatic-aromatic polyamides. In most cases,
this approach was sufficient to eliminate any observable
crystallinity.

Effect of Chain Length. The initial part of this study consisted
of determining the effect of aliphatic chain length on the
permeation properties of the polyamides. A series of isophthalamides (n-I) was prepared where the aliphatic chain length was
systematically altered from 2 to 10 methylenes (5). Crystalline
melting points were observed by DSC for 2-I and 3-I, so permeation
data was measured only for 4-I through 10-I. The thermal,
density, and oxygen permeation data for this series are contained
in Table I.

Table I.  Data for n-I Polyamide Series

| Polymer | OPV* (dry) | OPV* (80% RH) | Density (g/mL) | Tg (°C) | Wet Tg (°C) | 1/SFV (g/mL) |
|---------|------------|---------------|----------------|---------|-------------|--------------|
| 4-I | 0.4 | 0.5 | 1.25 | 141 | --- | 11.91 |
| 5-I | 1.2 | 0.9 | 1.23 | 129 | 46 | 11.12 |
| 6-I | 1.9 | 1.2 | 1.19 | 123 | 42 | 10.51 |
| 7-I | 3.9 | 2.9 | 1.18 | 113 | 41 | 10.02 |
| 8-I | 7.0 | 4.1 | 1.15 | 114 | 46 | 9.62 |
| 9-I | 11.3 | 7.8 | 1.13 | 105 | 53 | 9.28 |
| 10-I | 12.8 | 11.1 | 1.11 | 97 | 53 | 8.99 |

*cc-mil/(100 sq.in.-day-atm)

It is apparent from the OPV data in Table I that with each
additional methylene group in the polymer backbone, OPV at both
0% and 80% RH significantly increases. This trend can also be
discerned in a series of copolyesters in which 8-16% of the
terephthalic acid portion of poly(ethylene terephthalate) (PET)
is replaced by aliphatic diacids of various lengths (6).
Another significant feature of the OPV data in Table I is the
variance in the effect of RH on OPV. The OPV at 80% RH is greater
than the OPV at 0% RH only when n=4. When n=5, the dry OPV is
slightly greater than the OPV at 80% RH, but it becomes
significantly greater than the OPV at 80% RH as n increases. The
effect of RH upon OPV of the majority of these isophthalamides is,
therefore, similar to that observed for 6-I/T. As n increases,
the isophthalamide structure will approach linear polyethylene,
and the effect of RH upon OPV should become negligible. The amide
density of the isophthalamides examined here is still too high,
however, for confirmation of this prediction.

Factors Affecting Polyamide OPV. In order to determine the
polymer properties which affect polyamide permeation properties,
the n-I series was studied in more detail. Figure 3 shows the

effect of chain length upon the glass transition temperature (Tg) for this series. In the dry state, increasing the aliphatic chain length leads to lower Tg (also observed in other polyamide series (7,8)). The polymers, however, are all glassy at the permeation test temperature of 30°C. It is, therefore, impossible to attribute the observed dependence of OPV upon n to a transition of the polyamide from a glass to a rubber. As shown in Table I, the wet Tg of the polymer is still above the test temperature when n is greater than or equal to 5. This means that the 80% RH OPV data is obtained from polymers which are still in the glassy state. It was not possible to observe a wet Tg for 4-I in the DSC, which may indicate that it dropped below room temperature. If this is the case, the 80% RH OPV of the 4-I might be expected to be higher than the dry OPV due to an increase in rubbery character at high RH.

The free volume in a polymer is considered to be a very important parameter affecting the amount of gas permeation. Unfortunately, this is a very difficult parameter to quantify. One approach that has been used is to compare the densities of two polymers and infer that the denser polymer has a lesser amount of free volume and thus lower gas permeation rates (9,10). This approach, however, has been abused in that it has been used to compare the free volumes of structurally dissimilar polymers.

Since the polymers in this series are homologues, there is some justification for using a density comparison to determine relative amounts of free volume. Figure 4 shows that as n increases, the polyisophthalamide density decreases. Similar trends were observed by Ridgway in other polyamide series (11). This trend indicates that free volume is increasing with n and that permeation would be expected to increase, which is what in fact is observed.

A more direct method to determine the free volume differences in the series is to calculate them using the method of Lee (12), which uses a group contribution approach. Table I contains the values for 1/SFV (specific free volume) which were calculated using the additive molar volumes provided by Van Krevelen (13). Figure 5 shows a plot of the log of the dry OPV for the n-I series against 1/SFV. A linear relationship, which is what would be expected if free volume is a determining factor in oxygen permeation, is obtained in this plot.

Subglass motions are postulated to aid in the transport of gases through glassy polymers (14-16). These transitions in the n-I series were examined using dielectric spectroscopy. The relaxation data will be reported in greater detail elsewhere (Coburn, J. C.; Krizan, T. D., Du Pont, unpublished data). A plot of the dielectric loss of 6-I/T is provided in Figure 6. The magnitude of the large subglass transition (beta) is much less than that of the glass transition due to the hydrogen bonding which effectively reduces local segmental motion. This behavior is not observed in other thermoplastic polymers such as PET where the magnitude of the subglass transition is comparable to that of the glass transition (17).

Figure 7 shows a plot of the temperature of the beta transition at 10 kHz against n. This transition occurs close to room temperature in 4-I and shifts to lower temperatures as the number of methylene groups increase. This means that the amount

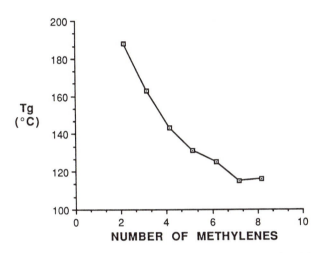

Figure 3. Effect of Chain Length Upon Isophthalamide Tg.

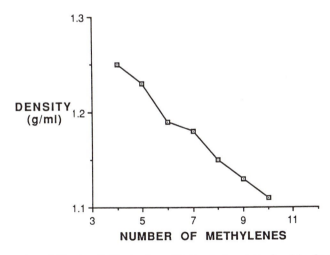

Figure 4. Effect of Chain Length Upon Isophthalamide Density.

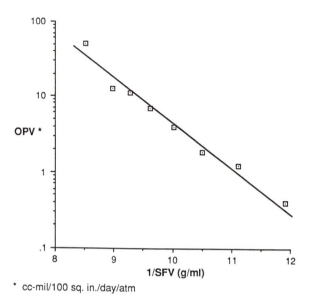

* cc-mil/100 sq. in./day/atm

Figure 5. Correlation of Isophthalamide OPV to Specific Free Volume.

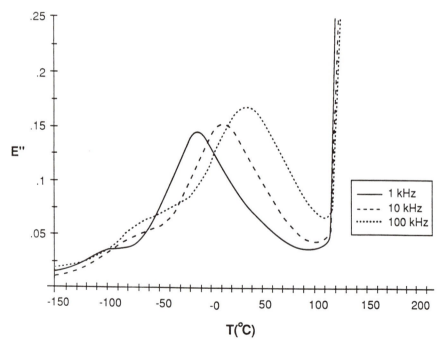

Figure 6. Dielectric Loss of 6-I/T Polyamide.

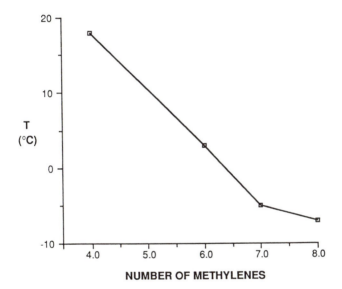

Figure 7. Effect of Chain Length Upon Beta Transition
Temperature at 10 kHz.

of segmental motion occurring at room temperature increases with
increasing methylene content. The gamma transition, which occurs
near $-100°C$, is observed when there are six or more methylenes in
the aliphatic chain. The magnitude of this transition increases
with increasing methylene content. It is attributed to motion
involving the methylene groups and occurs in the same temperature
range as the gamma process in aliphatic polyamides (18) and
polyethylene (19). A minimum of four consecutive methylene groups
is usually required to observe this transition (20).

The trends observed in both the beta transition (increased
motion at room temperature with increasing methylene content) and
the gamma transition (increased magnitude of the transition with
increasing methylene content) are consistent with the observed
effects of n on OPV. It is likely, however, that the amount of
free volume in the polymer is the dominant factor in determining
the OPV for the n-I series (although free volume and subglass
motion are not completely independent properties). This
hypothesis is based on the linearity of the log OPV against 1/SFV
line plot in Figure 5, and the magnitude of the subglass
transitions relative to the glass transitions. A strong
dependence of polyester OPV on the amount of subglass motion has
been reported (15), but in that class of polymers, the magnitude
of the beta transition is comparable to that of the glass
transition. It is therefore reasonable to expect that subglass
motion will have greater importance in determining the permeation
properties of that series than for the amorphous polyamide series.

Other Modifications of Polymer Structure. In order to more fully
determine the effects of structural change upon polyamide OPV,
several other modifications of the basic aliphatic-aromatic
backbone were performed. One simple modification is to reverse
the direction of the amide linkage. The data in Table II indicate
that for at least the shorter chain polyisophthalamides, amide
reversal has no measurable effect upon the OPV at either 0% or 80%
RH. It is surprising to note that the amide direction does not
seem to affect either density or Tg in addition to OPV. Furthermore, Morgan and Kwolek reported that the amide direction had
little, if any, effect upon the melting points on aliphatic
terephthalamides and their analogs (21). It is due to the
indifference of polyamide properties to amide direction that data
for MPD-14 is included in Figures 3 and 5.

Table II.  Effect of Amide Reversal on Polyamide Properties

| Polymer | OPV* (dry) | OPV* (80% RH) | Density (g/mL) | Tg (°C) |
|---------|-----------|---------------|----------------|---------|
| 4-I     | 0.4       | 0.5           | 1.25           | 141     |
| MPD-6   | 0.4       | 0.5           | 1.26           | 144     |
| 6-I     | 1.9       | 1.2           | 1.19           | 123     |
| MPD-8   | 2.1       | 1.2           | 1.19           | 129     |

*cc-mil/(100 sq.in.-day-atm)

Incorporation of N,N'-dialkyldiamines into the polymer chain would disrupt the normal hydrogen bonding since the repeat units would have no available amide hydrogens. Comparison of the OPV data in Table III for DMe6/6-I (25/75) to that for unsubstituted 6-I indicates that the dry OPV is increased much more dramatically than the 80% RH OPV. The reason for this observation may be due to the two different effects of the N-methyl groups. Not only do the methyl groups obliterate 25% of the hydrogen bonds relative to 6-I, but they also increase the free volume by physically increasing the interchain distance. At 0% RH, both effects are operative and the combination leads to a large increase in permeation. At 80% RH, the hydrogen bonding disruption imparted by the methyl groups is inconsequential as there is more than enough water present to provide the same disruption. The only observed difference in OPV at 80% RH is due to the steric effects of the methyl groups, which is slight compared to the effects of hydrogen bond disruption. The chain alkylation data described below confirm this conclusion.

Table III. Oxygen Permeation Data for Modified Polyamides

| Polymer | OPV* (dry) | OPV* (80% RH) |
|---|---|---|
| DMe6/6-I  (25/75) | 3.4 | 1.5 |
| 2Me5-I | 1.4 | 0.9 |
| 6-I/T  (from nylon salt) | 3.5 | 1.8 |
| MPD/5ClMPD-8  (50/50) | 5.2 | 2.3 |
| Pip/6-I  (20/80) | 2.8 | 1.8 |
| PACM/6-I/T  (50/50-70/30) | 5.6 | 3.7 |
| 6-2,6Pyr/I  (50/50) | 3.3 | 2.6 |

*cc-mil/(100 sq.in./day/atm)

Table III contains data for both 5-I and the chain alkylated 2Me5-I. In this case, the methyl group leads to a slight increase in OPV at 0% RH and a negligible difference at 80% RH. Additional chain alkyl groups lead to even greater increases in permeation. For example, Trogamid T, an amorphous polyamide made by Dynamit Nobel from two trimethylhexamethylenediamine isomers and terephthalic acid, has an OPV of 5.1 cc-mil/(100 sq.in.-day-atm) at 80% RH. This OPV is nearly three times greater than the OPV of unsubstituted 6-I/T (Table III) made in the same manner. (The difference in OPV between 6-I and 6-I/T reported here is not due to differences in the I/T ratio as in other polymer classes (see, for example: Schmidhauser, J. C.; Longley, K. L., this volume). It is instead related to differences in synthetic method (more branching in the polymer prepared from the nylon salt) or processing (more unrelaxed free volume in films which are cast through an extruder die and onto a quench roll)).

The effect of placing substituents on the aromatic ring was studied. Several polymers containing substituted meta-phenylene-diamines and isophthalic acids were prepared. Great difficulty was generally encountered in preparing polymers in high molecular weight from these monomers, so 6-I or MPD-8 copolymers containing 10-50 mol% of the monomer of interest were prepared. Although rings substituted in the 5-position with sulfonate, alkyl, nitro, and carboxamide groups were examined, the MPD/5ClMPD-8 example in Table III serves as a representative due to the relatively high level of monomer incorporation. As in the previous examples of chain substitution, substitution of the rings, at least in the 5-position, leads to increased OPV. This is probably due to increased interchain distance which leads to increased free volume.

It is difficult to evaluate the effect of incorporating cycloaliphatic groups due to the presence of other complicating factors in commonly available monomers of this class. The Pip/6-I polymer listed in Table III will serve as an example. In this case, the piperazine contains only two carbons between amide nitrogens. As demonstrated earlier, this would tend to lower OPV. On the other hand, the lack of hydrogen bonding imparted by the piperazine moieties would be expected to increase OPV. In fact, the presence of the piperazine increases permeation.

Another example of cycloaliphatic group incorporation in Table III is the data for PACM/6-I/T (made from the nylon salt) (Vassallo, D. A., DuPont, unpublished results). In this case, hydrogen bonding is possible, but the distance between amide nitrogens has increased. It is difficult from these examples to delineate the effect of the aliphatic rings on polyamide OPV, although it is likely that their presence increases interchain distance much as a chain substituent would.

A final modification of the basic polymer structure examined was the substitution of a heterocycle for the benzene ring. Table III contains data for 6-2,6Pyr/I. The pyridine ring increases permeation relative to isophthalic acid, which may be due to increased oxygen solubility imparted by the pyridine nitrogen.

Effect of RH on OPV. It was also of interest to determine the factors which lead to a decrease in OPV with increasing RH in amorphous polyamides. As noted above, this behavior is unique for commercial oxygen barrier materials. This phenomena, however, appears to be general for amorphous polyamides, so the discussion which follows will assume that the OPV decrease is caused by the same effect in all cases.

Although the OPV decrease with increasing RH is unique for barrier materials, decreased gas transmission rates in membrane materials in the presence of moisture have been previously noted. For example, workers at DuPont found significant reductions in the permeability of hydrogen and methane through polyimide films in the presence of water vapor (22). Koros and coworkers also observed reductions in the permeability of carbon dioxide through Kapton polyimide films in the presence of moisture (23). Both groups proposed that the gas permeability decrease was due to competition between the gases and the water vapor for the excess free volume in the polymer matrix. Because of this competition, the pathways available for diffusion are effectively reduced. The data obtained so far for the amorphous polyamide series indicates that the same effect is operative.

For the above explanation to be valid, there must be unrelaxed free volume in the polymer matrix after exposure to moisture. This means that the temperature at which the permeation is tested must be below the polymer Tg at high RH. As shown in

Table I, the wet Tg for every polyamide tested with more than five carbons in the aliphatic chain meets this requirement (OPV measured at 30°C). The Tgs of wholly aliphatic polyamides such as nylon 66 drop below this temperature at high RH (18), so it is not surprising that these polyamides exhibit increases in permeation with increasing RH.

Figure 8 shows a detailed dependence of the OPV of 6–I/T on RH. The majority of the permeation decrease occurs at low RH. This drop is clearly not due to the hydrogen bond disruption which accounts for the drop in polyamide Tg with increasing RH (24). Starkweather found that the Tg drop of the structurally similar 6–I nylon is much more linear with increasing RH (Starkweather, H. W., DuPont, unpublished data). The dissimilarity in behavior between Tg drop and OPV drop with increasing RH is indicative that the hydrogen bond disruption induced by water does not play a dominant role in the OPV drop observed. As discussed earlier, the fact that DMe6/6–I has a higher dry OPV than 6–I also indicates that hydrogen bond disruption in and of itself leads to an increase, and not a decrease, in OPV.

A final piece of evidence deals with the effect of moisture upon polyamide density. Sorption of water in excess free volume should lead to a increase in density while sorption with concurrent swelling should result in the additivity of volumes (25,26). In the case of 6–I/T, the density of a dry film sample is 1.178 g/mL while the density of a sample after immersion in water is 1.189 g/mL. A likely explanation for the observed increase is filling of the excess free volume of 6–I/T by water, which must dominate the effects of the concurrent plasticization by the water.

Conclusions

Through systematic modification of the polymer backbone, the effects of chemical structure upon the oxygen permeation properties of aliphatic–aromatic amorphous polyamides were determined. In this class of polymers, the greatest effects were obtained by alteration of the chain length and disruption of the amide hydrogen bonding by N–alkylation. It is remarkable that reversal of the amide linkage has no effect whatsoever on the permeation properties of the examples studied.

In an attempt to determine the factors which determine the barrier properties of this polyamide series, it was found that the permeation results were consistent with both the relative levels of subglass motion as measured by dielectric spectroscopy and the relative levels of free volume as calculated using a group contribution approach. It appears that free volume is the dominant effect in determining the OPV due to the relatively small magnitude of the subglass transitions as compared to the glass transition.

The substantial decline in OPV as relative humidity increases, which is unique for an oxygen barrier resin, was studied. It was concluded that this decline is due to water occupying the excess free volume through which the oxygen would otherwise travel.

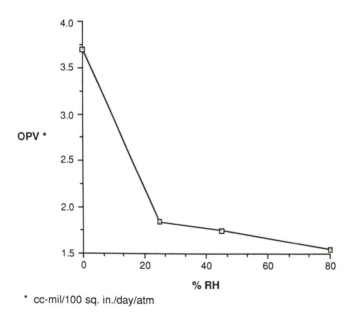

* cc-mil/100 sq. in./day/atm

Figure 8.  Effect of Relative Humidity on OPV of 6-I/T Polyamide.

Acknowledgments

We would like to thank DuPont Polymer Products Department for supporting this work and allowing its publication. We thank Gerald Horack and Robert Tomczak for performing the OPV tests and Michael Panco for obtaining the dielectric relaxation data. D. A. Vassallo is acknowledged for many helpful discussions in the early days of this program.

Literature Cited

1.  Long, F. A.; Thompson, L. J.  J. Polym. Sci. 1954, 14, 321.
2.  Shashoua, V. E.; Eareckson, W. M.  J. Polym. Sci. 1959, 40, 343.
3.  Yamazaki, N.; Higashi, F.; Kawabata, J.  J. Polym. Sci. Polym. Chem. Ed. 1974, 12, 2149.
4.  Michael, A. S.; Parker, R. B.  J. Polym. Sci. 1959, 41, 53.
5.  Gorton, B. S.  J. Appl. Polym. Sci. 1965, 9, 3753.
6.  Weemes, D. A.; Seymour, R. W.; Wicker, T. H.  U.S. Patent 4 401 805, 1983.
7.  Saotome, K.; Komoto, H.  J. Polym. Sci. A-1 1966, 4, 1463.
8.  Ridgway, J. S.  J. Polym. Sci. A-1 1970, 8, 3089.
9.  Pilato, L. A.; Litz, L. M.; Hargitay, R. C.; Osborne, A. G.; Farnham, A.; Kawakami, J. H.; Fritze, P. E.; McGrath, J. E. Polym. Prepr. 1975, 16(2), 42.
10. Nakagawa, T.; Fujiwara, Y.; Minoura, N.  J. Membr. Sci. 1984, 18, 111.
11. Ridgway, J. S.  J. Polym. Sci. Polym. Chem. Ed. 1974, 12, 2005.
12. Lee, W. M.  Polym. Eng. Sci. 1980, 20, 65.
13. Van Krevelen, D. W.  Properties of Polymers; Elsevier: New York, 1972; pp 574–581.
14. Chern, R. T.; Koros, W. J.; Hopfenberg, H. B.; Stannett, V. T. In Materials Science of Synthetic Membranes; Lloyd, D. R., Ed.; ACS Symposium Series No. 269; American Chemical Society: Washington, DC, 1984; pp 25–46.
15. Light, R. R.; Seymour, R. W.  Polym. Eng. Sci. 1982, 22, 857.
16. O'Brien, K. C.; Koros, W. J.; Husk, G. R.  J. Membr. Sci. 1988, 35, 217.
17. Coburn, J. C.; Boyd, R. H.  Macromolecules 1986, 19, 2238.
18. Starkweather, H. W.  In Nylon Plastics; Kohan, M. I., Ed.; Wiley: New York, 1973; pp 307–325.
19. McCrum, B.; Read, B.; Williams, G.  Anelastic and Dielectric Effects in Polymeric Solids; Wiley: New York, 1967; p 180.
20. Willbourn, A. H.  Trans. Faraday Soc. 1958, 54, 717.
21. Morgan, P. W.; Kwolek, S. L.  Macromolecules 1975, 8, 104.
22. Pye, D. G.; Hoehn, H. H.; Panar, M.  J. Appl. Polym. Sci. 1976, 20, 287.
23. Chern, R. T.; Koros, W. J.; Sanders, E. S.; Yui, R.  J. Membr. Sci. 1983, 15, 157.
24. Reimschuessel, H. K.  J. Polym. Sci. Polym. Chem. Ed. 1978, 16, 1229.
25. Bueche, F.  J. Polym. Sci. 1954, 14, 414.
26. Turner, D. T.  Polymer 1982, 23, 197.

RECEIVED October 17, 1989

Chapter 6

# Transport of Penetrant Molecules Through Copolymers of Vinylidene Chloride and Vinyl Chloride

J. Bicerano, A. F. Burmester, P. T. DeLassus, and R. A. Wessling

Materials Science and Development Laboratory, Central Research, The Dow Chemical Company, 1702 Building, Midland, MI 48674

The initial results of a systematic study of the transport of penetrant molecules through copolmers of vinylidene chloride (VDC) and vinyl chloride (VC) will be summarized. A synergistic combination of computational and experimental techniques is being utilized. The computational techniques include (i) the free volume theory of Vrentas and Duda, to present an overall "global" physical perspective; (ii) the statistical mechanical models of Pace and Datyner, to provide an "intermediate" perspective on the scale of parameters describing short chain segments and their interactions; and (iii) study of local unoccupied volume distributions and the dynamics of polymer and penetrant motions, to build understanding on a true molecular level, i.e., gain a "local" perspective. The experimental techniques consist of characterizations of several VDC/VC copolymers, and measurements of the diffusion coefficients of several penetrants through thin films made of each one, at several temperatures.

The study of the transport of penetrant molecules through polymers [1,2] is important in many areas of technology. There are two types of industrially important polymeric systems for which such transport phenomena are crucial:

1. Barrier plastics, used in food and beverage packaging applications, which have high resistance to permeation of gas and flavor-aroma molecules.

2. Separation membranes, used for the purification of mixtures of gases or liquids flowing through them. These materials have (i) high selectivities, and (ii) permeabilities that are sufficiently high to allow reasonable recovery of a product of the desired purity.

Barrier plastics will be used as examples in this paper. Barrier and selectivity are two sides of the same coin. Both properties are determined by the same types of transport phenomena. [3] Similar techniques can, therefore, also be applied to study separation membranes. The same approximations are valid if the mixture flowing through the membrane is sufficiently dilute that (i) it does not significantly affect the structure and properties of the membrane, and (ii) the components of the mixture can be treated as independent penetrants. The same general approaches can also be applied to concentrated mixtures, but only provided that certain simplifying approximations are not made.

A promising combination of techniques to utilize in studying the transport of penetrant molecules in polymers is: (i) the free volume theory of Vrentas and Duda (V&D), [4-11] to provide an overall "global" physical perspective; (ii) statistical mechanical models, the best-developed of which are the models of Pace and Datyner (P&D), [12-17] to provide an "intermediate" perspective on the scale of parameters describing short chain segments and their interactions; and (iii) study of local unoccupied volume distributions and the dynamics of polymer and penetrant motions, to build understanding on a true molecular level, i.e., gain a "local" perspective. Each perspective can contribute directly to the construction of a unified physical model, as shown schematically in Figure 1. Conversely, as the model is improved by input from any one of these perspectives, the modifications might point out the revisions necessary in the interpretation of phenomena on the scale of the other two perspectives. In other words, the physical model forms the hub of a synergistic interaction between the three perspectives, which constitute a complete and systematic approach to transport phenomena.

Figure 2 shows how glass transition temperatures ($T_g$) obtained by dynamic mechanical spectroscopy (DMS), percent crystallinities obtained by wide angle x-ray scattering (WAXS) or differential scanning calorimetry (DSC), experimental diffusion coefficients, and information on tortuosity obtained by studies of morphology, can be useful in applying both the theory of V&D and the model of P&D. The Williams-Landel-Ferry (WLF) parameters [18] $c_1g$ and $c_2g$, which can be determined by DMS, are needed as additional input for the theory of V&D. Densities and thermal expansion coefficients are needed as additional input for the model of P&D.

Figure 3 shows how a sequence of semi-empirical quantum mechanical (QM), force field (FF) and molecular dynamics (MD) calculations can provide a detailed local perspective of the transport phenomena. The information that can be obtained includes the dynamics of the polymer in the absence of the penetrant, the local unoccupied volume

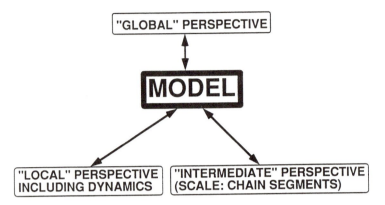

**Figure 1.** Schematic illustration of how the results of three different types of calculations, each one providing a perspective at a different scale, can be combined synergistically, to construct a unified physical model for the transport of penetrant molecules in plastics.

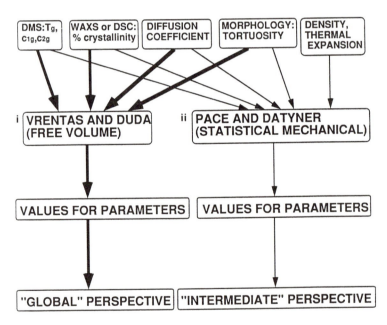

**Figure 2.** Flow chart on the use of (i) the free volume theory of Vrentas and Duda, to obtain a "global" perspective; and (ii) the statistical mechanical model of Pace and Datyner, to obtain an "intermediate" perspective on the scale of parameters describing polymer chain segments.

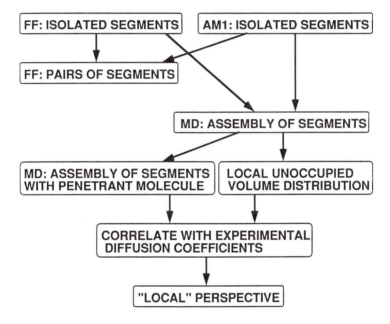

**Figure 3.** Flow chart on the use of force field (FF) and molecular dynamics (MD) calculations, to obtain a "local" perspective including the dynamics of the polymer and the penetrant.

distributions, and the dynamics (i.e., trajectory) of the penetrant as it diffuses through the polymer.

Sample copolymers of vinylidene chloride (VDC) [19] and vinyl chloride (VC) are being used in the experimental part of this study, which complements the calculations being performed on similar systems by a variety of techniques.

There are two major simplifying features of VDC/VC copolymers:

1. They are excellent barrier materials. Penetrants are likely to be present at concentrations of much less than 2%. Simpler forms of the V&D and P&D formalisms, valid in the low-concentration limit, can therefore be used.

2. They are above $T_g$ at room temperature, unless the fraction of VC is very high.

The major complicating feature of these copolymers is their semicrystallinity with a possibly substantial crystalline fraction, and all the concomitant structural, morphological, and rheological complexities:

1. The presence of crystallites, which are assumed to be impermeable, makes a fraction of the polymeric volume unavailable for transport.

2. The crystallites cause tortuosity in the diffusion pathway, i.e., they behave as randomly scattered obstacles which cause the diffusion pathway to become longer than it would have been otherwise. None of the existing simple general expressions for tortuosity form factors is always reliable. [20] There is no clearcut, foolproof way to extract tortuosity information from morphological studies either.

3. The amorphous transitions are attenuated by the presence of crystallites, [21] which can cause very significant changes in the rheological behavior. There are, consequently, many serious technical difficulties in the determination of reliable WLF parameters for the amorphous regions of semicrystalline polymers.

Starting with a maximally amorphous VDC/VC sample copolymer (i.e., one of fairly high VC content) can result in major simplifications. Several different compositions were therefore synthesized, to allow the complexities introduced by semicrystallinity to become only gradually more important.

## Comparisons Between "Global" Free Volume and Statistical Mechanical Models

The calculations to be summarized below will be described in much greater detail in a future publication. [22] Global parameters, related to the total free volume, are defined in the V&D theory. [4-11] This theory can be used to correlate experimental diffusion data by selecting optimum values for its adjustable parameters. Reasonable correlations have been obtained for non-barrier polymers such as polystyrene (PS) and poly(vinyl acetate) (PVAc). Attempts to use this theory in a fully predictive mode have been less successful.

By contrast, all parameters (whether adjustable or uniquely determined) are expressed in terms of the molecular level structural features in statistical mechanical theories, such as the diffusion model of P&D. [12-17] These parameters are defined by the local structure of, and interactions in, the polymer-penetrant system. Most of them describe features at the scale of polymer chain segments. The trade-off for this apparently more molecular level perspective is that physical assumptions whose general applicability and validity are questionable have to be made in defining the model.

In the present section, general comparisons will be presented between these two theories. Both theories will be utilized in a correlative mode. The parameters for barrier polymers will be defined for idealized completely amorphous poly(vinylidene chloride) (PVDC) and a VDC/VC copolymer. It will be shown that physically significant qualitatitive differences exist between the results calculated by the two theories.

The diffusion coefficient D, at the limit of the diffusion of a trace amount of penetrant in a completely amorphous polymer, at temperatures (T) above the $T_g$ of the polymer, is given by the following expression in the theory of V&D:

$$D(V\&D) = D_{01}exp[-2.303c_1^g c_2^g \xi/(c_2^g+T-T_g)]. \tag{1}$$

In the preliminary calculations on idealized polymers, where the purpose is to obtain "ball park" estimates of major trends, Ferry's "universal" values ($c_1^g$=17.44 and $c_2^g$=51.6) [18] will be used for the WLF constants. The preexponential factor $D_{01}$ and $\xi$ are the only two adjustable parameters in Equation (1). $\xi$ is intended to denote the ratio of the critical molar volumes of two "jumping units": penetrant/polymer. In the formalism of V&D, the size of the jumping unit of the polymer is treated as a constant. Consequently, if $\xi$ has been determined for a pair of penetrants in a given polymer, the ratio of the $\xi$'s determined for the same pair of penetrants in any other polymer should, in principle, be equal to their ratio in the first polymer.

$\xi < 0.8$ for penetrants as large as ethylbenzene in PS. In barrier polymers, the chains are held together very tightly. The free volume elements (i.e., holes) are therefore likely to be much smaller than in PS. The critical molar volume of a "jumping unit" of the polymer, needed to fill an average hole, might therefore also be much smaller than in PS. $\xi$ could then become >1, even for small penetrants.

An apparent activation energy $(E_{app})$ can be defined by the following equation:

$$E_{app} = RT^2[\partial \ln(D)/\partial T], \tag{2}$$

where ln denotes the natural logarithm function and $\partial$ denotes a partial derivative.

The following expression is obtained for $E_{app}$ in the theory of V&D, again at the limit of the diffusion of a trace amount of penetrant in a completely amorphous polymer, with $T > T_g$, by substituting Equation (1) into (2):

$$E_{app}(V\&D) = 2.303 c_1 g c_2 g \xi RT^2 / [(c_2 g + T - T_g)^2]. \tag{3}$$

$E_{app}(V\&D)$ is not a constant, but a monotonically decreasing function of T, which asymptotically approaches the following limit at high temperatures:

$$E_{app}(V\&D, T \to \infty) = 2.303 c_1 g c_2 g \xi R. \tag{4}$$

The basic form of the model of P&D used in this work, which will be described elsewhere in detail, [22] treats the diffusion of small amounts of "simple" spherical penetrants, such as gas molecules, in "smooth-chained" polymers, such as poly(ethylene terephthalate) (PET) and cis-polyisoprene (natural rubber). [12] Whenever necessary, generalized equations are being used, for example for simple nonspherical penetrants [13] and for polymers which possess closely spaced, bulky side groups such as poly(vinyl chloride) (PVC). [14]

There are two possible ways to define $E_{app}$(P&D). The first way is to define $E_{app}$(P&D1) in terms of a "chain separation energy" $\Delta E$, as was done by P&D. [12] The second way is to define $E_{app}$(P&D2) by analogy to Equation (2), which was used for the theory of V&D. [22]

The following strategy was adopted to determine the general behavior of D and $E_{app}$ as functions of T and penetrant size: [22] (i) obtain D for VC mole fraction (x) values of 0.0, 0.5 and 1.0, via the model of P&D, with reasonable values for the parameters, either taken from or extrapolated from experimental data; (ii) derive a correlation between the penetrant diameter d used by P&D, and the adjustable parameter $\xi$ used by V&D as an indicator of the size of the penetrant; (iii) substitute this correlation into Equation (1) to express D(V&D) as a function of d; (iv) fit the expression for D(d) in the theory of V&D to the expression for D(d) in the model of P&D, to obtain values for $D_{01}$ and $\xi$; and (v) calculate $E_{app}$(V&D), $E_{app}$(P&D1) and $E_{app}$(P&D2) as functions of d and T.

It is found [22] that all of the derived quantities for amorphous PVDC, which are related to parameters describing chain segments of the polymer, fall into the same ranges as the values of the same parameters in non-barrier polymers. None of these parameters has a value so significantly different from its values in non-barrier polymers as to be, by itself, a plausible cause for the much better barrier performance of PVDC. It is their particular combination which results in a low D for PVDC. The only important parameter whose value for PVDC is very different from its values for the non-barrier polymers is not a molecular level parameter, but a "global" parameter, namely the density (used as an input parameter), on which most of the derived parameters [12] depend. The density of PVDC [19] is about 30% higher than the density of any of the homopolymers studied by P&D. [13,14] On the other hand, brominated polycarbonate has a higher density than PVDC, but a permeability similar to that of PS. [23] Therefore, it is not the density itself, but the packing efficiency (i.e., the fraction of the total volume of the polymer occupied by the van der Waals volumes of the atoms, as determined by the shape of the chain contour and the mobility of the chain segments) that makes the main difference. The importance of the density for the set of polymers examined by the model of P&D, both by P&D themselves and in the present work, is mainly its role as an indicator of the packing efficiency. The importance of the packing efficiency in determining the barrier performance of a polymer has also been demonstrated by the work of Lee on the prediction of the gas permeabilities of polymers from specific free volume considerations. [24]

The diffusion coefficients calculated at 298.15K are depicted as functions of the penetrant diameter d in Figure 4. The curve for PVC with d replaced by (d-0.12) corresponds to the results of P&D [14] for PVC. The curves in Figure 4 manifest most of the expected trends: (i) the D's are in the expected ranges, (ii) D increases with increasing fraction of VC, and (iii) the differences between the D's calculated at different compositions increase with increasing d. A trend which deviates from the expectations is the slight concavity of the curves. There is no evidence in the systematic experimental studies of the d dependence of D [25] for the concavity of D as a function of d. A concave D curve implies an unphysical gradual reduction in the rate at which D approaches zero with increasing d. As d becomes very large, D should probably become a slightly convex function of d because it should rapidly become more difficult to find any diffusion pathway, resulting in an acceleration of the rate at which D approaches zero with increasing d.

The penetrant diameter d used in the model of P&D [13] appears to have a relatively unambiguous physical significance. It is an effective hard sphere diameter estimated from viscosity data, whenever necessary modified to correct for the nonsphericity of the shape of the penetrant. Therefore, d has the dimensions of length. On the other hand, the penetrant size parameter $\xi$ used by V&D is somewhat *ad hoc*. It is described as the ratio of the molar volume of the "jumping unit" of the penetrant to the molar volume of the "jumping unit" of the polymer, where the latter is a rather ill-defined quantity. Since the molar volume of the jumping unit of the polymer is assumed to be constant, a set of $\xi$ values for a series of penetrants in the same polymer should ideally scale as $d^3$ if the jumping unit of the penetrant is being described by its molar volume. In practice, $\xi$ is most often used as an adjustable parameter. Its values are estimated by a variety of curve fitting procedures.

Vrentas, Liu and Duda [6] have provided estimates for $\xi$ for many penetrants in PS; and stated that the values for $H_2$, $CH_4$ and $C_2H_4$ are questionable. The application of correction factors to the values provided for $H_2$, $CH_4$ and $C_2H_4$ might result in more reliable values. [22] P&D [13] have provided values of d for many penetrants, nine of which are the same as nine of the penetrants for which $\xi$ values have been provided [6] in PS. The d values provided by P&D, [13] and the $\xi$ values provided by Vrentas, Liu and Duda [6] (where the values for $H_2$, $CH_4$ and $C_2H_4$ have been corrected) are shown in Figure 5. It is seen that the correlation between d and $\xi$ is poor, even though there are only nine data points to correlate. The best power law fit is:

$$\xi(\text{in PS}) \approx 0.63d^{1.24} \tag{5}$$

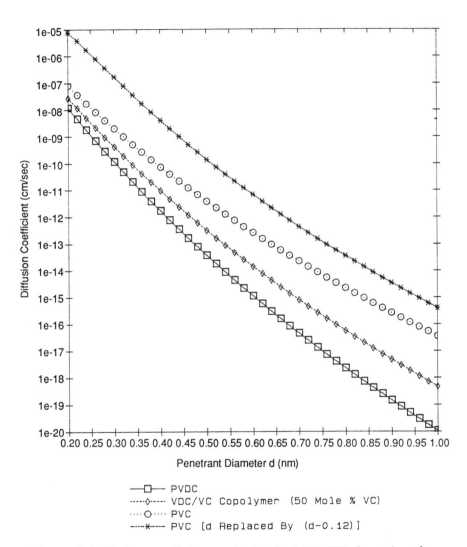

**Figure 4.** Diffusion coefficients calculated at 298.15K, by using the statistical mechanical diffusion model of Pace and Datyner for idealized completely amorphous polymers, as functions of the penetrant diameter d.

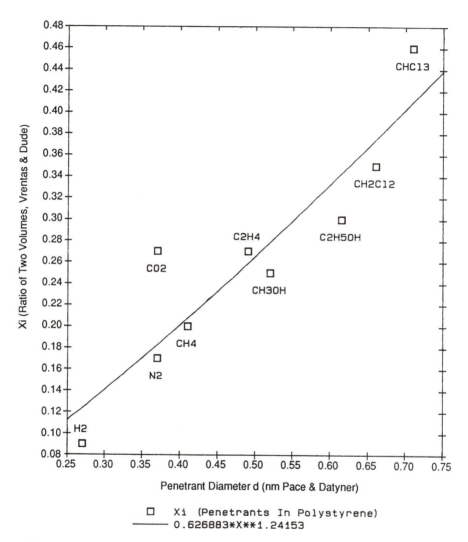

**Figure 5.** Correlation between penetrant diameter d used in the model of Pace and Datyner and size parameter $\xi$ used in the theory of Vrentas and Duda.

which is far from the cubic ($d^3$) correlation that a true volume parameter would be expected to show with a length parameter. Dimensional correlations between such different types of empirical molecular size parameters should probably not be taken too literally in any case.

The change of the type of polymer from PS to PVDC only changes the polymer jumping unit size. All penetrant jumping unit sizes remain the same. D(V&D) can therefore be expressed in terms of d by using the correlation given in Equation (5), and substituting Equation (6) into (1):

$$\xi(\text{in PVDC}) \approx 0.63cd^{1.24} \tag{6}$$

The constant c is an adjustable parameter equal to the quotient of the effective polymer jumping unit size in PS divided by the effective polymer jumping unit size in PVDC. In other words, c>1 would imply that a smaller polymer jumping unit is sufficient to fill an average hole in PVDC, and consequently that the size of an average hole in PVDC is smaller than in PS.

There is no systematic correlation between the preexponential factor $D_{01}$ and more direct measures of the size of a penetrant, such as its molar volume at 0K. [5] In the absence of such a correlation, and of suitable experimental data for diffusion in PVDC, $D_{01}$ was treated as a single adjustable parameter. Equations (1) and (6) were used to fit $\ln[D(V\&D,d)]$ to $\ln[D(P\&D,d)]$, at 298.15K, for the idealized completely amorphous PVDC system. The results of this fit are shown in Figure 6. Unlike D(P&D,d), D(V&D,d) has the expected slightly convex shape, allowing it to fall off increasingly rapidly with increasing d.

$D_{01} \approx 5.7 \times 10^{-8}$ cm$^2$/sec and c$\approx$2.252 are the values of the adjustable parameters which give the best fit. Since $\ln(D_{01}) \approx -16.7$, the average effective $D_{01}$ calculated for PVDC is much smaller than any of the values calculated in PS. {The lowest $\ln(D_{01})$ listed by VD [5] is -11.4.} It is possible to interpret the average effective $D_{01}$ calculated here as an analogue of exp(-0.115$\pi$), where $\pi$ is the Permachor value [26] of the polymer. The average effective $D_{01}$ then becomes an indicator of the intrinsic resistance of the polymer to permeation, independent of the penetrant used. The $\pi$ value calculated for PVDC in this manner is 145, which is substantially larger than the value of 87 [26] estimated by Salame. Note, also, that c$\approx$2.252>>1.0. The values of $\xi$ in PVDC are, therefore, larger than those in PS by a factor of 2.252. The average hole size in PVDC is,

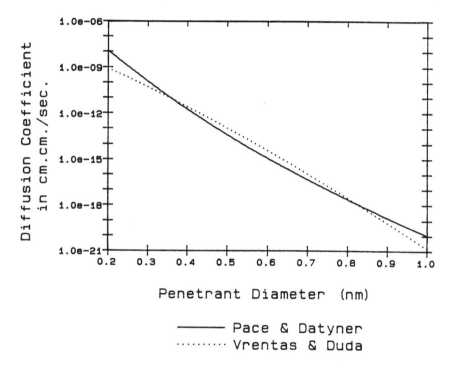

**Figure 6.** Best fit of the logarithm of the diffusion coefficient D calculated by the theory of Vrentas and Duda to the logarithm of D calculated by the model of Pace and Datyner, at 298.15K, for an idealized completely amorphous sample of PVDC, as a function of the penetrant diameter d.

consequently, much smaller than in PS. The penetrant has much less "space" to diffuse through, and therefore "appears" larger.

The drastic differences of $D_{01}$ and $\xi$ from the values observed in PS, and the observation that the density was the only parameter entering the model of P&D with a very different value for PVDC than for any of the non-barrier polymers studied, [13,14] indicate that the packing efficiency is the most important descriptive physical parameter both at the "global" and at the "intermediate" scales. The investigation of (i) the local distribution of unoccupied volume, and (ii) the MD trajectory of the penetrant molecule in the polymer matrix, will therefore be very useful. It must be kept in mind, however, that chemical effects (strong interactions between the penetrant and the polymer), as observed, for example, in (i) the high moisture sensitivity of many polar polymers, (ii) the high solubility of some penetrants in some polymers, and (iii) the plasticization of the polymer matrix by a penetrant, can cause the behavior of a polymer-penetrant system to become quite different from what would be predicted on the basis of physical effects (such as packing efficiency) alone.

The apparent activation energies are shown as functions of the penetrant diameter d in Figure 7. $E_{app}$(V&D) is a function of T whose dependence on T becomes weaker as T increases. Its limiting form as T→∞ is given by equation (4). By contrast, $E_{app}$(P&D1) and $E_{app}$(P&D2) are both almost independent of T. This fundamental qualitative difference in the dependence of $E_{app}$ on T reflects the very different interpretations of the authors of these two models, [7,12] of the meaning of $E_{app}$ in a non-Arrhenius system, i.e., in a system where the diffusion coefficient is not a simple exponentially decreasing function of a temperature-independent activation energy.

$E_{app}$(V&D) is slightly concave as a function of d, reflecting an acceleration in its rate of increase with d. $E_{app}$(P&D1) and $E_{app}$(P&D2) are slightly convex as functions of d, reflecting a deceleration in their rates of increase with increasing d. As with the diffusion coefficients, and for the same reason, the trend in the rate of change of $E_{app}$(V&D) as a function of d appears more reasonable than the trend in the rate of change of $E_{app}$(P&D1) and $E_{app}$(P&D2). In fact, it has been shown [27,28] that a second power dependence is often found between the Lennard-Jones diameters of penetrant molecules and the activation energy. Such a second power dependence, probably related to sensitivity to the cross sectional area of the penetrant, would obviously be manifested as a

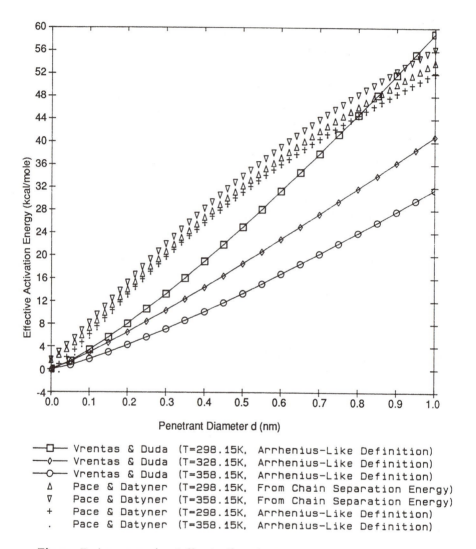

**Figure 7.** Apparent (or "effective") activation energies as functions of temperature T and penetrant diameter d.

concavity in $E_{app}$ when plotted as a function of d with linear x and y axes. See [22] for more detailed discussion of the activation energies.

A few final remarks are necessary concerning the dependence of D and of the experimental activation energy $E_a$ for diffusion on penetrant size. Some experimental results [29-31] on PVC and on rubbery polymers suggest that, contrary to the arguments made and the references provided in the discussion above, D may sometimes decrease less rapidly, and $E_a$ may sometimes increase less rapidly, with increasing penetrant size. Three possible sources can be identified for such apparent discrepancies:

1. The difference in dimensions between molecular diameter (d), cross-sectional area (proportional to $d^2$), and van der Waals (vdW) volume (proportional to $d^3$), each one of which has been used as an indicator of penetrant size by different authors. For example, if the rate of increase of $E_a$ were proportional to the cross-sectional area of the penetrant, $E_a$ would increase much faster than linearly if plotted against d, but much slower than linearly if plotted against the vdW volume.

2. For nonspherical penetrants, such as the linear alkanes or the flat aromatic molecules, the orientation of the penetrant in diffusing through the polymer results in an effective diffusional dimension related to the cross-sectional area of the molecule. The increase of this effective diffusional dimension in a series of such molecules is much slower than the increase of the van der Waals volume. Furthermore, even among small penetrant molecules, some (such as $CO_2$) are much less spherical in shape than others (such as $O_2$).

3. The possibility that an asymptotic limit might really exist for D and/or for $E_a$ as a function of increasing penetrant size can nonetheless not be ruled out on the basis of existing data. For example, the $E_a$ of penetrant molecules in rubber appears to approach an asymptotic limit as a function of molecular size, similar in magnitude to the $E_a$ for viscous flow of rubber. [30,31] Very large (and especially relatively linear and/or flexible) penetrants may well be behaving like short polymer chains, and moving through the rubbery matrix via reptational motions of the type commonly encountered in the dynamics of polymer chains. [32] Further experiments are needed to study this possibility that the diffusion process might asymptotically approach polymer self-diffusion with increasing penetrant size for penetrants of suitable shape and flexibility.

## Semi-Empirical Calculations on Model Molecules

**General Remarks and Notation.** This section is a brief summary of previously published material. The reader is encouraged to refer to the original papers [33,34] for more details, including extensive discussions intended to place such calculations in the general content of the study of the conformations and interactions of model molecules and polymer chain segments, and comparisons with experimental results and with similar calculations.

Standard notation will be used to describe the conformations defined by the values of dihedral angles $\Phi$ about successive C-C bonds in chains of polymers and model molecules containing a backbone of tetravalent carbon atoms: T=trans ($\Phi=180^{\circ}$), G=gauche ($\Phi=60^{\circ}$), C=cis ($\Phi=0^{\circ}$), X=($180^{\circ}>\Phi>0^{\circ}$, but $\Phi$ not equal to $60^{\circ}$), G'=-G ($\Phi=-60^{\circ}$), and X'=-X.

The letter Y will be used to denote a halogen atom, since the more commonly used letter X has already been used in describing the conformations. VF, VC and VB denote vinyl fluoride, vinyl chloride and vinyl bromide, respectively. In addition to PVDC and VDC/VC copolymers, PVDF and VDF/VF copolymers (Y=F), and PVDB and VDB/VB copolymers (PVDB denotes that Y=Br) will also be studied. Such a study of the entire isoelectronic series (Y=F, Cl and Br) can provide a more complete understanding of the effects determining the barriers of VDC/VC copolymers.

**Calculations on Model Molecules Simulating Isolated Chain Segments.** Semi-empirical calculations [35] were first carried out [33,34] on model molecules simulating isolated chain segments of PVDY and of VDY/VY copolymers. The geometries of the model molecules were optimized [33] by the Molecular Mechanics 2 (MM2) option in CHEMLAB-II, which is a general-purpose molecular modeling software package developed and owned by Chemlab Incorporated, and marketed by Molecular Design Limited. MM2 is an FF technique for calculating molecular geometries by minimizing the total steric energy E. A lower E implies a thermodynamically more stable molecular structure. The geometries of twelve model molecules of four different types were optimized. [33] Each one of the three molecules of any given type contains a different halogen atom (Y=F, Cl or Br):

1. $C_{20}H_{22}Y_{20}$ with only head-to-tail bonding between successive pairs of monomers. These molecules are oligomers of ten VDY monomers, terminated by an extra H atom at each end.

2. $C_{20}H_{22}Y_{20}$ with one head-to-head bond (at the tenth and eleventh C atoms), where a "head" has been defined as a C atom bonded to H atoms, rather than to Y atoms as in the more commonly used convention.

3. $C_{20}H_{22}Y_{20}$ with one tail-to-tail bond (at the tenth and eleventh C atoms), where a "tail"has been defined as a C atom bonded to Y atoms.

4. $C_{20}H_{23}Y_{19}$ with head-to-tail bonding between successive pairs of monomers. These molecules are oligomers of nine VDY and one VY monomers, terminated  by an extra H atom  at each end. The sixth monomer is the VY unit.

The optimized molecular geometries for Y=F and Cl are shown in space-filling illustrations in Figures 8 and 9 respectively. [33] To avoid confusion, note that many atoms are hidden from view. The optimized geometries for Y=Br are very similar to those for Y=Cl, and they are therefore not shown. The E's of the four types of model molecules are shown as functions of the halogen atom Y (1=F, 2=Cl and 3=Br) in Figure 10.

The optimized geometry of $C_{20}H_{22}F_{20}$ was very close to having TGTG' symmetry ($\Phi \approx 60.5^0$ as an average value for G), in agreement with the interpretation of the crystal structure of that polymorph of PVDF most closely resembling PVDC. A geometry very similar to TXTX' was obtained for $C_{20}H_{22}Cl_{20}$ ($\Phi \approx 48^0$ for X). The geometry calculated for $C_{20}H_{22}Br_{20}$ ($\Phi \approx 46^0$ for X) was isomorphous to the geometry computed for $C_{20}H_{22}Cl_{20}$. These results show that steric energy considerations are crucial factors in determining the chain conformations.

Conformations with a head-to-head bonding defect resembled the conformations of the standard structures. The only exception occurred in the vicinity of the two C atoms, tenth and eleventh along the chain, where the defect was incorporated. The bonding around the defect site became trans, interrupting the TGTG' or TXTX' pattern. A succession of three trans $\Phi$'s (TTT) resulted, as in the preferred TTTT conformation of PE. The winding pattern then returned to TGTG' or TXTX', with an equal likelihood of the "handedness" of the helix being reversed or remaining the same as before the defect. This type of defect might therefore serve as a site where the helix loses all "memory" of its direction of winding, and can either continue in the same direction or reverse directions.

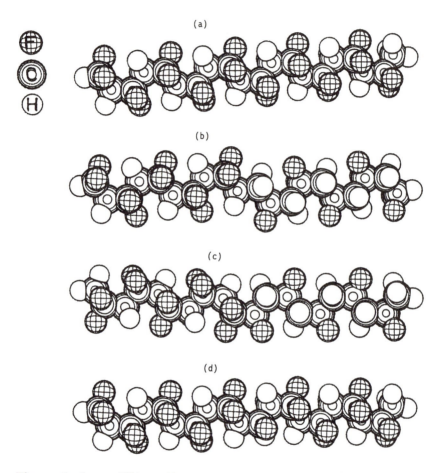

**Figure 8.** Space-filling illustrations: (a) standard geometry for $C_{20}H_{22}F_{20}$; (b) $C_{20}H_{22}F_{20}$ with a head-to-head bond at the tenth and eleventh C atoms; (c) $C_{20}H_{22}F_{20}$ with a tail-to-tail bond at the tenth and eleventh C atoms; and (d) $C_{20}H_{23}F_{19}$ with the eleventh C atom bonded to both an H and an F.

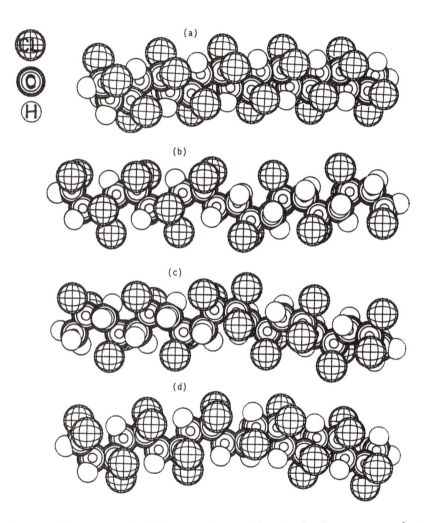

**Figure 9.** Space-filling illustrations: (a) standard geometry for $C_{20}H_{22}Cl_{20}$; (b) $C_{20}H_{22}Cl_{20}$ with a head-to-head bond at the tenth and eleventh C atoms; (c) $C_{20}H_{22}Cl_{20}$ with a tail-to-tail bond at the tenth and eleventh C atoms; and (d) $C_{20}H_{23}Cl_{19}$ with the eleventh C atom bonded to both an H and a Cl.

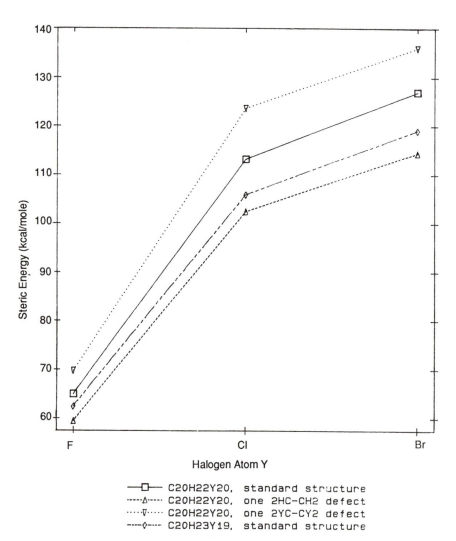

**Figure 10.** MM2 steric energy as a function of the halogen atom.

With a tail-to-tail bonding defect, the TGTG' or TXTX' winding pattern was retained everywhere along the chain. Around the defect, the $\Phi$ corresponding to X was somewhat larger than the range observed in the standard structures. This defect site $\Phi$ is $70^0$ for Y=F, in comparison with $60.5^0$; $68^0$ for Y=Cl, in comparison with $48^0$; and $67^0$ for Y=Br, in comparison with $46^0$. The increase in $\Phi$ at the defect site enables some reduction of the steric repulsions between halogen atoms.

The geometry of each $C_{20}H_{23}Y_{19}$ was very similar to the geometry of the standard structure of $C_{20}H_{22}Y_{20}$. The largest change was observed for Y=Br; however, even with this largest halogen atom, the $\Phi$ corresponding to X, at the eleventh carbon atom along the chain, which is bonded to both an H and a Br, only increases to $52.5^0$, i.e., very slightly above the average value of $46^0$ calculated for $C_{20}H_{22}Br_{20}$. Adding a small percentage of VY comonomer should therefore not change the preferred chain conformation. A small number of VY units should be able to enter the crystallites, without distorting the chains away from their preferred conformations. Such defective crystallites may, however, have lower cohesive energies than perfect crystallites of PVDY, as discussed in the following section. As the percentage of VY is increased, the loss of stabilization resulting from the loss of polar interactions, as well as the increasingly imperfect packing, should, at some composition, cause a transition from a predominantly PVDY-like structure with VY defect sites, to a PVY-type structure with VDY defect sites.

It can be seen from Figure 10 that the stability of isolated chain segments decreases in the order Y=F>>Cl>Br. The increase of E is much larger from Y=F to Y=Cl than from Y=Cl to Y=Br. This result is consistent with the much larger percent difference between the atomic volumes of F and Cl, than the atomic volumes of Cl and Br. The heat of formation is negative for Y=F or Cl, and positive for Y=Br. [A negative (positive) heat of formation indicates that a molecule is thermodynamically stable (unstable) relative to its constituting elements.]

Each structure with a head-to-head bond has a lower E than the corresponding standard structure with only head-to-tail bonds. This effect becomes stronger with increasing size of the halogen atom. In amorphous regions, a chain with this type of defect can be expected to have lower thermodynamic energy than the standard structure. In crystalline regions, however, periodicity would be disrupted and the energy would go up. This defect would, therefore, be quite unlikely to be found in crystallites of PVDC and PVDB, because of the large disparity between the sizes of H and of Cl or Br. One of its preferred locations, especially for high percent crystallinity, might be the boundary between a crystallite and the amorphous region surrounding it. After one of these defects is incorporated in a chain, the occurrence of a second such defect in the same chain becomes unlikely. A much less favored tail-to-tail bond is required, to create a new chain end with H attached to the terminal C, and make a second head-to-head bond possible.

The E of each defective structure with a tail-to-tail bond is considerably higher than the E of the corresponding standard structure. This type of defect is, therefore, thermodynamically disfavored. It is also kinetically much more disfavored than a head-to-head bond, since its formation requires the bonding of two carbon atoms hindered by the bulky Y's. Replacement of one Y atom by a less bulky H atom results in a decrease in E, as shown in Figure 10. This effect becomes more pronounced with increasing size of Y.

The AM1 (Austin Model 1) Hamiltonian available in Version 3.10 of the MOPAC (Molecular Orbital PACkage) program was used to carry out QM calculations, for the purpose of estimating the partial atomic charges to be used in the electrostatic portion of the potential energy for the calculations on the interactions between pairs of model molecules. [34] The AM1 calculations also yielded quantum mechanical estimates of the heats of formation of the model molecules. These heats of formation generally followed similar trends to the MM2 steric energies.

**Calculations on Interacting Pairs of Model Molecules.** The geometries optimized by MM2 [33] were used for $C_{20}H_{22}Y_{20}$ and $C_{20}H_{23}Y_{19}$ in these calculations. [34] The total intermolecular energy ($E_2$) of a pair of molecules was calculated as a function of the relative positions of the two molecules. The PHBIMIN program in CHEMLAB-II was used to minimize $E_2$ as a function of the intermolecular configuration of pairs of rigid molecules, using both Lennard-Jones and electrostatic (coulombic) terms in the potential energy.

Such simple calculations on pairs of chain segments are mainly useful as bridges between the single-chain and the many-chain assembly levels of calculation. They provide information on the preferred interactions of two chain segments in the absence of other chain segments. Comparison of their results with the results for isolated chains and multi-chain assemblies can facilitate the identification of which effects are primarily caused by (i) intrachain factors, (ii) the intrinsically preferred patterns of interaction between pairs of chains, and (iii) constraints and/or superpositions of effects induced by large-scale packing. [34]

The optimized intermolecular configurations of the $C_{20}H_{22}F_{20}$-$C_{20}H_{22}F_{20}$ and $C_{20}H_{22}Cl_{20}$-$C_{20}H_{22}Cl_{20}$ pairs are shown in Figures 11 and 12, respectively. The intermolecular energies are plotted as a function of the period of the halogen atom Y in Figure 13.

Define $\delta E_2$ as the change in $E_2$ resulting from the replacement of $C_{20}H_{22}Y_{20}$ by $C_{20}H_{23}Y_{19}$ as the second molecule in a pair. $\delta E_2$ is a measure of the energetic effect of replacing 10% of the VDY by VY in one of the two model molecules. Define the structural dissimilarity index $\sigma$ as the root mean square deviation in a geometric match of corresponding $C_{20}H_{22}Y_{20}$ and $C_{20}H_{23}Y_{19}$ molecules. High $\sigma$ indicates poor match between the two structures. The magnitude of $\delta E_2$ ($-\delta E_2$ for Y=F and $+\delta E_2$ for Y=Cl or Br), is shown as a function of $\sigma$ in Figure 14.

The two molecules are antiparallel and perfectly aligned for Y=F. Figure 11b shows that the intermolecular configuration can be viewed from a perspective in which one molecule is behind the other, almost completely eclipsed by it.

The two molecules are antiparallel but at a slight oblique angle for Y=Cl or Br. Figure 12b shows that, in the perspective in which one molecule is behind the other, the molecule in the back is not almost completely eclipsed by the one in the front, and a considerable portion of it can still be seen. It is not obvious, from simple qualitative considerations, that the use of model molecules much longer than the twenty-carbon chains utilized in this work would necessarily result in greater parallelization for Y=Cl or Br. The presence of more than two chains might force additional parallelization, but the energetic cost of forcing pairs of chains into the less favorable fully aligned pairwise configurations could be quite high.

A greater propensity of the chains to parallelize when Y=F than when Y=Cl may favor a higher percent crystallinity in PVDF, but the incorporation of a much larger number of head-to-head bonding defect sites in chains of PVDF [36] may disfavor higher crystallinity in PVDF. Consequently, the crystallinities of PVDF and PVDC are comparable, with PVDC in general being slightly more crystalline than PVDF when prepared in the same manner.

The size disparity between Br and H is only slightly larger than the size disparity between Cl and H. Consequently, the crystal structures of PVDC and PVDB can be expected to be much more similar than the crystal

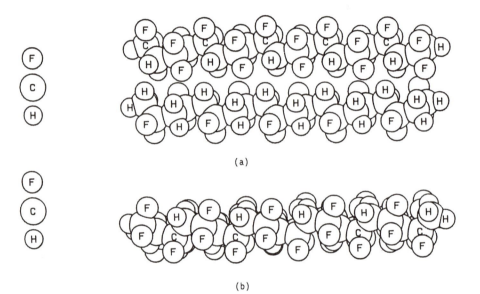

**Figure 11.** Space-filling illustrations of the $C_{20}H_{22}F_{20}$-$C_{20}H_{22}F_{20}$ pair: (a) a perspective showing both molecules; (b) an alternative perspective with one molecule behind the other, and almost completely eclipsed by it.

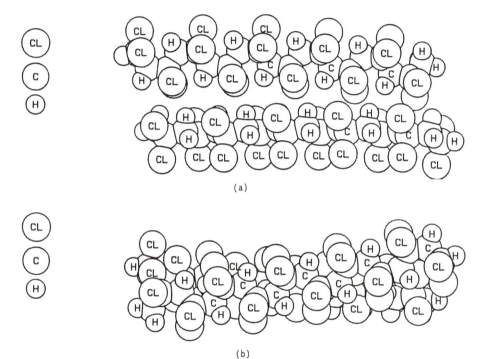

**Figure 12.** Space-filling illustrations of the $C_{20}H_{22}Cl_{20}$-$C_{20}H_{22}Cl_{20}$ pair: (a) a perspective showing both molecules; (b) an alternative perspective with one molecule behind the other, but only partially eclipsed by it because the two molecules are at an oblique angle.

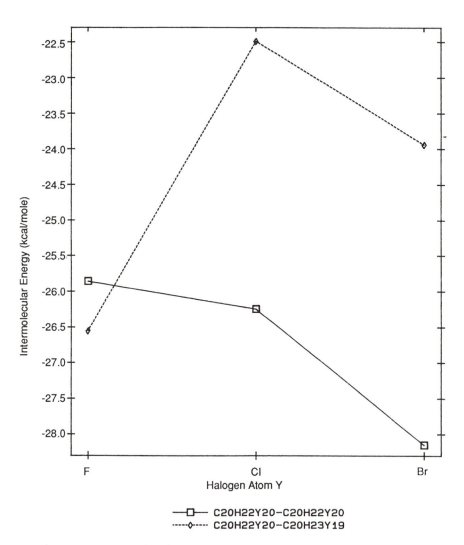

**Figure 13.** Intermolecular energy as a function of the halogen atom.

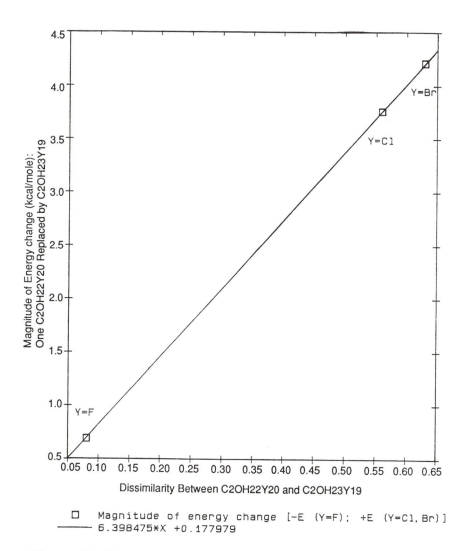

**Figure 14.** Magnitude of energy change ($\delta E_2$) resulting from replacement of one $C_{20}H_{22}Y_{20}$ molecule in a pair by $C_{20}H_{23}Y_{19}$, as a function of dissimilarity ($\sigma$) between $C_{20}H_{22}Y_{20}$ and $C_{20}H_{23}Y_{19}$.

structures of PVDC and PVDF. This expectation is in agreement with crystallographic data [37] on oriented fibers.

The replacement of a Y atom by an H atom in a non-terminal site did not cause any significant changes in the intermolecular configurations, just as it did not cause any significant changes in the conformations of isolated chains. Addition of a small percentage of VY comonomer (10% in these calculations) should therefore not cause major changes in the packing patterns.

As shown in Figure 13, $E_2$ was attractive (-26 to -28 kcal/mole) for all three $C_{20}H_{22}Y_{20}$-$C_{20}H_{22}Y_{20}$ pairs. There are twenty monomer units in two $C_{20}H_{22}Y_{20}$ molecules. The range of $E_2$'s is therefore equivalent to a stabilization energy of 1.3 to 1.4 kcal/mole per monomer unit. It can be concluded that the instability of PVDB is not caused by weak intermolecular binding, but by the instability of its individual chains. [34] This instability is manifested by the positive heat of formation of individual model molecules representing its isolated chain segments. [34]

The $E_2$ of the $C_{20}H_{22}Y_{20}$-$C_{20}H_{23}Y_{19}$ pair is slightly more negative for Y=F, and considerably less negative for Y=Cl or Br, than the $E_2$ of the corresponding $C_{20}H_{22}Y_{20}$-$C_{20}H_{22}Y_{20}$ pair. As shown in Figure 14, the magnitude of $\delta E_2$ is proportional to $\sigma$, i.e., to the index of mismatch.

The significant loss of stabilization energy caused by the replacement of one Cl or Br atom by an H atom shows that, although the intermolecular packing patterns might not change by the addition of a small percentage of comonomer, the intermolecular stabilization energy (and hence also the cohesive energy of the polymer) may decrease significantly.


Preliminary Experimental Results

Preliminary results on the oxygen permeabilities of thin films of four VDC/VC copolymers and a PVC homopolymer are shown in Figure 15.

The films were prepared by flash molding at about 10°C above the melting temperature of the resin. They were then heat-aged at 333K for several hours before testing. The oxygen permeabilities were determined with an OX-TRAN permeability tester. The permeabilities at several temperatures above 296K were measured, and the results were extrapolated to 296K. Note that the natural logarithm of the permeability increases linearly with the mole percent of VC in the VDC/VC copolymers, but that the data point for the PVC homopolymer lies significantly below what would be predicted by a linear extrapolation of

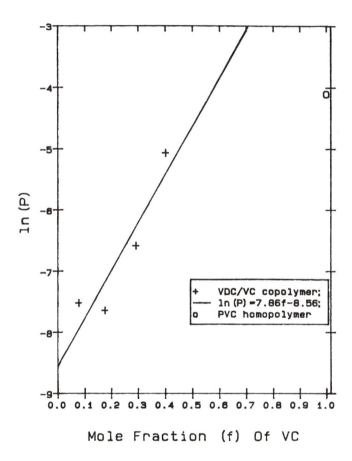

**Figure 15.** Preliminary experimental results, showing the natural logarithms of the oxygen permeabilities [i.e., ln(P)] of four VDC/VC copolymers and a PVC homopolymer, where P is in barrers. Data extrapolated to 296K from higher temperatures.

the copolymer data. It is quite possible that an enhanced packing efficiency in the homopolymer (perhaps, but not necessarily, also accompanied by an enhanced crystallinity) will provide an explanation for this poor extrapolation of the diffusion data to the PVC homopolymer. Such issues are among the types questions which will be addressed in future experimental and theoretical work.

## Summary and Conclusions

A study is in progress on the transport of penetrant molecules in barrier plastics, utilizing a synergistic combination of techniques.

The systematic use of a combination of different types of calculations can provide perspectives at different scales. These perspectives can be used to attempt to construct a unified working model for transport phenomena. Flow charts have been used to summarize how the different types of calculations and experiments fit together as a complete and coherent approach.

Copolymers of vinylidene chloride and vinyl chloride have been used as test cases. Permeation of penetrant molecules is generally believed to occur through the amorphous regions of these semicrystalline polymers. Their study is, however, complicated by the necessity to understand the effects of the presence of the crystallites in addition to the amorphous regions.

General comparisons have been presented between diffusion coefficients and apparent activation energies derivable from the "global" free volume theory of Vrentas and Duda, and the statistical mechanical model of Pace and Datyner, as an initial step in the utilization of these theories.

The results of semi-empirical calculations on isolated chain segments and on interacting pairs of chain segments of PVDY and VDY/VY copolymers (Y=F, Cl or Br), utilizing both force field and quantum mechanical techniques, have been summarized, and used to draw as many conclusions as possible, from such a limited set of calculations, about chain packing patterns and stabilities.

The main conclusion of the calculations summarized here is that the packing efficiency (as determined by the shape of the chain contour and the mobility of chain segments) is an extremely important physical factor in determining the permeability. This conclusion is also supported by positron annihilation studies of the microstructure of polymers in relation to their diffusional properties. These studies show that the diffusion coefficients of hydrogen and methane in a wide variety of amorphous polymers are determined by the free volume in the disordered regions of the polymer. [38] Finally, a new technique, based on the photoisomerization of photochromic and fluorescent probe molecules, which was recently developed for measuring the distribution of local free volume in glassy polymers, appears to be very promising for studying the packing in greater detail. [39,40]

## Acknowledgments

We thank H. A. Clark, I. R. Harrison, A. J. Hopfinger, D. J. Moll, J. K. Rieke, N. G. Rondan and J. S. Vrentas for helpful discussions.

## Literature Cited

1. Crank, J.; Park, G. S. *Diffusion in Polymers*, Academic Press: New York, 1968.
2. Stannett, V. *J. Membrane Sci.*, 1978, 3, 97.
3. Koros, W. J.; Story, B. J.; Jordan, S. M.; O'Brien, K.; Husk, G. R. *Polymer Engineering and Science*, 1987, 27, 603.
4. *Encyclopedia of Polymer Science & Technology*, the "Diffusional Transport" subsection in the article titled "Diffusion".
5. Vrentas, J. S.; Duda, J. L. *J. Appl. Polym. Sci.*, 1977, 21, 1715.
6. Vrentas, J. S.; Liu, H. T.; Duda, J. L. *J. Appl. Polym. Sci.*, 1980, 25, 1297.
7. Ju, S. T.; Duda, J. L.; Vrentas, J. S. *Ind. Eng. Chem. Prod. Res. Dev.*, 1981, 20, 330 .
8. Vrentas, J. S.; Duda, J. L.; Ling, H.-C. *J. Polym. Sci., Polym. Phys. Ed.*, 1985, 23, 275.
9. Vrentas, J. S.; Duda, J. L.; Ling, H.-C.; Hou, A.-C. *J. Polym. Sci. Polym. Phys. Ed.*, 1985, 23, 289.
10. Vrentas, J. S.; Duda, J. L.; Hou, A.-C. *J. Polym. Sci., Polym. Phys. Ed.*, 1985, 23, 2469.
11. Vrentas, J. S.; Duda, J. L.; Hou, A.-C. *J. Appl. Polym. Sci.*, 1987, 33, 2571.
12. Pace, R. J.; Datyner, A. *J. Polym. Sci., Polym. Phys. Ed.*, 1979, 17, 437.
13. Pace, R. J.; Datyner, A. *J. Polym. Sci., Polym. Phys. Ed.*, 1979, 17, 453.
14 Pace, R. J.; Datyner, A. *J. Polym. Sci., Polym. Phys. Ed.*, 1979, 17, 465.
15. Pace, R. J.; Datyner, A. *J. Polym. Sci., Polym. Phys. Ed.*, 1979, 17, 1675.
16. Pace, R. J.; Datyner, A. *J. Polym. Sci., Polym. Phys. Ed.*, 1979, 17, 1693.
17. Pace, R. J.; Datyner, A. *J. Polym. Sci., Polym. Phys. Ed.*, 1980, 18, 1103.
18. Ferry, J. D. *Viscoelastic Properties of Polymers*, 2nd edition, Wiley: New York, 1970.

19. Wessling, R. A. *Polyvinylidene Chloride*, Gordon and Breach Science Publishers: New York, 1977.
20. Manson J. A.; Sperling, L. H. *Polymer Blends and Composites*, Plenum Press: New York, 1976; p. 410.
21. Alfrey, Jr., T.; Boyer, R. F. *Molecular Basis of Transitions and Relaxations* (Meier, D. J., ed.), Gordon and Breach Science Publishers: New York, 1978; p. 193.
22. Bicerano, J.; in *Molecular Level Calculations of the Structures and Properties of Non-Crystalline Polymers* (Bicerano, J., ed.), Marcel Dekker: New York (to be published).
23. Muruganandam, N.; Koros, W. J.; Paul, D. R. *J. Polym. Sci., Polym. Phys. Ed.*, 1987, 25, 1999.
24. Lee, W. M. *Polymer Engineering and Science*, 1980, 20, 65.
25. Berens, A. R.; Hopfenberg, H. B. *J. Membrane Sci.*, 1982, 10, 283.
26 Salame, M. *Polymer Preprints*, 1967, 8, 137.
27. Allen, S. M.; Stannett, V.; Hopfenberg, H. B. *Polymer*, 1981, 22, 912.
28. Aharoni, S. M. *J. Appl. Polym. Sci.*, 1979, 23, 223.
29. Chern, R. T.; Koros, W. J.; Hopfenberg, H. B.; Stannett, V. T. *ACS Symposium Series*, 1985, 269, 25.
30. Van Amerongen, G. J. *J. Polym. Sci.*, 1950, 5, 307.
31. Van Amerongen, G. J. *Rubber Chemistry & Technology*, 1951, 24, 109.
32. Doi, M.; Edwards, S. F. *The Theory of Polymer Dynamics*, Clarendon Press: Oxford, 1986.
33. Bicerano, J. *Macromolecules*, 1989, 22, 1408.
34. Bicerano, J. *Macromolecules*, 1989, 22, 1413.
35. See Clark, T. *A Handbook of Computational Chemistry*, John Wiley and Sons: New York, 1985, for detailed discussions of computational methods.
36. Elias, H.-G. *Macromolecules, Volume 1: Structure and Properties*, 2nd edition, Plenum Press: New York, 1984.
37. Narita, S.; Okuda, K. *J. Polym. Sci.*, 1959, 38, 270.
38. V. V. Volkov, A. V. Gol'danskii, S. G. Dur'garyan, V. A. Onishchuk, V. P. Shantorovich and Yu. P. Yampol'skii, *Polym. Sci. USSR*, 29: 217 (1987).
39. Victor, J. G.; Torkelson, J. M. *Macromolecules*, 1987, 20, 2241.
40. Victor, J. G.; Torkelson, J. M. *Macromolecules*, 1987, 20, 2951.

RECEIVED October 27, 1989

Chapter 7

# Gas Transport Through Bisphenol-Containing Polymers

John C. Schmidhauser and Kathryn L. Longley

Corporate Research and Development, General Electric Company, Schenectady, NY 12301

*The gas transport properties of bisphenol containing polymers were investigated in order to determine which structural features lead to enhanced polymer barrier properties. Permeability measurements on a series of structurally different aromatic polycarbonates indicate a strong relationship between monomer structure and polymer permeability, with gas transport rates varying by a factor of 200 between the most and least permeable materials. Unexpectedly, polycarbonates prepared from 3,3' - dimethyl substituted bisphenol-A's were found to possess enhanced barrier properties compared to polymers prepared from either bisphenol A or monomers bearing larger substitutents. Measurements performed on series of aromatic polyestercarbonates, polyesters and polyetherimides were used to assess the relative importance of the monomer structure in the repeat unit versus the presence of carbonate/ester/etherimide linkages in determining a polymer's gas permeability. A study of aromatic polyesters showed that, when the rest of the repeat unit is the same, polymers containing predominantly isophthaloyl units exhibit a lower permeability than polymers which contain predominantly terephthaloyl units. The ability of several current structure-permeability models to reproduce this new polymer permeability data was explored.*

Gas transport through polymers is an area of growing interest as materials with unique transport properties continue to find use in new, specialized applications ranging from extended life tennis balls (1) to natural gas separation systems (2). Concurrent with this increased interest is the desire to understand on a molecular level what determines the gas permeability properties of a particular material. The ability to better relate polymer molecular structure to gas transport properties is crucial in any attempt to rationally design materials for specific permeability applications such as gas barriers.

0097–6156/90/0423–0159$06.00/0
© 1990 American Chemical Society

Historically, the availability of experimental permeability data has been limited for the most part to common/commercial polymers (3). While this information demonstrated that gas transport rates of polymers may vary by many orders of magnitude (Figure 1), little effort was made towards relating the permeability or diffusion rates of gases to polymer molecular structure. The slow development of such a structure-property model can be traced to some extent to the type of transport data available, since the readily available, commercial polymers included few examples within any particular polymer class. It is therefore not surprising that early generalities were drawn from this limited data which emphasized the type of functionality of the polymer, either as its "link" or as a substituent, as the predominant factor influencing its gas transport rate. For example, statements such as ". . . In order to be a truly good barrier polymer the material must have some degree of polarity such as contributed by the nitrile, ester, chlorine, fluorine, or acrylic functional groups . . ." (4) reflect the observation that a particular member of each class exhibits high to moderate gas barrier properties.

Recent evidence indicates that the influence of molecular structure on gas permeation through polymers is complex. For example, reports investigating series of structurally varied polyimides (5-7), polyacetylenes (8), polystyrenes (9) and silicone polymers (10) show that gas transport rates within a particular polymer class can vary dramatically depending upon the structure of the monomer present. These observations on materials where the monomer changes while the functional "link" remains constant suggest that structural factors other than the polymer class are significant in determing gas transport properties.

Glassy polymers are attractive candidates for probing the effect of molecular structure on gas transport properties. Permeabilities of rubbery polymers (11) tend to be less responsive to structural change, while a molecular level interpretation of transport data from crystalline polymers can be complicated by the effects of sample morphology and orientation (12). The theory of gas transport through glassy polymers is well advanced (13-17), due in no small part to the fundamental studies performed on polycarbonate (18,19) and polysulfone (20) formed from bisphenol A. Scattered reports (21,22) have indicated some dependence of polycarbonate gas transport properties on polymer structure. In addition, detailed studies of the sorption and transport properties (23) of three tetrasubstituted bisphenol polycarbonates and a polysulfone (24) have recently been described. Finally, comparisons of gas transport properties of bisphenol A based polycarbonate and polyester (25), as well as the sorption (26) and permeability properties (27) of these and several other bisphenol A based polymers have appeared. Despite these advances, there is clearly a need for more systematic investigation of the influence of molecular structure on the transport behavior of glassy polymers.

To further investigate the question of polymer structure-permeability relationships, this study reports oxygen permeability measurements on a group of structurally varied bisphenol based polymers. In addition to representing commercially important classes of engineering thermoplastics, polycarbonates, polyarylates and polyetherimides can be easily prepared from a common set of bisphenol monomers. The availability of this systematic set of data can be used to address a number of structure-property related questions. First of all, varying the structure of the monomer segment between a constant functional "link", most extensively done for polycarbonates (Schmidhauser, J. C.; Longley, K. L. J. Appl. Polym. Sci., in press) helps to define the permeability limits of this polymer class and offers insight into how this behavior correlates with monomer structure. Extending these measurements to polyarylates and polyetherimides made from the same bisphenols explores whether these structure-property correlations hold for other classes of polymers which have the same monomer segments in the repeat units. In this way, general questions concerning the relative importance of monomer structure versus type of polymer link in determining gas transport properties can be addressed. Finally, this new permeability data offers an opportunity to evaluate how current structure-permeability models agree with experimental results.

Experimental

Literature procedures were followed for the preparation of polycarbonates (21), polyarylates (28) and polyetherimides (29) from different bisphenol monomers. Glass transition temperatures were determined using a Perkin Elmer DSC-7 differential scanning calorimeter. Densities of the substituted polymer films were deter-mined by floatation in potassium iodide gradient columns at 23 °C.
   Polymer film samples were prepared by solution casting. Typically a 8-10 wt. % methylene chloride solution of polymer was prepared and passed through a 0.5 μm filter onto a clean, dried soda-lime glass plate fitted with a glass casting ring. Solvent was allowed to diffuse through a lightly plugged, inverted funnel over 48 h. Drying for 96 h in a vacuum oven (70 °C, 15 torr) effectively removed residual solvent and gave film samples which, with few exceptions, were clear and ductile.
   Oxygen permeability measurements were made using a Modern Controls Inc. Oxtran 100 or 1000 analyzer. Film samples, 5-10 mil thick, were mounted between aluminum foil masks containing 5 $cm^2$ holes. One side of the film was flushed with a 1 atm stream of oxygen, while the oxygen concentration in the nitrogen stream purging the other side of the film was monitored. Once the oxygen concentration was constant over a 12 h period, it was assumed that a steady state flux of oxygen through the film was achieved and the permeability value was recorded. Control experiments show that reproducibility of the permeability reading had a precision of ± 5 %. In this report, permeability values are in units of Barrers, where 1 Barrer = $10^{-10}$ $cm^3$ (STP) · cm / $cm^2$ · s · cm Hg (30).

Results and Discussion

Effect Of Molecular Structure on Polycarbonate Oxygen Permeability. Permeability measurements on a series of structurally different aromatic polycarbonates indicate a strong relationship between monomer structure and polymer permeability. A major part of this study investigated properties of polycarbonates which were prepared from variously substituted bisphenols. The polycarbonate samples are based upon bisphenol monomers which differ from 2,2-bis(4-hydroxy-

phenyl)propane (bisphenol A) by substitution at the "central"
aliphatic carbon atom, substitution at the 3,3'-positions of the
aromatic rings, and by substitution at both positions. A summary of
transport data for these three sets of polycarbonates illustrates the
sensitivity of permeation rates to relatively minor polymer
structural changes.

A variety of groups were substituted at, or in place of, the
central carbon atom of the bisphenols, giving correspondingly
different effects on the resultant polycarbonate's permeability
(Table I). For example, replacement of the geminal dimethyl groups
of bisphenol A based polymer 1 with longer alkyl substituents gave
polymers 2-4 which had oxygen transport rates which increased with
the size of the alkyl groups. Replacement of the geminal methyl
groups by other bulky groups 5-6 raised the permeabilities in a
similar fashion. On the other hand, polymers based upon monomers
where the central, methylene carbon is contained in a medium-sized
aliphatic ring 7-9 have permeabilities lower than the bisphenol A
based polymer reference point. This triad of polycarbonates
illustrates the sensitivity of gas permeabilities to small structural
changes, with the 6-member ring-containing polymer 8 having a lower
permeability than either the 5- or 7-member ring-containing polymers.
The observation that the mid-sized ring containing polycarbonate has
the lowest permeability suggests that the well known ability of the
cyclohexane ring to adopt a sterically compact conformation is
helpful in reducing the polymers' gas transmission rate. Finally, it
is interesting to note that the nitrile containing polycarbonate 10
exhibits a much lower permeability than the hydrocarbon analog 2.
Therefore, while many of the permeability results can be
qualitatively rationalized by steric arguments, other factors such as
polarity and interchain attraction forces can clearly be involved.

Polycarbonates prepared from bisphenol monomers substituted on
the aromatic rings at the 3,3'-positions exhibit a range of gas
transport rates (Table II). Placement of methyl groups at these
positions gives a polymer 11 with one fifth the permeability of the
parent polymer 1. However, increasing the size of the alkyl groups
from methyl to ethyl or isopropyl (polymers 11, 12 and 13,
respectively) increases the permeability of the polymer to approach
or exceed that of 1. Furthermore, the polymer containing
symmetrically substitued 3,3',5,5'-tetramethylbisphenol-A 14
exhibits a much higher permeability rate than either the
unsubstituted 1 or dimethyl substituted 2 polycarbonate, as had been
previously reported (23).

The unexpected enhancement in gas barrier properties caused by
3,3'-dimethyl substitution warranted further investigation. To test
the generality of this effect, permeability measurements were
performed on polycarbonates which had various substituents at the
central carbon atom in addition to methyl groups at the aromatic
3,3'-positions (Table III). The comparison between data from these
materials and the analogous polycarbonates with unsubstituted
aromatic rings shows that in every case the dimethyl substituents
lower the permeability rate. However the magnitude of this permeability drop ranges from five-fold (1 to 11) to a factor of only
1.4 (10 to 16). Thus, while these results suggest some degree of
additivity of monomer structural effects towards polymer
permeability, it is qualitative at best. Finally, it should be noted

610      Polydimethylsiloxane

100.0

        Polyphenylene oxide
10.0
        LD Polyethylene

        Polyarylate
1.0     Polycarbonate

        Polyvinyl acetate
        Polyetherimide

0.1     PBT

        Nylon 6
        PET
0.01
        Polyvinylidene chloride

0.001

0.0003  Polyacrylonitrile

Figure 1.   Oxygen permeabilities of selected commercial polymers.

Table I. Oxygen Permeabilities of Polycarbonates Containing Bisphenols Substituted at the Central Carbon Atom

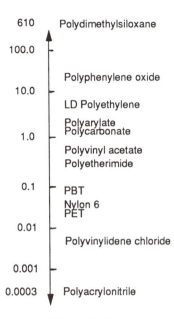

| Polymer No. | Bisphenol Composition | $PrO_2$ | Polymer No. | Bisphenol Composition | $PrO_2$ |
|---|---|---|---|---|---|
| 1 | | 1.50 | 6 | | 6.90 |
| 2 | | 1.48 | 7 | | 1.35 |
| 3 | | 2.35 | 8 | | 0.57 |
| 4 | | 3.41 | 9 | | 0.76 |
| 5 | | 2.02 | 10 | | 0.33 |

Table II. Oxygen Permeabilities of Polycarbonates Substituted at the 3,3'-Positions
of the Bisphenol Aromatic Rings

| Polymer No. | Bisphenol Compostion | $PrO_2$ |
|:-----------:|:--------------------:|:-------:|
| 1 | HO—⬡—⊦—⬡—OH | 1.50 |
| 11 | HO—⬡—⊦—⬡—OH | 0.30 |
| 12 | HO—⬡—⊦—⬡—OH | 0.72 |
| 13 | HO—⬡—⊦—⬡—OH | 1.83 |
| 14 | HO—⬡—⊦—⬡—OH | 5.94 |

Table III. Oxygen Permeabilities of Polycarbonates Containing Bisphenols With and Without
Methyl Groups at the 3,3'-Positions of the Aromatic Rings

| Polymer No. | Bisphenol Compositon | $PrO_2$ With Methyls | (Without Methyls) |
|:-----------:|:--------------------:|:--------------------:|:-----------------:|
| 15 | HO—⬡—⊦—⬡—OH | 1.26 | (2.35) |
| 11 | HO—⬡—⊦—⬡—OH | 0.30 | (1.50) |
| 16 | HO—⬡—⊦—⬡—OH (CN) | 0.21 | (0.33) |
| 17 | HO—⬡—◯—⬡—OH | 0.11 | (0.57) |

the polymer based upon 1,1-bis(4-hydroxy-3-methylphenyl)cyclohexane
17 exhibits one of the lowest oxygen permeabilities reported for a
bisphenol based polycarbonate (31).

3,3'-Disubstituted Bisphenol Containing Polycarbonates. The
intriguing and specific barrier enhancement brought about by
3,3'-dimethyl substitution of the bisphenol polymers serves as an
excellent case study for probing the relationship between monomer
(repeat unit) structure and oxygen permeability.  On a molecular
level, differences in the permeabilities of polymers are typically
rationalized by citing differences in structural factors such as
polarity, hydrogen bonding, cohesive energy density, crystallinity,
steric hindrance, chain flexibility and free volume (32).  Polarity,
hydrogen bonding and crystallinity effects are not expected to
contribute to differences in the observed gas transport properties of
polycarbonates 1 and 11-14, as they differ only by the structure of
the hydrocarbon substituents.  Nevertheless, oxygen permeabilities
within this series of polycarbonates differ by as much as a factor of
twenty.  Therefore, it remains to rationalize these gas transport
differences on the basis of polymer chain flexibility (segmental
mobility) and polymer chain packing.
    Considering chain mobility first, the predominate segmental
motion of unsubstituted polycarbonate 1 in its glassy state is a
partial rotation, or $\pi$-flip, of the phenylene group about its 1-4
axis (33).  A large volume element change has been assigned to this
motion by dynamic mechanical techniques.  On steric grounds, ortho
substitution would be expected to slow this motion and, indeed, this
has been found to occur experimentally by dynamic mechanical (33) and
NMR techniques (34).  Additionally, MINDO calculations modeling the
energy for rotation about the 1-4 axis of suitable diphenyl carbonate
models indicate an increase in this barrier in going from the
hydrogen to methyl subsituted system (2.5 to 8.0 Kcal/mole).  There
is only a small increase in the rotational barrier in going from the
methyl to the ethyl or isopropyl substituted models, while the
tetramethyl carbonate simply shows a twofold barrier to rotation.  In
summary, the experimental and computational data for the polymers
indicates that the ordering of segmental motion is expected to be:
1>11>12>13>14.  This order is not observed for the permeabilities
of the polycarbonates, implying that this particular motion is not by
itself the controlling factor for gas transport rates.
    The glassy morphology of the 3,3'-disubstituted polycabonates
precludes direct, crystallographic measurement of polymer chain
packing.  However, the densities of these polymers decreases as the
substituent size increases.  Thus this crude measure of packing
density does not correlate with the observed oxygen permability data.
To gain a more accurate view of polymer chain packing, segmental
models of these polycarbonates were prepared and studied by single
crystal diffraction methods.  A summary of these more detailed
examinations of effects of polymer segmental mobility and packing on
the permeability of substituted polycarbonates will be reported
shortly (Bendler, J. T.; Grabauskas, M. F.; Schmidhauser, J. C.
General Electric Co., unpublished data).

Permeability of Polymers Containing a Common Bisphenol.  To
investigate the importance of monomer structure versus polymer class

(i.e., functional group link) in determining a polymer's gas transport rate, it is necessary to obtain permeability data for several classes of polymer, the members of which are based upon monomers with a common structure. To this end, oxygen permeabilities measured for several polyarylates (PAr) and polyetherimides (PEI) are compared in Table IV with the previously listed transport data for polycarbonates (PC) based upon the same bisphenol monomers. (For comparison purposes, only polyarylates containing a 1 : 1 molar ratio of iso- and terephthalic acids are considered at this point.) Analysis of this transport data leads to several conclusions concerning the relationship between polymer structure and permeability.

Within a particular polymer class, oxygen permeability rates vary widely depending upon the different monomer structure present. The strong dependence of the gas permeabilities of polycarbonates on the structure of their repeat unit (rates vary by a factor of 63 for these examples) has been previously noted. The similarly wide range of rates observed for the polyarylates (factor of 17) and polyetherimides (factor of 10) indicates that this strong structural dependence is general for a number of classes of bisphenol derived polymers. In addition, it is found that the same ordering of oxygen permeabilities is followed by members within each polymer class, so that the spirobiindane bisphenol containing materials always have the highest transmission rates while the 1,1-bis(4-hydroxy-3-methyl-phenyl)cyclohexane based materials always exhibit the lowest rate. Both the wide range and the similiar ordering of permeabilities within these classes of polymers are evidence for the importance of monomer structure in determining polymer gas transport properties.

An alternative analysis of this gas transport data can be made by focusing on classes of polymers which were made from the same bisphenol monomers. This comparison of the oxygen permeabilities reveals the following general order: polyarylate > polycarbonate > polyetherimide. These results agree with previously reported observations on bisphenol A based polymers (27). As might be expected, a bisphenol A based polyestercarbonate (GE Lexan PPC, containing 20 % carbonate links and 80 % ester links) exhibits an oxygen permeability (1.77 Barrers) somewhere between that of the pure PC or PAr. The polyarylates have only slightly higher oxygen permeabilities than the structurally related polycarbonates, while the analogous polyetherimides have much lower permeabilites then either material. Overall, the range of transport rates exhibited by these groups of polymers with the same-monomer/different-link is smaller, i.e., never more than a factor of five (in the case of bisphenol A), than that displayed by the just discussed subsets of polymers with different-monomer/same-link. Therefore, it is concluded that for these sets of bisphenol containing polymers, monomer (repeat unit) structural factors, not functional group links, play a dominant role in controlling polymer permeability properties.

Are Polyesters Better Gas Barriers Than Polycarbonates? Commercial alkyl terephthalate polyesters exhibit lower gas permeabilities than commercial bisphenol based polycarbonates. On this basis, polyesters are generally classified as good barrier polymers while polycarbonates are typically not regarded as such. Furthermore, the ester functionality has been assigned higher barrier values than the

Table IV. Oxygen Permeabilities of Polymers Containing a Common Bisphenol

| Polymer No. | Bisphenol | Polycarbonate $PrO_2$ | Polyester $PrO_2$ | Polyetherimide $PrO_2$ |
|---|---|---|---|---|
| 6 | | 6.90 | | |
| 18 | | | 7.44 | |
| 23 | | | | 2.82 |
| 4 | | 3.41 | | |
| 19b | | | 3.03 | |
| 1 | | 1.50 | | |
| 20b | | | 1.81 | |
| 24 | | | | 0.33 |
| 15 | | 1.26 | | |
| 21 | | | 1.32 | |
| 17 | | 0.11 | | |
| 22b | | | 0.44 | |
| 25 | | | | 0.30 |

carbonate group in a model which attempts to assign barrier values to segmental increments of polymers (35). However, the gas transport data listed in Table IV shows that in four of five examples, the bisphenol containing polyesters exhibits the same or slightly higher oxygen permeability than the structurally related polycarbonates. These results indicate that a polyester based upon a "polycarbonate-like" monomer (i.e., a bisphenol A derivative) exhibits a "polycarbonate-like" gas transport rate.

A similiar comparison can be made between permeability data of polyesters and polycarbonates containing "polyester-like" monomers. For example, polycarbonates 26 and 27, which contain predominantly bisphenols which are structurally similiar to the repeat unit (circled) of an alkyl terephthalate polymer (PBT), possess essentially the same low permeability as that polyester (Table V). Furthermore, there has been a recent announcement (36) that aliphatic polycarbonates, such as polypropylene and polyethylenecarbonate, are being developed as potential gas barrier materials. In summary these results suggest that the relative low permeabilities of commercial polyesters and high permeabilities of commerical polycarbonates are not a direct consequence of the ester or carbonate links, but are due instead to the structure of the monomers they are prepared from, i.e., the aliphatic diol and aromatic bisphenol, respectively.

Effect of Diacid Structure on Polyarylate Permeability. On first inspection, it would be tempting to ascribe the relatively high permeability of the polyarylates to the presence of isophthalate units which could lead to ". . . interuption of the [polymer] chain packing caused by the random "kinks" . . . " (26). To test this hypothesis, the effect of diacid structure on the permeability of poylarylates was explored by preparing series of polymers containing different molar ratios of iso- and terephthalic diacids (Table VI). Surprisingly, increasing the amount of isophthaloyl links lowered the observed oxygen transport rates, with an isophthaloyl-rich polyester having a permeability approximately one half that of a polymer based predominantly on the more symmetric terephthalate group.

Several rationales other than monomer structure were considered in attempts to explain the gas barrier enhancement imparted by the m-phenylene units (from isophthalic acid). For example, the opaque appearance of films prepared from isophthaloyl-rich BPA 20a and DMBPC 22a polyarylates suggested the possibility that increased crystallinity may account for the lowered permeability of these materials rather than any structural effect of the repeat units. While no detailed study of the effect of diacid ratio on polyarylate crystallinity has been reported, certain observations point against this being a determining factor for observed trends in the transport rates. First of all, an early survey of polyarylate properties states (37) that bisphenol A containing polyesters show increasing amounts of crystallinity when they contain either predominantly iso- or terephthalic acid. In addition, visual inspection and DSC analysis of the 5,5-bis-(4-hydroxyphenyl)nonane containing poly-arylates 19 a-c indicates that all three of these compositions are amorphous. This evidence indicates there is no general relationship between extent of crystallinity caused by the isomeric diacid composition and oxygen permeability of the polyarylates. In a similar manner, there is no correlation between the polyester's

Table V. Comparison of Polyester and Polyestercarbonate Permeabilities

| Polymer | Polymer Composition | $PrO_2$ |
|---------|---------------------|---------|
| PBT | | 0.072 |
| 26 | | 0.036 |
| 27 | (Copolymer with 10 % BPA) | 0.114 |

Table VI. Effect of Diacid Content on Polyarylate Permeability

| Polymer No. | Bisphenol | Acid Content | | $PrO_2$ |
|-------------|-----------|-----|-------|---------|
| | | % Iso | % Tere | |
| 19a | | 90 | 10 | 2.61 |
| 19b | | 50 | 50 | 3.03 |
| 19c | | 10 | 90 | 3.53 |
| 20a | | 75 | 25 | 1.41 |
| 20b | | 50 | 50 | 1.81 |
| 20c | | 25 | 75 | 2.52 |
| 22a | | 90 | 10 | 0.35 |
| 22b | | 50 | 50 | 0.44 |
| 22c | | 0 | 100 | 0.60 |

density (Table VII) and oxygen permeability, unlike the recently reported (38) findings of the properties of poly(phenolphthalein phthalate)s. Molecular structure, rather than polyester morphology, is therefore found to exert a controlling influence on the materials' gas transport properties.

Fitting of Transport Data to Structure-Permeability Models. Measurements of the oxygen permeability properties of the new substituted polyarylates, polyetherimides, and polycarbonates show that gas transport rates for a polymer class are not centered around a single value, but instead can span a large range. Given this broad distribution of possible properties, the ability to relate polymer structure to polymer transport rates would be most desirable. Of several possible strategies, the approach of relating polymer permeability properties to another polymer physical property is attractive due to its simplicity. For example a correlation between polymer glass transition temperature (Tg) and permeability has been proposed. Unfortunately, no such relationship can be drawn based upon the properties of the bisphenol containing polymers (Table VII). To illustrate this point, a plot of permeability versus Tg for the substituted polycarbonates *1-17* (Figure 2) shows no discernable trend. Evidently a more sophisticated approach to permeability modeling is needed.

The process of a permeate diffusing through a transient network of channels or microvoids is basic to most models of polymer diffusion. The varying tightness of the polymer chain packing is therefore strongly implicated as a factor in determining polymer permeability rates. However when trying to predict polymer chain packing, one is confronted by a problem that has plagued enzyme/ protein chemists for years, namely that there is currently no good model for relating monomer (primary) structure to polymer chain folding/packing (tertiary) structure. On a very simple level, this is illustrated by a lack of correlation between monomer symmetry and gas permeability of the derived polymer, as was observed for the methyl-substituted bisphenol containing polycarbonates and the iso-/terephthalic acid containing polyarylates. With no direct method presently available to calculate the extent of polymer chain packing, other approximation methods must be used.

The free volume approach has been an increasingly popular method to relate polymer structure to gas transport properties. The basic premise of this technique is that a polymer with an open, poorly packed structure will have a large unoccupied free volume through which a gas can diffuse with ease. In a typical model, set forth by Lee (39) a specific free volume, SFV, is derived from the difference between the molar volume, Vm, (determined from the experimental density of a polymer) and the occupied volume, Vo, (calculated using a group additive method, in this case, that of Bondi) (40).

A plot of the logarithim of the oxygen permeability coefficients versus the reciprocal of the SFVs, as proposed by Lee, shows a fair correlation between these values. Data for the polyetherimides and polyarylates (Figure 3) is meager, but does allow some general conclusions to be drawn. The model does apparently overestimate the permeability for the bisphenol A based PEI *24*. In addition, the fit of the polyarylate data with the calculated model is favorable except for an overestimation of the permeability of the bis(4-hydroxy-

Table VII. Physical Properties of Polymers Prepared from Bisphenols

| Polymer No. | Glass Transistion Temperature (°C) | Density (g/cc) | SFV |
|---|---|---|---|
| 1 | 150 | 1.199 | 0.138 |
| 2 | 137 | 1.159 | 0.141 |
| 3 | 130 | 1.116 | 0.154 |
| 4 | 113 | 1.098 | 0.152 |
| 5 | 179 | 1.208 | 0.129 |
| 6 | 228 | 1.147 | 0.131 |
| 7 | 155 | 1.222 | 0.124 |
| 8 | 145 | 1.212 | 0.119 |
| 9 | 155 | 1.203 | 0.114 |
| 10 | 150 | 1.238 | 0.109 |
| 11 | 98 | 1.167 | 0.127 |
| 12 | 65 | 1.128 | 0.137 |
| 13 | 80 | 1.081 | 0.159 |
| 14 | 191 | 1.086 | 0.163 |
| 15 | 101 | 1.093 | 0.149 |
| 16 | 117 | 1.192 | 0.111 |
| 17 | 131 | 1.184 | 0.096 |
| 18 | 278 | 1.127 | 0.175 |
| 19a | 160 | 1.126 | 0.147 |
| 19b | 156 | 1.128 | 0.145 |
| 19c | 188 | 1.124 | 0.148 |
| 20a | 186 | 1.208 | 0.135 |
| 20b | 195 | 1.205 | 0.137 |
| 20c | 212 | 1.223 | 0.125 |
| 21 | 156 | 1.121 | 0.145 |
| 22a | 181 | 1.208 | 0.105 |
| 22b | 193 | 1.177 | 0.127 |
| 22c | 232 | 1.183 | 0.123 |
| 23 | 258 | 1.226 | 0.131 |
| 24 | 222 | 1.280 | 0.112 |
| 25 | 214 | 1.253 | 0.107 |

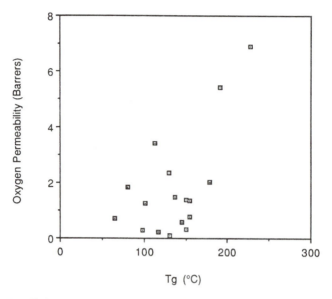

Figure 2. Relationship between glass transistion temperature and oxygen permeability of substituted polycarbonates.

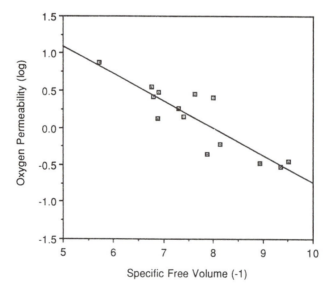

Figure 3. Correlation between polyarylate and polyetherimide oxygen permeability versus reciprocal of specific free volume.

3-methylphenyl)heptane based PAr **21**. One shortcoming of the group additive method is that the tables used (**41**) do not distinguish between isomeric disubstituted benzenes, with the result that the same occupied volumes are calculated for polyarylates containing different ratios of iso- and terephthalic acid.

The success of this model in analyzing the polycarbonate permeability data is mixed (Figure 4). At its best, this model correctly predicts the ordering of the permeabilities of the 3,3'-disubstituted polycarbonates **11 - 14**. On the other hand, the order of permeabilities predicted for the medium-sized ring containing bisphenol PCs **7 - 9** is wrong. In addition there is substantial scatter in the rest of the data, with an extreme example being SBI based PC **6** having a much greater permeability than is calculated. Therefore, a free volume approach to modeling polymer permeability, while having merit, also has room for refinement and work continues towards that end.

Conclusions

Oxygen permeabilities have been measured for polycarbonate, polyarylate and polyetherimide polymers made from a common set of structurally varied bisphenol monomers. Within each polymer class, the transport rates were found to vary dramatically depending upon which monomer was used, with the same relative order of rates being observed in each set of materials. By comparison, the difference in permeabilities among members of different polymer classes based on the same monomer was found to be smaller. These results can be interpreted as showing that monomer structure has a dominant role in determining the transport rates of these bisphenol containing polymers, while the type of functional group link plays a secondary role in this regard.

A more general comparison of oxygen permeability data for polycarbonates and polyesters shows that, given a similar monomer structure, a polycarbonate exhibits the lower gas transport rate. This observation is different than one would expect based upon available literature permeability data, pointing out the difficulty in drawing conclusions about polymer structure-permeability relationships using data from a relatively small number of polymers. Throughout these studies, synthesis of new polymers containing specifically varied monomers played a crucial role in detailing the complex relationship between polymer structure and gas permeability.

The finding that polymers within a particular class possess a wide range of transport rates should make it possible to synthesize materials whch exhibit a better match between permeability and some other property needed for a particular application, such as thermal stability. The challenge comes in attempting to develop a model for predicting polymer transport properties based on the structure of the repeat unit. This study shows that the presence of both the functional group link and the structure of the monomer must be considered when attempting to model polymer permeability. Furthermore, examples were found where the symmetry of a monomer's structure does not by itself correlate with polymer permeability (e.g., the iso-/terephthalate containing polyarylates). Additionally, no simple relationship between polymer permeability and other physical properties was found.

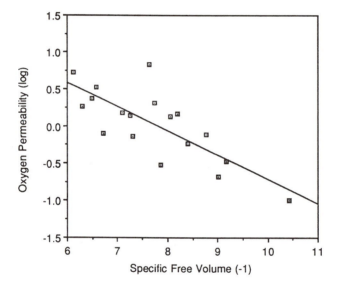

Figure 4.   Correlation between polycarbonate oxygen permeability versus reciprocal of specific free volume.

A detailed analysis of the 3,3'-disubstituted bisphenol containing polycarbonates provided an explanation for their observed oxygen permeabilities. However such an approach is not of general utility. Using an existing free-volume approach, a qualitative relationship between monomer structure and the experimental polymer permeabilities was found to exist. Therefore, these and other results are presently being used to refine an excess-free volume type method in hopes of achieving better estimated gas transport properties based on molecular structure.

Acknowledgments

We thank John Bendler for performing the molecular modeling calculations and for many helpful discussions. Monty Alger and Lorraine Rogers provided some of the permeability data, while John Campbell, Dave Dardaris, Gary Faler, Ed Fewkes, Paul Howson, Rick Joyce, Jerry Lynch and John Maxam contributed to the results described in this report.

Literature Cited

1.  Reed, T. F.; Thomas, G. B. Chemtech 1988, 18, 48-53.
2.  "New Generation of Membranes Developed for Industrial Separations", Chem. Eng. News, June 6, 1988, pp. 7-16.
3.  Stannett, V. T. Polym. Sci. Eng. 1978, 18, 1129-1134.
4.  Steingiser, S.; Nemphos, S. P.; Salame, M. In Kirk-Othmer Encyclopedia of Chemical Technology 3rd Ed.; Grayson, M., Ed.; Interscience: New York, 1978; Vol. 3, pp.480-502.
5.  Pye, D. G.; Hoehn, H. H.; Panar, M. J. Appl. Polym. Sci. 1976, 20, 287-301.
6.  Sykes, G. F.; St. Clair, A. K. J. Appl. Polym. Sci. 1986, 32, 3725-3735.
7.  Stern, S. A.; Mi, Y.; Yamamoto, H.; St. Clair, A. K. J. Polym. Sci. Polym. Phys. Ed. 1989, 27, 1887-1909.
8.  Takada, K.; Matsuya, H.; Masuda, T.; Hiyashimura, T. J. Appl. Polym. Sci. 1985, 30, 1605-1616.
9.  Kawakami, Y.; Kamiya, H.; Toda, H.; Yamashita, Y. J. Polym. Sci. Polym. Chem. Ed. 1987, 25, 3191-3204.
10. Stern, S. A.; Shah, V. M.; Hardy, B. J. J. Polym Sci. Polym. Phys. Ed. 1987, 25, 1263-1298.
11. Cassidy, P. E.; Aminabhavi, T. M.; Thompson, C. M. Rubber Chem. Tech. 1983, 56, 594-618 .
12. Perkins, W. Polym. Bull. 1988, 19, 397-401.
13. Meares, P. J. Amer. Chem. Soc. 1954, 76, 3415-3422.
14. Vieth, W. R.; Sladek, K. J. J. Colloid Sci. 1965, 20, 1014-1033.
15. Paul, D. R.; Koros, W. J. J. Polym. Sci. Polym. Phys. Ed. 1976, 14, 675-685.
16. Fredrickson, G. H.; Helfand, E. Macromolecules 1985, 18, 2201-2207.
17. Petropoulos, J. H. J. Polym. Phys. Ed. 1988, 26, 1009-1020.
18. Koros, W. J.; Chan, A. H.; Paul, D. R. J. Membrane Sci. 1977, 2, 165-190.
19. Koros, W. J . Ph. D. Dissertation; Univ. of Texas, Austin, 1977.
20. Erb, A. J. M. S. Thesis, Univ. Of Texas, Austin, 1977.

21. Schnell, H. Chemistry and Physics of Polycarbonates; Inter-
    science: New York, 1969.
22. Pilato, L.; Litz, L. M.; Hargitay, B.; Osbourne, R. C.; Farnham,
    A. G.; Kawakami, J.; Fritze, P. E.; McGrath, J. E. Polym. Prepr.
    1975, 16, 42-44.
23. Muruganandam, N.; Koros, W. J.; Paul, D. R. J. Polym. Sci.
    Polym. Phys. Ed. 1987, 25, 1999-2026.
24. Moe, M. B.; Koros, W. J.; Paul, D. R. J. Polym. Sci. Polym.
    Phys. Ed. 1988, 26, 1931-1945.
25. Sheu, F. R.; Chern, R. T.; Stannett, V. T.; Hopfenberg, H. B. J.
    Polym. Sci. Polym. Phys. Ed. 1988, 26, 883-892.
26. Barbari, T. A.; Koros, W. J.; Paul, D. R. J. Polym. Sci. Polym.
    Phys. Ed. 1988, 26, 709-727.
27. Barbari, T. A.; Koros, W. J.; Paul, D. R. J. Polym. Sci. Polym.
    Phys. Ed. 1988, 26, 729-744.
28. Tsia, H. B.; Lee, Y.-D. J. Polym. Sci. Polym. Phys. Ed. 1987,
    25, 3405-3412.
29. Wirth, J. G.; Heath, D. R. U. S. Patent 3 875 116, 1974.
30. Huglin, M. B.; Zakaria, M. B. Angew. Makromol. Chem. 1983, 117,
    1-13.
31. Mark, V.; Hedges, C. U. S. Patent 4 304 899, 1974.
32. Crank, J.; Park, G. S. Diffusion in Polymers, Academic Press:
    London, 1968.
33. Yee, A. F.; Smith S. A. Macromolecules 1968, 14, 54-73.
34. Ratto, J. A.; Inglefield, P. T.; Rutowski, R. A.; Li, K.-L.;
    Jones, A. A.; Roy, A. K. J. Polym. Sci. Polym. Phys. Ed. 1987,
    25, 1419-1430.
35. Salame, M. Polym. Eng. Sci. 1986, 26, 1543-1546.
36. Chem. Engineering November 24, 1986, p. 11.
37. Eareckson, W. M. J. Polym. Sci. 1959, 40, 399-406.
38. Sheu, F. R.; Chern, R. T. J. Polym. Sci. Polym. Phys. Ed. 1989,
    27, 1121-1133.
39. Lee, W. M. Polym. Eng. Sci. 1980, 20, 65-69.
40. Bondi, A. Physical Properties of Molecular Crystals, Liquids and
    Gases; Wiley: New York, 1968.
41. VanKrevelen, D. W.; Hoftyzer, P. J. Properties of Polymers;
    Eisevier: New York 1972; pp. 410-420.

RECEIVED December 11, 1989

# Chapter 8

# Water Transport Through Polymers

## Requirements and Designs in Food Packaging

Leonard E. Gerlowski

Polymeric Materials Department, Shell Development Company,
Houston, TX 77251−1380

The plastics packaging industry needs to understand the
transport of water through polymers in order to
properly utilize these materials in food containers.
The primary goal of any food container is to protect
the quality of the packaged food. To meet this goal,
and account for water-polymer interactions which can
affect the food, the transport of water must be
properly taken into account in material selection and
design of plastic containers. For example, the
abundance of water in the food packaging environment
for sterilization, sealing, etc., can affect the
polymer properties and limit the package to only
certain uses. Also, interactions of water with
polymers used in food packaging can affect the
organoleptic properties (taste characteristics) of
food-polymer interactions. On this basis, the
transport characteristics of water in polymers are
reviewed from a physico-chemical basis. These transport
characteristics are then included in design equations
and three design calculations are presented; i. an
equation to predict proper wall thickness; ii. a
heuristic rule to insure the polymer is transport
limiting; and iii. an example of a water mass balance
to predict the effect of water sorption on transport
properties common to the food packaging environment.

The many benefits offered by plastics over conventional materials
has led to increased use over the last decade (1). Some of these
benefits include light weight, ease of use and manufacture,
aesthetics, economics and many others. However, one deficiency of
plastics is that all polymers allow small molecules to dissolve into
and diffuse through this matrix at some rate. This process has been
termed permeation and is of interest to protect the food from

0097−6156/90/0423−0177$06.00/0
© 1990 American Chemical Society

external or internal permeants which can alter the food chemistry. Permeation of water is of particular interest in food packaging due to its abundance and importance to the food (2). Water is used at high temperatures and pressures in the food packaging environment for sterilization, sealing, etc. (3). Such processes can affect the initial package properties (2). Also, most foods require a certain water content to provide proper taste, consistency and other properties. It is the responsibility of the package design to insure the package can provide such food properties over a given shelf life. The objective of this work is to briefly review water transport in polymers and provide heuristic rules to aid in proper container design.

The interest of the plastics packaging industry in designing containers which limit water transport is two-fold. First, the quality of the packaged food must be maintained. In many cases, this requires keeping water out of dried foods or in wet foods (4). To design containers to properly meet these requirements, the water transport characteristics must be well understood. These transport properties can be highly non-linear and may require taking the water content of the food into account. With a proper understanding of these transport properties, a container can be designed which properly maintains the water content of a packaged good. Secondly, the permeability of gases other than water in many hydrophilic polymers can be affected by water content. The second case is of particular interest where food undergoes a high temperature, high water (steam) sterilization process following packaging. The high water activity in the barrier polymer resulting from sterilization can reside with the container for long periods of time. With a proper understanding of the water transport characteristics of the container, this long term effect can be properly handled in the design. Methodologies to predict the response of plastic package designs in such instances are also discussed in the container design section.

In order to use design methodologies, the basic data used in such calculations (transport properties) should be well understood. Simple gases such as oxygen and carbon dioxide, under most normal food packaging conditions, follow Fickian transport mechanisms. However water, due to its hydrogen bonding nature and other polar interactions with the polymer backbone, tends to follow different transport mechanisms under transient conditions. These water-polymer interactions have been generically quantified and are reviewed briefly in the transport mechanism section. Also, examples of alterations to the polymer backbone to reduce water transport are presented in the polymer structural characteristics section. These examples are presented as guidelines to follow in designing or altering polymer chemical structure to meet water transport requirements.

TRANSPORT MECHANISM

Water and polymer interactions have been recognized since the early developments of polymeric materials. These interactions have led to characterizing polymers as hydrophobic (e.g. polyolefins, etc.) and hydrophilic (e.g., polyvinylalcohol, nylons, etc.). The way in which polymers interact with water can affect the transport mechanism of water through polymers. Table I contains the transport parameters of several polymers of commercial food packaging interest. The water permeabilities of these polymers vary several orders of magnitude. Typically, the hydrophilic polymers sorption values are also much larger than the hydrophobic polymer values. This trend indicates that water sorption plays a role in determining the water transport characteristics.

The way in which water permeates through polymers is much the same as other vapors and has been well characterized ($\underline{5},\underline{6}$). As is the case with non-condensible gases, the permeability is a function of the sorption level and the diffusivity of the permeating substance:

$$P = D \times S \qquad (1)$$

Typically, in hydrophobic polymers, the sorption levels are relatively low, and the sorption follows Henry's Law. In these cases, the permeability is relatively unaffected by the surrounding water (water activity or relative humidity). The transport mechanism in these cases follows Fick's Law.

On the other hand, the water sorption level can be extremely high in hydrophilic polymers. In these cases, the permeability is greatly affected by the amount of water present. The state of water adsorbed can be considered: i.) as homogeneously distributed throughout the polymer; ii.) to reside primarily in localized areas, or iii.) to exist in clusters ($\underline{5}$). These high levels of water content can greatly affect the transport mechanism and dominate the transport process (as opposed to diffusion limited). In these cases, the water can be thought to move as a uniform front through the polymer [often referred to as Case II sorption ($\underline{7}$)] and the dynamics of water uptake is typically linear with time. The characteristic water uptake equations for the three states mentioned above are reviewed briefly in the following.

Homogeneous sorption occurs in cases where the interaction between the polymer and water is uniform. Flory ($\underline{17}$) described this case as the polymer and water being randomly distributed. The distribution can be described by:

$$p_0/p = V_{H2O} \exp (V_p + X V_p^2) \qquad (2)$$

where $p_0$ is the partial pressure of water in the environment, p is the total pressure, X is the interaction parameter which can be a function of the water activity, and $V_i$'s are the volume fractions polymer and water. Other more complicated forms have been postulated to include crosslinked systems and crystalline systems ($\underline{5}$). This interaction mechanism has been applied to cellulose, nylons, and polyvinylalcohol over limited water activity ($\underline{8\text{-}12}$).

Localized sorption occurs when adsorbed water is tightly bound to the polymer molecules. In these cases, two populations of water are considered to exist in the polymer: a bound population and a liquid like population. This sorption isotherm is described by:

Table I.  Water Transport Properties

| | P $100°F$ 90% $RH_2$ (gm-mil/100in$^2$-d) | S* Polymer (gm/100gmpolymer) | Reference |
|---|---|---|---|
| PVDC (coatings) | 0.02 - 0.06 | < 0.1 | 36,37 |
| (films) | 0.08 - 0.20 | -- | 36,37 |
| high density polyethylene | 0.38 | -- | 38 |
| polypropylene | 0.66 | 0.0071 | 38 |
| low density polyethylene | 1.4 | 0.0062 | 38 |
| PET | 1.8 | 0.8 | 38 |
| Polyvinylchloride | 2.2 | 1.5 | 36,38 |
| PAN | 2.1** | 3.6** | 35,18 |
| EVAL-F | 3.8 | 3.0 | 30 |
| -H | 2.1 | | 30 |
| -E | 1.4 | | 30 |
| Nylon-6/6,11 | 3.8 | 4.0 | 36 |
| Acrylonitrile/ Styrene (BAREX) | 6.1 | | 35 |
| PS | 7.4 | 0.048 | 36,38 |
| Nylon-6 | 10.9 | -- | 36 |
| Polycarbonate | 11.4 | -- | 35 |
| Nylon-12 | 63.5 | 4.0 | 36 |
| Cellulose acetate | 76.0 | -- | 36 |
| Hydroxy ethyl cellulose | 110 | -- | 36 |

*   at 25 C, 100% RH
**  at 30 C

$$\frac{A}{B} = \frac{C\ p_0}{(1 - p_0)(1 - p_0 + C\ p_0)} \qquad (3)$$

Where A molecules are distributed over B sites and C represents the partitioning of bound and liquid water. This "BET" like absorption model has been modified to more simplified forms for saturable sorption models. For low water activity, this sorption model has been applied to several hydrophilic polymers including epoxies (13) and nylons (14).

Clustering accounts for the association of water molecules in the polymer. In clusters water molecules can associate with other water molecules, but not necessarily polymer molecules. NMR evidence of these types of associations have been found in the literature. In this case, the sorption isotherms can be described as:

$$\frac{G}{v_{H2O}} = -v_p\ [\frac{\partial(a_{H2O}/v_{H2O})}{\partial a_{H2O}}]\ P,T\ -\ 1 \qquad (4)$$

Where $v_i$ is the partial molar volume, G is the clustering integral defined such that when $G/v$ is greater than -1 there is a probability for clustering to occur. This sorption isotherm has been applied to high water adsorbing polymers [nylon (15), polyurethanes (16), and polyacrylonitrile (18)] at high water activities (5).

For most applications for food packaging, sorption of water in most hydrophilic polymers can be fit to a saturation isotherm, such as Equation 3. A simplified form would be:

$$S = \frac{Ap_0}{B\ +\ p_0} \qquad (5)$$

where A and B are empirically determined constants at a given temperature. Although, this empirical approach may not provide physico-chemical insight to polymer/water interactions, it provides adequate information for design of plastic packages. Sorption of water in most hydrophobic polymers can be fit to simple Henry's law linear sorption:

$$S = p_0\ k \qquad (6)$$

where k is the sorption coefficient.

The diffusivity can be affected by concentration of water, but in most instances the transport of water is sorption limited. The temperature and surrounding relative humidity effects on uptake kinetics are usually more important than the water content effects.

POLYMER STRUCTURAL CHARACTERISTICS

As was seen from Table I for most hydrophilic polymers, the sorption of water appears to be the controlling element of the transport equation (1). With this in mind, there are several means to characterize this process. The polarity of the polymer backbone is a good indication as to the level of sorption and water transport.

This indication is not as straight forward as is the case for
non-interacting transporting gases (such as oxygen and carbon
dioxide). The structural permeability relationships for
non-interacting gases have been established with empirical
mechanisms relating permeability to a backbone interaction
parameter. One method develops a "permachor" and has been
established by Salame for a variety of polymers (19,20). The second
method is based on amorphous free volume (21). Based on this
approach, the more specific free volume in the polymer, the higher
the permeability or diffusion coefficient. Both of these methods
are based on polymer-polymer interactions and predict gas transport
properties of non-interacting systems well. It appears that in
highly adsorbing water-polymer systems, the transport is dependent
on polymer-water interactions and these structural prediction
methods do not apply at high RH.

        With this in mind, we consider the functional groups on a
polymer backbone (Figure 1). In the case where pendant groups have
no free electrons for interaction with other groups (hydrogen), the
water sorption and permeability are extremely low. In cases where
the pendant groups are highly hydrogen bonding groups and can easily
attach to water, the sorption and permeability are both high and
water activity dependant. In cases where the pendant groups are
very strong interactors - for example polyvinylidene chloride and
polyvinylidene fluoride - pendant groups interact with each other
more strongly than with water. In these cases, the water sorption
and permeability are extremely low.

        Van Krevelen (22) provides a correlation of water sorption for
various backbone and pendant groups at several relative humidities.
For the hydrophobic olefin backbone groups, the contribution is 2-5
orders of magnitude below that for the hydrophilic backbone groups.
In ranging water activity from 0.3 to 1.0, the increase in group
contribution from the olefin structures is of the order of 5.
Whereas, in the case of hydrophilic backbone groups, the increase
over this water activity range is an order of magnitude. For
pendant groups of halogen molecules or olefin groups such as
methane, the contribution is similar to that from an olefin
backbone. However, in the case of hydrophilic pendant groups such as
alcohols, the contribution is greater than the hydrophilic backbone
groups by an order of magnitude.

        For cases of new polymer design, many times it may be important
to increase or decrease water transport in a given polymer backbone
for some application. Choices of chemical groups to alter the
backbone can be based on heuristic guidelines from Van Krevelen
(22). As was previously discussed, adding olefinic or halogen
nature to the backbone can reduce transport and adding hydrophilic
groups can increase transport. In an example of structural
alterations to reduce water sorption, Barrie et al. (23) considered
halogenation of epoxy resins. They added bromine, chlorine, and
trifluoromethyl groups to the benzene rings in MY720. At 40, 50,
and 60°C they found lowering of sorption levels with the largest
reduction with the fluorinated system. The reduction was almost 3X
based on a dual mode model with the fluorinated system. Little
effect was seen on the diffusion coefficient. The non-Fickian

```
        H X1
        | |
    --(-C-C-)--
        | |
        H X2
```

| X1 | X2 | P | S |
|------|-------|------|-------|
| H | H | 1.4 | 0.006 |
| H | $CH_3$ | 0.66 | 0.007 |
| H | Cl | 2.2 | 1.5 |
| Cl | Cl | 0.06 | <0.1 |
| H | F | 0.04 | 0.01 |
| F | F | 0.06 | 0.04 |
| F,F | F,F | 0.02 | -0- |
| Cl,F | F,F | 0.03 | -0- |
| F,F | $CF_3$,F | 0.02 | 0.01 |

------------------
P = units of gm-mil/100sqin-d

S = gm per 100 gm polymer

Data taken from Reference 39

Figure 1.  Structure Relationship:  WVTR

transport characteristics of the base epoxy were also evident in the halogenated compounds.

## CONTAINER DESIGN

The primary objective of the food container is to protect the packaged food from spoilage. For many foods this protection includes prohibiting water transport into the container (for dry foods) or out of the container (for wet foods), and to protect against oxygen transport in or carbon dioxide out. This next section outlines a method of container design which properly accounts for the water related transport characteristics of the package. The first two characteristics relate to the transport of water itself: to insure the polymer is rate limiting, and to design for adequate wall thickness to protect food from water loss. The third characteristic relates to the effect of water on other transport properties. The complete design of a container includes gas transport, material strength, economics, etc. Inclusion of all of these issues is complex and beyond the scope of this text. In what follows, the focus will be on the water transport implications of container design.

To properly design a container to limit the water transport, the food water transport characteristics must also be accounted for. To provide guidance in design, Taoukis, et al. (24), have quantified the water transport in a food and container. A dimensionless value (L) can be defined based on the internal (food) to external (polymer) transport characteristics of the system:

$$L = \frac{\beta/a}{P/l} \tag{7}$$

where $\beta/a$ is the permeability of the food normalized with the half thickness of the food (a), and $P/l$ is the permeability of the polymer normalized for a wall thickness (l). This value is somewhat analogous to a Sherwood number in mass transport. At high L values, the transport is limited by the polymer. At low L values, the food water sorption isotherm (as part of the food permeability $\beta$) dominates the transport. Taoukis, et al. (24), found that predictions of water content were generally below the measured values when the food transport was the limiting step. Khanna and Peppas (25,26) applied a rigorous mathematical analysis to combined oxygen and water transport in package design. Their results also indicated that the polymer should be the limiting transport mechanism. If the polymer is not rate controlling, the non-linear effects of the food sorption isotherm can greatly affect the results of the analysis. A heuristic rule from these works is to insure water transport is dominated by the package to reliably design package performance. This constraint can be met by choosing container wall thickness (l) or a polymer of low permeability (P) which results in an L value which is large (greater than four). Taoukis, et al. state that L can be as low as 0.2 without greatly affecting the ability to predict moisture transport. However, when L is less than 0.2, predictions were quite below measurements. For conservative design, an L value of greater than 4 is recommended.

For most rigid packaging applications with thick polyolefin layers, this requirement is easily met. In cases of flexible packages (film wraps, etc.), the transport limitations should be checked. Once the water transport limitations have been designed to be polymer dependent for a given container, the amount of polymer necessary to perform this task can be determined. For most containers, this implies estimating a wall thickness. For a polymer with water transport characteristics not highly dependent on environmental water activity, Labuza (27,28) has determined the design equation to be:

$$m(t) = m_e - (m_e - m_i) \exp\{(P/l)(A/W)(p_0/b)t\} \tag{8}$$

where $m(t)$ = moisture content at time t (gm water/gm food), $m_e$ = equilibrium moisture content of the food in contact with external relative humidity, $m_i$ = initial moisture content of the food, P = water permeability of package wall (gm-mil/100sqin-d-mmHg), l = wall thickness (mil), A = area of transmission (square inches), W = mass of food in gm, $p_0/b$ = water vapor pressure of pure water a function of temperature (mmHg) divided by the slope of the moisture sorption isotherm (MSI) curve from m to $m_i$. This method was able to predict data for cheese food packaged in three different package designs over a one month time frame (27). For a polymer with water transport properties which are dependent on surrounding water activity, Equation 8 needs to be applied over limited ranges of constant environmental humidity and solved piecewise.

Hence, there are two criteria to properly design a container for water transport. The first is to insure the polymer is rate limiting (Equation 7), and the second is to insure a minimum weight loss over the shelf life. To properly apply these equations, the wall thickness required for a given polymer is found from Equation 8. This wall thickness then is the minimum value to use in Equation 7. That is, apply Equation 7 with the P/l from Equation 8. If the L from Equation 8 is less than 4, then increase the wall thickness with the criteria that Equation 7 must be greater than or equal to 4. If these conditions cannot be met practically or economically, the designer will need to iterate on polymer choice to insure these requirements can be met.

The third characteristic which must be considered is the effect of water on transport properties of other permeants. Certain food processing steps such as sterilization, sealing, etc. require the use of steam. As an example of such a processing step, we will consider the food sterilization (or retorting), although other processes such as sealing (3) or storage (24) may indeed play a role. Typically, this sterilization step is performed on the packaged food to insure against microbial growth. Maximum conditions for sterilizations are $250°F$ for one hour in a pressurized aqueous environment (hot water or steam). At this temperature, the water permeability of most polymers increases drastically [approaching 100 X over ambient conditions based on typical Arrhenius relationship (5)]. As was discussed in the structure/property section, this increased water activity can also affect other transport properties of hydrophilic polymers. Both of these effects can be properly designed for with an understanding of

the transport characteristics of the polymers used in the container
construction.

To account for such effects on hydrophilic polymers used in
container design, the container structure needs to be properly
understood. Many food packages are formed in a composite laminate
structure with a barrier layer for water and a barrier layer for
oxygen. In containers manufactured with this technology, the gas
barrier is typically a very hydrophilic polymer (for example,
polyvinyl alcohol copolymers, nylons, etc.)(29). Due to the high
water activity during processing, the water transport characteristics of the gas barrier polymer must be well understood to
properly design such a container. Of primary importance, and as has
been well documented, the oxygen permeability of the gas barrier
layer can increase over an order of magnitude in going from the dry
state to a high water activity (30). In a typical multilayer
package, the structure wall consists of a thick polyolefin layer, a
thin adhesive layer, a thin layer of the barrier polymer, a thin
adhesive layer, and a thick polyolefin layer. In this symmetric
configuration, the water available to the barrier layer is limited
by the sorption and diffusion through the thicker polyolefin layers.
This configuration has been shown to have an oxygen permeability,
which at ambient humidity, is capable of maintaining adequate shelf
life of many foods (29). During sterilization, however, the barrier
layer can become saturated with water. Once the barrier layer is
saturated and the container is cooled to room temperature, the
barrier polymer needs to readily lose the water it gained during
sterilization. Thus, in cases where sterilization is necessary, the
transient transport of water from the gas barrier layer back through
the water barrier layer must be adequately understood.

With a proper understanding of the transport properties in such
multilayered structures, the containers for use in sterilization
processes can be appropriately designed. As an example in designing
for such a situation, a mass balance on water permeating into or out
of a multilayered structure can be made to estimate the amount of
water which can affect the hydrophilic polymer. With this analysis,
the time required to dry the water sensitive layer and return the
container to ambient oxygen permeability values can be estimated.
Assuming the transport of water in the multilayer configuration is
limited by the outer layers, the water content (M) of the inner
layer in a prescribed container can be described by:

$$\frac{dM}{dt} = \frac{P_{OL}}{L_{OL}} (a_{w,IL} - a_{w,o}) \qquad (9)$$

where $P_{OL}$ is the water vapor permeability of the outer layer, $L_{OL}$ is
the thickness of the outer layer, $a_i$ is the water activity, and $t$ is
time. As an example, the inner layer was assumed to be EVOH and the
outer layer was assumed to be polypropylene. The water activity
absorption curve was determined from EVALCA data (30) and fit to a
saturation isotherm as Equation 5. The water permeability of the
outer layers was taken from the open literature for polypropylene.
The EVOH was assumed to be saturated at the beginning of the
"drying" cycle and "dry" when the water content was reduced to 5% of
the initial value. This analysis implies the water content of the

EVOH layer will be reduced in an exponential manner with time. Consequently, the EVOH layer will return to nearly its original barrier level following the "drying" period. To compare several design configurations, Equation 9 was solved implicitly to provide a time required to "dry" the EVOH. Three multilayer configurations were considered and are summarized in Table II. One configuration has an outer layer of polypropylene of 5.6 mils. This calculation was made to compare with a similar experiment reported in the literature (31). The calculation estimates approximately 12 days are necessary to reduce the water content to 5% of the saturation level. The reported experiments indicate that approximately 10 days were necessary to reduce the EVOH oxygen barrier to original levels. The measurement of oxygen barrier indicates that the barrier layer has lost most its water, or to a point we assumed to be 5% of the saturation value. This agreement between independent experiments and calculations indicates Equation 9 can be used as a quantitative tool to screen various container designs. Caution should be exercised to never allow water content to rise to a level above which irreversible transport changes may occur (Wachtel, J. A. and B. C. Tsai, to be published in this symposium series). Two other extreme cases are also presented. One with a very thick outer layer of (25 mils) and one with a very thin outer layer (1 mil). As these cases show, by increasing and decreasing the thickness of the outer layer, the time required to reach the initial oxygen barrier levels can be affected. Of course, these two extreme cases are used as illustrative examples. In the very thick PP case it could be argued that the barrier layer may not reach the same water sorption level as the other cases. And this may indeed be the case as the multilayer container structure is finding its way into the retortable food packaging market with thicker barrier layers (1). These calculations are included to compare various design scenarios and should be appropriately applied. The objective of the design of such containers is to reduce the time needed to regain the original properties of the barrier polymer. Thus, by reducing the time at the higher permeability levels, the cumulative oxygen uptake by these containers can be minimized.

LARGE MOLECULE EFFECTS

Food deterioration can occur by means other than transport of gases into the food container or water into or out of the container. The loss of food characteristics in plastics packaging may also occur from direct interaction of the food with the polymer. The measurement of these types of transport is often termed organoleptic analysis because they may involve human taste tests. The tainting of food by transport of substances from the polymer into the food is termed migration. The loss of flavor constituents from the food into the polymer is termed scalping (32). Water content can affect these transport characteristics as well.

The transport mechanisms of larger molecules is similar to that of smaller molecules, where the molecule is first sorbed into the polymer and then is free to diffuse throughout the polymer. In non-interacting hydrophobic systems, the transport mechanism can be described by Fick's law and the presence of water has little effect.

Table II.  CALCULATED RECOVERY TIME OF EVOH LAYER

```
|                |          |          |
|                |          |          |
|     Layer 1    |          | Barrier  |
|                |          | Polymer  |
|<----L1------>|          |          |
|                |          |          |
```

| Layer 1 | L1(mils) | $\theta*$ | $\theta*_{exp}$ |
|---------|----------|-----------|-----------------|
| Low WVTR<br>(0.3 gm-mil/<br>100sqin-d) | 1.0 | 4 | - |
| Low WVTR<br>(0.3 gm-mil/<br>100sqin-d) | 5.6 | 12  days | 10 days |
| Low WVTR<br>(0.3 gm-mil/<br>100sqin-d) | 25 | 21 | - |

$\theta*$  =  time required to reach 5% of saturation level of EVOH
       layer
$\theta*_{exp}$  =  data from ref 31 - time for structure to return to
       original oxygen permeability.

In these cases, the absorption or migration can be described by the equations developed for transport into a uniform slab (33). The amount of substance which migrates or is scalped can be typified by a square root of time dependence. In hydrophilic polymers, the presence of water tends to complicate this mechanism. Water can swell the polymer and/or "drag" other dissolved substances along with it into the polymer. Little work has been done in characterizing hydrophilic polymer with combined water and larger molecule transport. As guidance characterizing such systems, Chang, et al. (34) have quantified the uptake of n-octadecane in polypropylene. N-octadecane is a larger molecule which also swells the polymer (polypropylene). They described the polymer matrix to contain two transport mechanisms, one for the non-swollen portion and one for the swollen portion behind the advancing front. They also assumed the swollen portion to grow linearly with time. With this type of analysis, they were able to compare with experimental results of n-octadecane in polypropylene. Although this analysis was applied to organic systems, the same rationale can be applied to highly water adsorbing polymers and the loss of taste components from a food to the polymer in a high water environment.

Salame (35) has considered the scalping of flavor constituents into polymers by dilute solution absorption. He found sorption in hydrophilic polymers to be dependent on surrounding water activity (Table III). In the most hydrophilic polymers such as nylon and EVOH, there was an increase both in non-polar organics and polar organics in these two polymers in the presence of water. The increase in non-polar organics could be attributed to increased free volume of the polymer in the sorbed polymer state, while the increase in polar organics could be attributed to additional interactions available for the polymer molecule.

Table III.   FLAVOR ADSORPTION OF HYDROPHILIC PACKAGING POLYMERS

|  |  | PERMACHOR | AVERAGE FLAVOR ADSORPTION (%) (50°C) | |
|---|---|---|---|---|
|  |  |  | Non-Polar Organics | Polar Organics |
| NYLON | DRY | 80 | 1.1 | 2.8 |
|  | WET | 64 | 3.3 | 6.2 |
| EVOH | DRY | 110 | <1.0 | 1.4 |
|  | WET | 70 | 2.3 | 8.4 |

Reprinted from ref. 34. Copyright 1988 American Chemical Society.

## CONCLUSIONS

The use and design of polymers in the food packaging industry requires an understanding of the water transport properties of the

polymer and the effects of water on other transport properties of
the package. Water can play a role in shelf-life, polymer
properties, and polymer-food interactions. Along these lines,
several issues have been organized to provide guidance in material
selection and container design. First, water transport through
hydrophilic polymers is primarily controlled by the level of
sorption. Hence in choosing polymers for low water transport the
polar, dipole, and ionic interactions should be either limited or
the inter-chain interactions need to be large. Secondly, the design
equations for water transport through containers are available.
Several rules have been outlined: i.) the design equation should
take the food sorption characteristics into account as per
Equation 8; ii.) once a chosen polymer is designed for proper wall
thickness, the polymer should be the transport limiting step as in
Equation 7; iii.) in cases where multilayer structures are
sterilized, the dynamics of the container following retort can be
properly determined from a mass balance of water in the system.
Finally, it must be emphasized that the water content of the polymer
can affect the organoleptics of the package. This may be
particularly true when food processing conditions may result in high
water content in hydrophilic polymers. With these transport
properties accounted for, containers can be designed which maintain
the properties of packaged foods.

LITERATURE CITED

1. Kreisher, K.R. Modern Plastics, August, 1989, 31-33.
2. Duckworth, R.B. Water Relations of Foods, Proceedings of an International Symposium, Glasgow, Sept., 1974, Academic Press, NY, NY, 1975.
3. Lopez, A. A Complete Course in Canning and Related Processes, 12th ed., The Canning Trade, Inc., Baltimore, MD, 1987.
4. Lazarides, H.N.; Goldsmith, S.M.; Labuza, T.P. Chem. Eng. Prog., May, 1988, 46.
5. Barrie, J.A. In Diffusion in Polymers, Ch. 8, Crank and Park, Eds., Academic Press, NY, 1968.
6. Hopfenberg, H.B., Ed.; In Polymer Science and Technology, 6, Plenum Press, NY, 1974.
7. Cromyn, Ed. Polymer Permeability, Elsevier Applied Science Publishers, Essex, 1985.
8. Cutler, J.A.; Mclaren, A.D. J. Polym. Sci., 3, 1948, 792.
9. Rowen, J.W.; Simha, R. J. Phys. Coll. Chem., 53, 1949, 921.
10. Kawai, T. J. Polym. Sci., 37, 1959, 181.
11. Starkweather, H.W. J. Appl. Polym. Sci., 2, 1959, 129.
12. Taylor, N.W.; Cluskey, J.E.; Senti, F.R. J. Phys. Chem. Wash., 65, 1961, 1810.
13. Slabaugh, W.H.; Sparlaris, G.N.; Holbroke, L.E. J. Appl. Polym. Sci., 2, 1959, 241.
14. Kawaski, K.; Sekita, Y. J. Polym. Sci., A, 2, 1964, 2437.
15. Starkweather, H.W. J. Polym. Sci., B, 1, 1963, 133.
16. Barrie, J.A.; Nunn, A.; Sheer, A. In Ref. 5, p. 167.
17. Flory, P.J. Principles of Polymer Chemistry, Cornell University Press, Ithaca, NY, 1953.

18. Stannett, V.; Haider, M.; Koros, W.J.; Hopfenberg, H.B.  Polym. Eng. Sci., 20 (4), 300, 1980.
19. Salame, M.  Polym. Eng. Sci., 26 (22), Dec., 1986, 1543.
20. Salame, M.  Am. Chem. Soc., Div. Polym., Chem. Polym. Prepr., 8 (1), 1967.
21. Lee., W.M.  Polym. Eng. Sci., 20 (1), Jan., 1980, 65.
22. Van Krevelen  Properties of Polymers, Their Estimation and Correlation with Chemical Structure, Elsevier, NY, 1976.
23. Barrie, J.A.; Sagoo, P.S.; Johncock, P.  Polymer, 26, August, 1985, 1167.
24. Taoukis, P.S.; Meskine, A.E.I.; Labuza, T.P.  In Food and Packaging Interactions, ACS, 1988, 243.
25. Khanna, R.; Peppas, N.A.  Food Process Engineering, AIChE Symp. Ser., H.G. Schwatrzberg, D. Lund, J.L. Bomben, eds., AICHE, v. 78, No. 218, 1982, 185.
26. Peppas, N.A., Khanna, R.  Polym. Eng. Sci., 20, 1980, 1147.
27. Labuza, T.P.; Contieras-Medellin, R.  Cereal Foods World, 26,7 (1981).
28. Lazarides, H.N.; Goldsmith, S.M.; Labuza, T.P.  Chem. Eng. Prog., May, 46, 1988.
29. Drennan, W.  Plastics Packaging, July/August, 1(4), 1988, 1.
30. Iwanami, T.; Hirai, Y.  TAPPI J., 66 (10), Oct., 1983, 85.
31. Rohn, B.A.  Current Technologies in Flexible Packaging, ASTM STP 912, M.L. Troedel, Ed., A.S.T.M., Philadelphia, 1986, 13.
32. Gray, J.I.; Harte, B.R.; Miltz, J., Eds.  Food Product-Package Compatibility, Proceedings of a Seminar at Michigan State School of Packaging, July, 1986, Technimic Publishing Company, Lancaster, PA, 1987.
33. Crank  The Mathematics of Diffusion, Clarendon Press, Oxford, 1956.
34. Chang, S.S.; Guttaman, C.M.; Sanches, I.C.; Smith, L.E.  In Food and Packaging Interactions, ACS, 1988, 106.
35. Salame, M.  Plastics Packaging, 1 (4), 1988, 28.
36. Blackwell, A.  Proc. of Eleventh Int. Ryder Conf. on Beverage Packaging, March 23-25, Atlanta, Georgia, 1987.
37. SOLVAY, Inc., Technical Bulletin.
38. Wessling, R.A.; Edwards, F.G.  In POLYMER ENCYCLOPEDIA.
39. Sweeting, O.J., Ed.  The Science and Technology of Polymer Films, Vol. II, Wiley-Interscience, NY, 1971.

RECEIVED November 14, 1989

# Chapter 9

# Barrier Properties of Ethylene–Vinyl Alcohol Copolymer in Retorted Plastic Food Containers

Boh C. Tsai[1] and James A. Wachtel

American National Can Company, 433 North Northwest Highway, Barrington, IL 60010

Ethylene vinyl alcohol copolymer (EVOH) is used in multilayer containers for packaging shelf-stable foods which are retorted to achieve commercial sterility. EVOH is an excellent oxygen barrier when dry, but suffers a serious decrease in barrier properties when exposed to moisture. Retorting increases oxygen permeability both by moisture plasticization of the EVOH, and by irreversible changes in the EVOH structure, resulting from increased free volume. A previous study developed a method for predicting oxygen permeability, including the irreversible increase in steady state permeation due to retorting. This paper discusses the prolonged transient state after retorting, over six months to reach steady state oxygen permeation, and demonstrates the beneficial effects of incorporating desiccant in the structure in achieving low oxygen permeability after retorting.

Multilayer plastic food containers made with an ethylene vinyl alcohol copolymer (EVOH) oxygen–barrier layer are packed on commercial filling, can-closing, and thermal-processing equipment [1]. The development is remarkable because EVOH, an excellent oxygen-barrier material under dry conditions, but inherently moisture sensitive, becomes increasingly permeable on absorbing moisture. In fact, permeability increases by a factor of one hundred when containers containing an EVOH barrier layer are subjected to the food industry's means of commercial sterilization, which involves exposing containers to steam or pressurized water at temperatures of 205 to 275°F. To put it into perspective, the EVOH

[1]Current address: Amoco Research Center, Amoco Chemical Company, P.O. Box 400, Naperville, IL 60566

0097–6156/90/0423–0192$06.00/0
© 1990 American Chemical Society

layer absorbs more moisture during this single event than it does in months on the shelf at ambient temperatures.

Steam retorting of ethylene vinyl alcohol copolymer (EVOH) not only causes a sharp increase in initial $O_2$ permeability, but also results in higher steady state permeability at a given relative humidity. This latter increase is attributed to the creation of excess free volume in EVOH during retorting, a path dependence that cannot be eliminated under normal storage conditions. Experimental results are presented showing that the permeability increase due to retort path dependence is very significant for multilayer containers. The incorporation of desiccant [2] in the adhesive layers, however, overcomes the short term permeability increase of a retorted container due to moisture plasticization and also greatly reduces the permanent increase due to retort path dependence.

The effect of retorting on steady state permeability has been treated in a previous paper [3]. The abnormally slow rate at which steady state is established after retorting is treated in this paper and further examples and explanations of the beneficial effects of desiccants in EVOH multilayer structures are included.

## Experimental

The permeability test samples were five-layer polyolefin plastic cylindrical containers with EVOH as the barrier material, modified adhesive layers on either side of the EVOH, and polyolefin layers as outer protection. The EVOH used was Evalca's EPF which has 32 mole percent ethylene. The layer thicknesses in the container sidewalls were 14/1/2/1/14 mils.

The containers had a net capacity of 250 $cm^3$ and a surface area (excluding the metal end of 26.5 $in^2$). The containers had an integral plastic body (bottom and sidewall) and were closed with double seamed metal ends. Double seaming is the technique used to close metal and plastic cans for commercial use.

Permeability measurements started with conditioning the containers. They were filled with water and closed with a double seamed metal end, fitted with two rubber grommets, and thermally sterilized in steam at 240°F or 250°F for pre-determined time periods. After retorting, the containers were prepared for permeability testing by pulling the grommets, removing the water, placing 10 cm of water in the container to maintain 100% relative humidity, replacing the grommets and then flushing the interior of the container with nitrogen. This was accomplished by inserting a hypodermic needle through each grommet and connecting one to a nitrogen cylinder and the other to a vacuum line. The containers were then maintained for two days at room temperature to outgas the inside polyolefin layer. Oxygen concentration was then measured by sampling the gas space with a hypodermic syringe (inserted through a grommet) and injecting the gas into a MoCon Toray LC700F Oxygen Analyzer. Following this, the containers were stored at various temperatures and humidities under one atmosphere of oxygen. Oxygen concentration in the containers was measured periodically.

The permeability coefficients (P* in units of cc · mil/100 $in^2$ · atm · day) were calculated from the above data. In this paper, P* is an incremental permeability coefficient for a given time period. This provides a better understanding of the time required to reach steady state permeation.

By using Incremental-P*, more information about permeation can be deduced than with the more widely used secant P* which gives the average permeation between time 0 and each testing time.

The Quasi-Isostatic technique has several advantages over an isostatic technique, such as would be used with a Modern Control Oxtran, which is limited in the number of samples (ten) that can be measured at one time. Because of the very long time required to achieve steady state oxygen permeation for high barrier packages, the Quasi-Isostatic method is preferred since thousands of containers can be under test at the same time. Also, with the Quasi-Isostatic technique, the storage temperature and humidity can be controlled since the Oxygen Analyzer and containers are not coupled throughout the test period.

## Results

By plotting oxygen permeability as a function of storage time, both the short term and long term effects of retorting can be seen. The short term effect reflects high moisture content following retorting, which plasticizes the EVOH, and the long term effect reflects excess free volume created during retorting which was not eliminated during normal storage.

In the early days of storage, the expected sensitivity of EVOH permeability to moisture plasticization is seen. Figure 1 shows that the permeability immediately following a 100 minute retort at 240°F is about 25 times higher than the permeability when the container is not retorted. As retort time increases, the increased moisture ingress into the EVOH layer causes higher permeability.

The change in EVOH water activity with retort time or storage time can be described by equations published in a previous paper [3].

Early investigators [4] modeled oxygen permeability, assuming that retorting did not affect permeability. For a specific grade of EVOH, oxygen permeability was thought to be affected by water activity in the EVOH and by storage temperature; its permeability was thought to be predictable by an equation having only two variables, e.g.:

$$P_{O2} = f \text{ (EVOH water activity, storage temperature)} \qquad (1)$$

This work cannot explain the fact that the three curves in Figure 1 do not coalesce into a single curve at long storage times. This apparent anomaly is explained by a semi-empirical equation, published earlier [3], describing the relationship between EVOH permeability and water content after retorting, water activity during storage, and temperature during storage:

$$P_{O2} = (A+Bw) \cdot \exp[C \cdot (a-a_{ref})] \cdot \exp(-E_a/RT) \qquad (2)$$

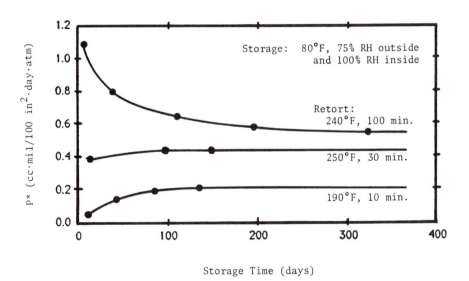

Figure 1.  Effect of retort processing on oxygen permeability of EVOH in a multilayer container as a function of storage time.

Equation 2 has the new term "Bw" which corrects for the dependence of oxygen permeability on the thermal—moisture history of the sample, sometimes called the Retort Shock, A and B are constants and w is the concentration of water in the EVOH layer following retort (as calculated from equations published in reference [3] or determined experimentally).

Figure 2 shows the excellent agreement between measured and predicted oxygen permeability extending over more than one order of magnitude. While there is a 2.7X increase in oxygen permeation as a result of retorting (as shown in Figure 1), which was not explainable in the early work (application of Equation 1), application of Equation 2 provides an excellent fit between experimental and calculated values. This figure presents data for containers both retorted and not, stored at three different relative humidities. Two retort temperatures and three retort times were used.

The steady state water content of EVOH is affected by its thermal—moisture history. The water content for containers stored at 75% RH/80°F (with 100% RH inside) is 7.7%, 7.3%, and 6.4% for containers processed at 240°F/100 minutes, 250°F/30 minutes, and 190°F/100 minutes, respectively.

In addition to undergoing Retort Shock, the oxygen barrier properties of EVOH are unusual because the lag time for reaching steady state oxygen permeability is more than 200 days. During retorting, moisture permeates through the polyolefin layers and the moisture content in the EVOH layer can reach as high as 7-14%; for the containers in this study, the moisture content in the EVOH layer is about 9.4%. This is above the steady state water content that the containers reach during storage (7.7%, Figure 3). The curve in Figure 3 is calculated using the methodology published earlier [3], the data points in the figure were determined experimentally. While the moisture content approaches steady state asymptotically, practically steady state is reached at about 60 days. The surprising result is that the oxygen permeability coefficient does not reach steady state until after 200 days (Figure 4). This is long after moisture steady state is reached. The calculated curve in Figure 4 is obtained by converting water content in EVOH (as shown in Figure 3) to water activity in EVOH (a) and then to the corresponding permeability coefficient through Equation 2. Comparing the experimental data with the calculated curve, it is clear that the oxygen barrier recovery process is slower than the moisture equilibration process.

The water content of the EVOH for the containers of Figure 1 was also determined experimentally at the beginning and at the end of permeability testing. The water content change was found to be −1.7%, +0.9%, and +2.8% for containers processed at 240°F/100 minutes, 250°F/30 minutes, and 190°F/10 minutes, respectively. Even though the water content change of the 240°F/100 minutes variable, −1.7%, is less than that of the 190°F/10 minutes variable, +2.8%, the time to reach oxygen barrier steady state is much longer for the 240°F/100 minutes variables as shown in Figure 1. The slower barrier recovery process is clearly demonstrated. In practical applications, the slow barrier recovery

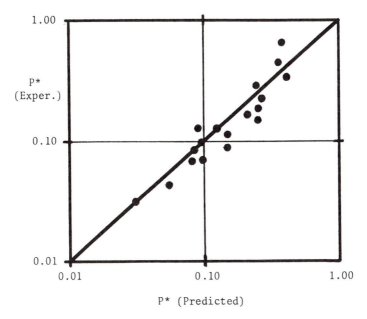

Figure 2. Correlation between experimental data and predicted values for oxygen permeability of EVOH as calculated from Equation 2.

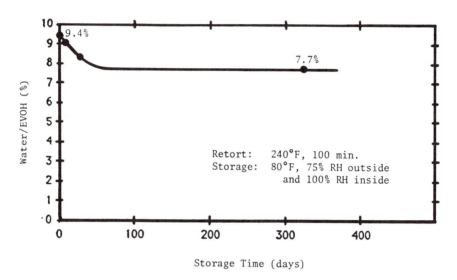

Figure 3. The moisture content of the EVOH as a function of
storage time.

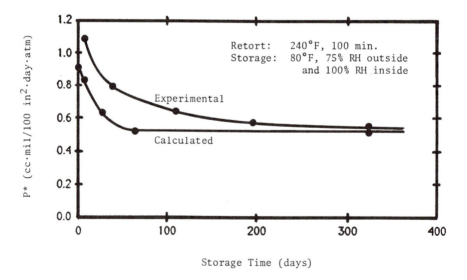

Figure 4. Comparison of experimental permeability and calculated
permeability as a function of storage time.

process can result in substantial additional oxygen permeation into a retorted container.

Incorporation of a desiccant into a multilayer structure overcomes the adverse effects of retorting on the barrier properties of EVOH reported above. While the desiccant used commercially is proprietary and cannot be disclosed, we can state that a desiccant system has been developed which meets the requirements for commercial use. Three criteria are paramount when selecting a drying agent for multilayer containers: effectiveness in protecting the EVOH layer, applicability to the container manufacturing process, and public safety. The drying agent must have a high capacity for holding water in the range of 30 to 80 percent relative humidity. A good drying agent should be able to hold one to two times its own weight of water, and hold much more water at 80 percent relative humidity than at 30 percent relative humidity.

Applicability to the manufacturing process means that the drying agent must be commercially available at a reasonable price, grindable to a fine particle size for dispersion in the resin, dispersible in commercial compounding equipment, and sufficiently thermally stable for can manufacture.

Desiccants are available which meet the above criteria. Effectiveness in protecting barrier properties is shown in Figure 5. This same desiccant is processible on plastic processing equipment including multilayer injection blow molding, multilayer extrusion blow molding, sheet coextrusion, and thermoforming. This same desiccant has a favorable FDA status and is in compliance with FDA Regulations for direct food contact.

The incorporation of desiccant in the structure significantly reduces the increase in permeability due to retorting. The desiccant lowers the concentration of water in the EVOH immediately after retorting and thereby lowers the permeability in the short term and greatly reduces or eliminates Retort Shock. This is in addition to the desiccant absorbing water at room temperature to protect the EVOH from reaching moisture steady state with its surroundings for about 18 months, delaying moisture plasticization. The permeability at steady state is reduced because the desiccating function greatly reduces the excess free volume after retorting. Figure 5 compares the permeability of containers with desiccant incorporated into the adhesive layers against the permeability of containers without desiccant. Immediately after retorting, the oxygen permeation rate (cc/day) of the container without desiccant is about 100 times higher. Even after over 500 days of storage, the container without desiccant has about 20 times the volume of permeated oxygen and over six times the permeation rate of the container with desiccant in the adhesive layers.

With an adequate amount of desiccant in the structure, the EVOH is not prone to Retort Shock and does not suffer from a prolonged lag time in approaching steady state. Oxygen permeation through EVOH barrier/desiccant containers is controlled by the partition of water between the desiccant and the EVOH. Simply said, the desiccant keeps the EVOH dry.

Containers with desiccant in the structure have much lower
steady state oxygen permeabilities than is predicted by
Equation 2. The difference lies in the Retort Shock term "Bw".
Moisture is present in the EVOH layer at the end of retorting, but
with desiccant, there is minimum Retort Shock. It appears that
desiccant serves to quickly remove moisture from the EVOH avoiding
the creation of excess free volume in the EVOH. As seen in
Figure 6, the relationship between permeation and storage time is
predicted by Equation 2 using B = 0, which is not true for the
non-desiccant containers, where the Retort Shock effect is large
and Bw is a significant factor in predicting permeability. For
containers with desiccant, B is much smaller than in the absence
of desiccant and can approach zero.

Not only is there negligible Retort Shock, but with desiccant
containers, a lag time to reestablish oxygen barrier in the EVOH is
not detectable. In fact, a very low permeability coefficient of
0.02 is established within seven days of retorting, the minimum
time period at which permeability can be measured. The high
initial oxygen permeation rate, that is seen immediately after
retorting with non-desiccant containers, is never seen with
desiccant containers.

## Discussion

In the absence of a desiccant, the steady state oxygen permeation
of retorted EVOH is determined by the moisture content of the EVOH
immediately after retorting and by the steady state moisture
content of the EVOH. Immediately after retorting, the oxygen
permeation rate can be ten times greater than the steady state, and
the time for barrier recovery takes over six months for commercial
containers. The oxygen permeation rate of EVOH after retorting
takes much longer than the moisture content of EVOH to reach steady
state. The length of the lag time depends primarily on the
severity (temperature and time) of the retort process.

Retorting EVOH, in the absence of desiccant, results in
substantial changes: a higher moisture content at a given water
activity, a higher steady state oxygen permeability (the Retort
Shock effect) and a long lag time before reaching its steady state
oxygen permeation rate. These are attributable to a change in EVOH
morphology, i.e., increased free volume. It has been well
established that permeation studies are excellent tools to assess
polymer morphology; the current work confirms this.

Putting desiccant in a layer of multilayer containers made
with the same EVOH polymer results in much less oxygen permeation.
The Retort Shock and slow barrier recovery effects do not take
place. Moreover, this is true even when the desiccant is separated
from the EVOH by a thin polymeric layer so that the EVOH and the
desiccant cannot interact chemically. Retorted containers with
desiccant behave more like unretorted containers without desiccant;
oxygen permeation is controlled almost exclusively by the water
activity of the EVOH layer. Morphological changes in the EVOH
layer after retorting appear to be very small when desiccant is
present. We attribute this to rapid transport of excess moisture
from the EVOH to the desiccant which results in healing the

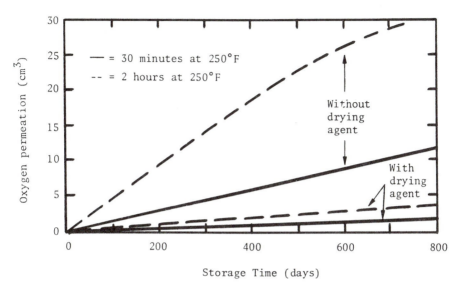

Figure 5. Oxygen permeation as a function of storage time (70°F, 100% RH inside and 75% RH outside) for EVOH containers with and without drying agents and for different thermal sterilization conditions.

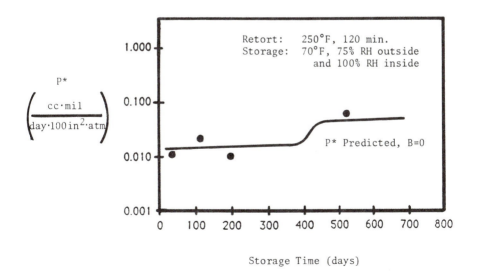

Figure 6. Oxygen permeation after retorting, EVOH multilayer container with desiccant.

structure, eliminating the free volume increase which occurs in the absence of desiccant. Containers with desiccant initially have approximately a hundred-fold improvement in oxygen permeation rate over those without desiccant and, at long storage time, the improvement is about six-fold.

## Legend of Symbols

A            Experimentally Determined Constant,
             $8.2x10^8$ ($cc \cdot mil/100$ $in^2 \cdot day \cdot atm$)

a            Water Activity

$a_{ref}$    Reference EVOH Water Activity, 0.77

B            Experimentally Determined Constant,
             $2x10^{10}$ ($cc \cdot mil/100$ $in^2 \cdot day \cdot atm$)

C            Experimentally Determined Constant, 9.9

$E_a$        Activation Energy, $14x10^3$ $cal/g$-mole, for EVOH

P*           Oxygen Permeation Coefficient (Incremental),
             ($cc \cdot mil/100$ $in^2 \cdot day \cdot atm$)

R            Gas Constant, 1.987 ($cal/g$-mole$\cdot °K$)

T            Absolute Temperature $°K$

w            Water Concentration in EVOH After Retorting

## Acknowledgments

We want to acknowledge the contributions of our colleagues, Ms. Julieta Cruz, Mr. Ward Edwards, Mr. Geno Nicholas, and Ms. Becky Jenkins.

## Literature Cited

1.   J. A. Wachtel, B. C. Tsai, and C. J. Farrell, Plastics Engineering, 41(2), 41 (1985).
2.   U.S. Patent No. 4,407,897.
3.   B. C. Tsai and B. J. Jenkins, J. Plastic Film & Sheeting, 4, 63 (1988).
4.   K. Ikari and Y. Motoishi, Packaging Japan, 5(24), 43 (1984).

RECEIVED October 17, 1989

# Chapter 10

# Retortable Food Packages Containing Water-Sensitive Oxygen Barrier

## M. M. Alger, T. J. Stanley, and J. Day

Process Engineering and Chemistry Laboratory, General Electric Company, Corporate Research and Development Center, 1 River Road, Schenectady, NY 12301

Multilayer ethylene-vinyl alcohol (EvOH) retortable food packages are steam sterilized at temperatures of 220-270°F. Water permeates through the outer package layers and into the center EvOH layer. The wet EvOH offers a reduced resistance to oxygen permeation immediately following retort. As the EvOH dries to a steady-state moisture profile, the barrier properties substantially recover, but a significant amount of oxygen can pass through the package wall during the drying stage. The long-term oxygen barrier performance of a retortable package is a very complex function of EvOH properties and history, layer thickness, EvOH layer placement, retort time, retort temperature, and storage conditions. This study was undertaken to develop a procedure for predicting and measuring barrier performance of multilayer retortable food packages. A model that includes major variables for calculating water and oxygen permeation through the life of a retortable package is presented. Oxygen ingression tests were done on food packages to test the model. The results showed that dissolved oxygen in the package wall can make a significant contribution to oxygen gain. Package tests also suggest that there is an irreversible loss of EvOH barrier effectiveness following retort. The exact mechanism of the barrier loss is unclear; water and oxygen permeation experiments were done on thin coextruded films and provide some insight into mechanisms of transport in multilayer structures.

There are two polymers used in a majority of high barrier food packages: copolymers containing polyvinylidene chloride (Saran ) and ethylene-vinyl alcohol copolymer (Eval , Selar-OH ). Packages in which either polyvinylidene chloride or ethylene-vinyl alcohol copolymer (EvOH) provide the barrier are invariably multilayer since neither of these materials offers the properties required to make a good monolayer structure. There are advantages and disadvantages associated with each of these materials, and the

0097–6156/90/0423–0203$06.50/0
© 1990 American Chemical Society

proper choice depends upon the application. For cases where EvOH is the more attractive alternative, a significant concern is the sensitivity of the barrier properties to moisture.

Most multilayer barrier packages are either hot-filled or retorted. In hot fill the package is charged with food typically between 180 - 200°F. Retortable food packages are steam retorted typically from 220 to 270°F to sterilize the contents. The latter retort operation greatly stresses the package and subjects it to a high water activity. Since EvOH has a very high water affinity it has been found that a significant quantity of water can be absorbed during the retort process (2-6). In addition, the generally excellent EvOH oxygen barrier effectiveness is seriously reduced at high humidity (1-9). Therefore, immediately following retort, the wet EvOH offers a reduced resistance to oxygen permeation. As the EvOH dries to a "steady-state" profile, the barrier properties substantially recover, but a significant amount of oxygen can pass through the package wall during the dry-out stage.

Because many food packages must be shelf-stable for many months to years, package testing becomes both expensive and time-consuming. It is necessary to make some estimate of the performance of various candidate structures to assess their barrier potential over the expected shelf life of the product. There have been numerous studies on the retortable food packages (1-8); however, a single consistent theory has yet to emerge and experimental details are limited.

The long range goal of our work is to develop a consistent theory for predicting the barrier performance of any multilayer food package. At present quantitative models do not exist for EvOH-containing structures due to poorly understood gas transport processes in EvOH under conditions found in food package applications. In this paper there are two generic package structures which considered, "PEP": Food/PP/EvOH/PP/ambient and "LEP": food/PP/EvOH/PC/ambient [PC = polycarbonate, PP = polypropylene].

This paper is divided into several sections which cover retort modelling, package testing, and thin-film testing. First, the basic equations are given for describing the performance of a retortable food package. With the model it is possible to capture the essential features of the package's performance during and following retort. More importantly, the model provides a good basis for organizing the many variables which directly influence the final package performance: EvOH properties and history, layer thicknesses, EvOH layer placement, EvOH thickness, retort time, retort temperature, storage conditions, and package geometry. Second, we discuss results on package testing that we have done to test model predictions. An oxygen headspace analysis was used to measure accumulated oxygen in packages as a function of time. We show that dissolved oxygen in very thick polycarbonate walls can contribute to the total oxygen gain following retort. A loss of EvOH steady-state oxygen barrier effectiveness following retort has been observed and correlates roughly with severity of retort.

In the final section water and oxygen permeation experiments were done on thin coextruded films to aid in understanding possible transport mechanisms in multilayer structures. An important conclusion is that water transport is important and must be understood before quantitative calculation of oxygen permeation will be possible.

## EvOH Properties

EvOH is an excellent gas barrier and as such its use in the packaging market has increased markedly in the last several years. EvOH is commercially

available in several grades with different ratios of ethylene/vinyl alcohol in
the chain (1). The major limitation of EvOH is that at elevated humidity,
>70%, its oxygen and water barrier effectiveness is severely reduced because of
disruption of the polymer-polymer hydrogen bonds (10-12). Also, because of
the hydroxyl groups on the polymer backbone, water tends to be extremely
soluble in EvOH. Figure 1 gives the water solubility in EvOH-32[1] as a
function of humidity at 20°C and 120°C. We see that the water solubility
increases markedly as humidity is raised above 65-75%. Measurements made by
Hopfenberg et al. (10) show that the maximum water solubility at 100% humidity
is a very strong function of the vinyl alcohol content in the polymer.

   Figure 2 shows the oxygen permeability coefficient for EvOH-32 as a
function of water content at 20C. The oxygen barrier effectiveness of EvOH-
32 decreases markedly with the absorption of water because of plasticization of
the EvOH-32 matrix.

PRINCIPLES of an EvOH RETORTABLE FOOD PACKAGE

Retortable food packages using EvOH as the oxygen barrier are multilayer
structures which are formed by coextrusion. The number of different layers in
a the structure can vary; a representative five-layer structure is shown in
Figure 3. Often one or more layers are added to reuse scrap and regrind in
the package. There is a wide range of possible combinations of layer
thicknesses which can be used in this type of structure.

   In retort the package is sterilized at high temperature, 220-270°F, for
one to two hours. During this period of time the relative humidity on both
the inside and outside package walls is 100%. As a result, water permeates
through the outer layers and into the center EvOH layer. The solubility of
water in EvOH at high temperatures is remarkably high as shown in Figure 1.
Since the EvOH layer is protected from the water by the outer layers, the
total amount of water absorbed in the EvOH layer during retort is limited by
water permeation through these outer layers. If very long retort times are
used then the amount of water absorbed in the EvOH layer can be quite
significant (4). Water solubility at 120 °C is about 35 to 40 g/100 g EvOH
(4, 5).

   After the retort cycle is complete the package is placed in "storage".
While the inside humidity of the package remains at 100%, the outside humidity
can vary over a wide range depending upon the exact storage conditions.
Changes in temperature are very likely to occur during storage. If we assume
that there is no irreversible loss of EvOH barrier effectiveness following retort
then the primary variable of importance is the total water absorbed by the
EvOH. The water content is an initial condition for the drying period
following retort.

   During storage the excess water in the EvOH layer desorbs and the
package approaches a steady-state. During this latter phase the oxygen
permeability decreases as water is desorbed from the EvOH layer in accordance
with the relationship given in Figure 2. The drying stage can last from weeks
to many months depending on the particular package layer distribution and the
total water absorbed during the retort process. After excess water has
permeated out of the EvOH layer, the package reaches a steady-state condition

---

   [1] EvOH-32 refers to ethylene-vinyl alcohol copolymer with 32 mol %
ethylene. All subsequent references to EvOH-32 will be for Evalca Eval-F
grade material.

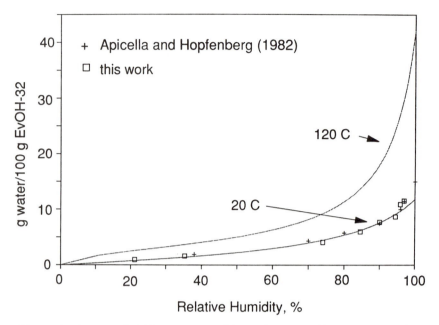

Figure 1. Water Solubility in EvOH-32 as a function of humidity. Results at 25°C were taken from ref 10, 11, and 26. Results at 120°C are from ref 5.

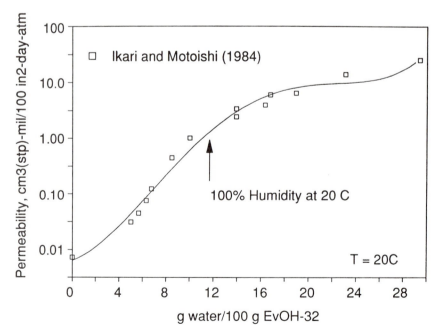

Figure 2. Oxygen Permeability in EvOH-32 as a function of water regain at 20°C (5).

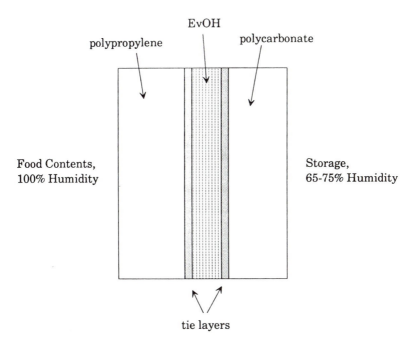

Figure 3. Five-Layer Structure with PP/tie/EvOH/tie/PC.

during which the water and oxygen permeation rates reach a constant value. While somewhat of an idealization, because storage conditions are not fixed in reality, the important aspects of the performance of a retortable package are captured using this simplified model.

## MODEL of a RETORTABLE PACKAGE

To predict the performance of a retortable package both water and oxygen permeation must be included in the model. To illustrate a simplified calculation of the performance of a retortable package we consider the idealized five-layer structure shown in Figure 3. There are two distinct parts of the calculation: retort and storage. To a first approximation the retort calculation determines the total water absorbed by the EvOH layer during the retort cycle and can be thought of as an initial condition for the storage calculation.

RETORT CALCULATION During the retort process we assume that no oxygen is present so that only water transport is considered. At retort temperatures the diffusion coefficient of water in materials of interest is sufficiently high so that a pseudo-steady-state model, in which linear profiles are assumed across all structural layers, can be used. The EvOH layer is assumed to be at a uniform water activity at a given time which is good assumption based on water permeability measurements in EvOH at humidities found in package applications (13).

Writing a mass balance for water in the structure gives the following equation for regain, $\Phi$ [g water/100 g EvOH]:

$$\frac{d\Phi}{dt} = \frac{P_w^{vp}(T)}{\rho_{EvOH}\, \delta_{EvOH}} \{\beta_{w_1}(RH_1 - RH_{EvOH}) - \beta_{w_2}(RH_{EvOH} - RH_2)\} \tag{1}$$

We use the term "regain" to refer to an amount of water in the EvOH layer at a given condition and "solubility" at a given water activity to refer to the amount of water absorbed when the EvOH is in equilibrium with the surrounding water activity.

In the water mass balance, effective water transmission rates, $\beta_w$, are defined for the layers between the food and EvOH layer

$$\beta_{w_1} = \{\sum_{i=1}^{k-1} (\frac{\delta_i}{P_{w_i}})\}^{-1} \tag{2}$$

and for the layers between ambient and the EvOH

$$\beta_{w_2} = \{\sum_{i=k+1}^{N} (\frac{\delta_i}{P_{w_i}})\}^{-1} \tag{3}$$

For the general case, the $k^{th}$ layer is EvOH and in this example k=3 and N=5. $\delta_i$ is the thickness and $P_{w_i}$ is the water permeability of layer i, respectively.

The temperature profile during retort can be specified as any general function of time; however, we have used only constant rise and fall rates and fixed retort temperature. By so doing we assume that heat transfer through the package and rate of heating the contents is very rapid. This assumption is reasonable; to relax it would serve only to complicate the calculation.

An important part of the retort simulation for LEP and PEP structures is the temperature dependence of the polypropylene and polycarbonate water permeabilities. Measurements of polypropylene and polycarbonate water permeabilities were made using a MOCON Permatran-W™ and it was found that polycarbonate has a water permeability which decreases slightly with temperature whereas polypropylene increases with temperature. Near retort temperature the water permeabilities of both materials are similar; at storage polycarbonate has a water permeability about ten times greater than polypropylene. Measurements shown in Figure 4 are in good agreement with previously reported results for polycarbonate (2,14-18) and polypropylene (5,19-21).

The retort simulation calculation was performed by integrating eqn 1 using the isotherm for EvOH-water in Figure 1 and the water permeabilitity for polypropylene and polycarbonate shown in Figure 4. Since volumetric measurements are not available, the EVOH-32 was assumed to have a constant density of 1.19 $g/cm^3$ in the calculation.

STORAGE CALCULATION During storage the temperature and outside package humidity are both lowered. Storage temperatures in the range of 10 to 40°C and humidities from 50 to 85% RH are common. Again, while it is possible to use a generalized time-dependent storage humidity and temperature, most of the interesting performance characteristics are realized with constant values. The inside package humidity was set to 100% for all conditions we considered for wet food applications. Modelling of variable package humidity coupled with oxygen permeation has been discussed by Howsmon and Peppas (22).

The water regain in the EvOH layer is calculated as a function of time using eqn 1. The initial loading of water is taken directly from the retort calculation; outside humidity is imposed only on the outside of the package. The food contents are assumed to remain at 100% humidity at all times.

With the water content of the EvOH available from integration of eqn 1, the oxygen transmission rate of the package, OxTr, is calculated from:

$$OxTr = \{ \sum_{i=1}^{N} (\frac{\delta_i}{P_{O_{2_i}}}) \}^{-1} \qquad (4)$$

where $P_{O2i}$ is the oxygen permeability of layer i. Normally all of the oxygen barrier is supplied by the EvOH layer. The OxTr can then be integrated with time to calculate accumulated oxygen:

$$cm^3(stp) = 0.21 \ A \int_0^t OxTr \ dt \qquad (5)$$

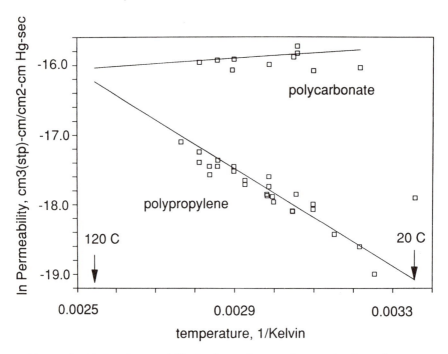

Figure 4. Water Permeability for Polypropylene and Polycarbonate.
Polypropylene measurements were made on 9.5 mil extruded sheet;
polycarbonate measurements were made on 10 and 20 mil extruded sheet.
Note that permeability has water pressure in the denominator which is
distinctly different from a WVTR (water vapor transmission rate).

where A is the surface area of the package and the factor of 0.21 is added for ambient oxygen pressure. Oxygen concentration is calculated from total $cm^3$(stp) found from eqn (5) and the volume of the package. An expression for the EvOH oxygen permeability as a function of humidity and temperature is required to integrate eqn (5).

Using any standard numerical routine, the above set of coupled equations, eqns 1-5, can be solved to simulate both the retort and storage of a multilayer package.

RESULTS of SIMULATED PACKAGE PERFORMANCE  Figure 5 shows the calculated ppm's of oxygen vs time for symmetrical LEP and PEP structures stored both at 65 and 75% RH. Figure 6 gives the water desorption in the EvOH layer as a function of time corresponding to the ppm's of oxygen in Figure 5.

From solution of the retort model equations we find that the use of a polycarbonate outside layer decreases the amount of oxygen which will permeate into a package for a fixed shelf life. The reason for this is the following. At retort conditions, the water permeability of polycarbonate and polypropylene are comparable, as shown in Figure 4, so that the amount of water allowed into the EvOH during retort is similar in the two structures. However, under typical storage conditions, polycarbonate has a water permeability which is approximately 10 times greater than that of polypropylene; thus, the EvOH dried out faster in the LEP structure and, therefore, less oxygen can diffuse through the LEP during dry out. Some early studies suggested that the ratio of polycarbonate to polypropylene water permeabilities was the same under retort conditions as under storage conditions which would result in a "flood" of water passing through the outer polycarbonate skin into the EvOH layer during retort. As seen in Figure 4, this assumption is a poor one.

While the above model certainly captures the performance of a retortable package, package tests have shown that complications arise from the retort process which lead to an irreversible change in EvOH barrier properties (4). Since the exact nature of this change is very poorly understood, it is necessary to develop a test methodology to provide insight into mechanisms operating in the package to make a truly quantitative model.

OXYGEN INGRESSION EXPERIMENTS on FOOD PACKAGES

An oxygen head-space analysis was used to measure oxygen ingression into packages as a function of time following retort. Packages were placed in an oxygen-free glove box, filled with 1-2 cc of oxygen-free water, and then sealed. The packages were retorted under predetermined conditions and then placed in storage at a fixed humidity. At various time intervals individual packages were selected and destructively tested using a MOCON/TORAY LC-700F™ headspace analyzer to measure oxygen concentration in the package.

The procedure can be modified by first retorting the package, flushing with nitrogen, and placing it in storage (8). While usually not a problem, this latter approach will 'miss' any oxygen which permeates into the package during the retort process. We have found that dissolved oxygen in the package material can diffuse in during retort. This is illustrated by Figure 7 which shows ppm oxygen vs time following retort for a LEP package. In Figure 7 a 2 ppm oxygen offset was observed immediately following retort. The offset was presumed to be a result of the very thick polycarbonate outer wall in which there was dissolved oxygen available for diffusion into the package during retort.

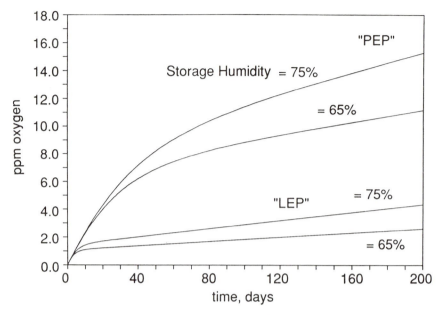

Figure 5. Calculated ppm's oxygen vs time for symmetrical LEP and PEP structures (10 mil PP/2 EvOH/10 PP or PC; Retort 90 minutes, 120°C).

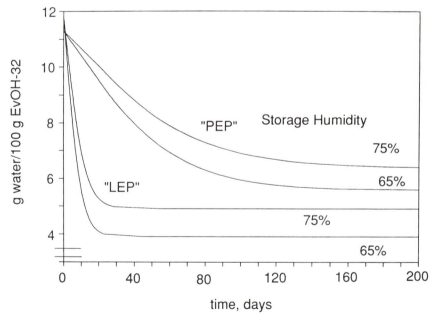

Figure 6. Calculated Water Desorption vs time for symmetrical LEP and PEP structures (10 mil PP/2 EvOH/10 PP or PC; Retort 90 minutes, 120°C).

The dissolved oxygen hypothesis was proven by filling 20 packages with nitrogen, as described above, and then dividing them into two lots. The first lot was stored in the nitrogen glove box before retort; the second lot was allowed to equilibrate with air for 1-2 days before retort. There was a substantial offset between these two groups: nitrogen saturated $0.21 \pm 0.12$ ppm (n=10); air saturated $1.3 \pm 0.28$ ppm (n=10). The dissolved oxygen occurred because of the particular layer construction of the packages and has not been found in structures with polycarbonate thicknesses of 5-10 mils on the outside of the package.

Since solubility effects are ignored in the simplified retort model it is not possible to predict this result. However, if a more general model is used, the dissolved oxygen can be included as an initial condition in the calculation.

Figure 8 gives results for "PEP" and "LEP" jars. The curves were calculated using the retort simulation model. While agreement between experiment and model seems reasonable, the steady-state permeabilities for the PEP and LEP packages are greater than expected using values estimated from film measurements. More importantly, it is difficult to decide what did or could cause loss of steady-state oxygen barrier effectiveness after retort. Loss of barrier properties has been observed previously in PEP packages (4) and more recent testing on LEP showed that the steady-state permeation rate also increased (23) with severity of retort.

The integrity of the EvOH following the retort process is essential to maintain adequate barrier during storage. It is possible to speculate on possible mechanisms which might cause an irreversible loss of barrier after retort. It may be due entirely to changes brought about to EvOH morphology and barrier properties or, on the other hand, by mechanical rupture of the layer integrity due to water absorption resulting in swelling or water condensation. Since it is difficult to make sensitive tests on food packages we have begun a series of experiments using thin coextruded LEP samples to understand better the dynamics of these structures in retort.

THIN FILM EXPERIMENTS

PrintPack, Inc. manufactures thin films for use in barrier applications. We used PrintPack EX429-325 coextruded film with the following layer distribution: 1.8 mil LDPE // 0.3 mil tie // 0.3 mil EvOH-32 // 0.3 mil tie // 0.8 mil polycarbonate. The tie layer was Plexar 3342 from Quantum Chemical (The chemical composition of this tie layer material is proprietary and has not been disclosed by the manufacturer).

Two questions have been studied using the thin film samples: does EvOH offer any resistance to water permeation in a LEP structure and can an irreversible change in EvOH oxygen permeability during retort be caused in controlled experiments? We consider these questions in separately.

WATER PERMEATION RATES A common design calculation for multilayer structures is to use an assumed steady-state water profile across the package and, from the resulting calculated EvOH humidity, to look-up the corresponding EvOH oxygen permeability on a graph. In this approach it is implicitly assumed that there is no resistance to water transport across the EvOH layer. However, published values of WVTR for PP, EvOH-32, and PC at 40 C, 90%-0% RH are, respectively, 0.70, 3.8, and 11.0 g-mil/100in$^2$-day (1). Because the resistance to water transport is given by the film thickness divided by the WVTR, the EvOH becomes the limiting resistance if the polycarbonate layer is

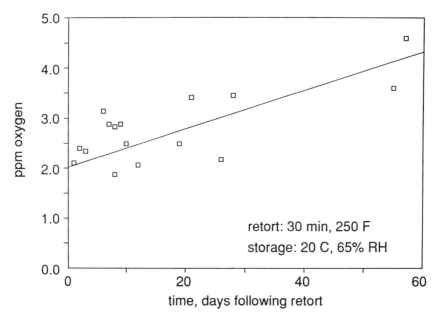

Figure 7. Measured ppm's oxygen vs Time for LEP Bottles (8.5 mil PP/1.3 Eval-F /45 PC; volume=150 cm$^3$, area=155 cm$^2$; Storage 20°C, 65% RH).

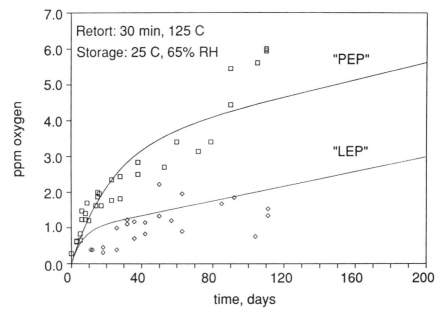

Figure 8. Measured ppm's oxygen vs Time for LEP and PEP Structures (PEP: 18 mil PP/2.4 Eval-F /6.2 PP; LEP: 9.2 PP/2.1 Eval-F /17 PC; volume=150$^3$, area=155 cm$^2$; Retort 30 min, 125°C; Storage 25°C, 65% RH).

of the same order thickness as the EvOH layer. Under these conditions, the benefit from the use of PC as the outside layer of the package is reduced because the EvOH layer can be a limiting water resistance. Several sorption experiments were done to assess the water barrier effectiveness of EvOH in coextruded samples.

Water sorption experiments were done with a Cahn RG balance on the PrintPack EX429-325 films. Constant water pressure was maintained using an MKS 252A pressure controller. The water source was contained in a round bottom flask and was flashed across a valve into the balance assembly. A low flow rate of water was maintained so that the pressure control valve could operate. Pressure was measured with an MKS 510A readout. The sample was suspended in a water-jacketed chamber at 20°C which was 4-5 °C cooler than the ambient so that the higher water vapor density in the sample tube minimized convection.

Figure 9a gives the step change of water pressure from 0 to 68% humidity. Figure 9b gives step change from 68 to 93% humidity. Figures 9c and 9d give the desorption curves from 93 to 68% and 68 to 0%, respectively. The curves in Figures 9a-d were calculated using the retort model. In the calculation it was assumed that the tie layers offered no resistance to water transport. Although including a water resistance for the tie layer shifts the curves slightly out to longer times, the relative position between calculated and measured remains unchanged.

We note that for the low humidity measurements the agreement between the calculated and measured weight change is poor relative to the high humidity experiments. We believe that this is because at low humidities EvOH offers a significant water transport resistance. The reason is that when a one point 90%-0% WVTR (Water Vapor Transmission Rate) measurement is made with the standard ASTM E96-E test method the measured water flux is actually averaged over all the humidity levels between the two sides of the film. Near 0% humidity EvOH is an excellent water barrier and as such most of the resistance to water transport comes from only a small section of the film. Corrections can be made to measured WVTR's to put them on a basis consistent with application within a package: 65-90% RH instead of 0%-90% RH. In doing so we find that the EvOH water permeability is a factor of 5 to 10 greater than the value measured in the ASTM test. Further, at conditions relevant for a food package, EvOH will offer very little resistance to water transport and the fact that the reported EvOH WVTR is lower than polycarbonate is an artifact of the ASTM test method.

It is important to be able to quantitatively predict the water profiles in the multilayer structure as a function of time. Failure to make this calculation results in incorrect oxygen flux calculations. Since there is very little detailed understanding of water transport in polymers over the range of temperatures and pressures of interest, a good quantitative model for water transport cannot be made for a general case at present.

LOSS of EvOH BARRIER PROPERTIES DURING RETORT Oxygen permeation measurements were made using a modified MOCON Ox-Tran 100 permeation cell which is a commercially available instrument. The experimental procedure has been described previously (24). Film samples were fixed in a machined brass block with a diffusion area of 50 cm$^2$.

We simulated a mild retort process by boiling the LEP films in water for predetermined time intervals. Following boiling the film was mounted in the MOCON Ox-Tran diffusion cell and 0% RH gas streams were flushed past both sides of the film.

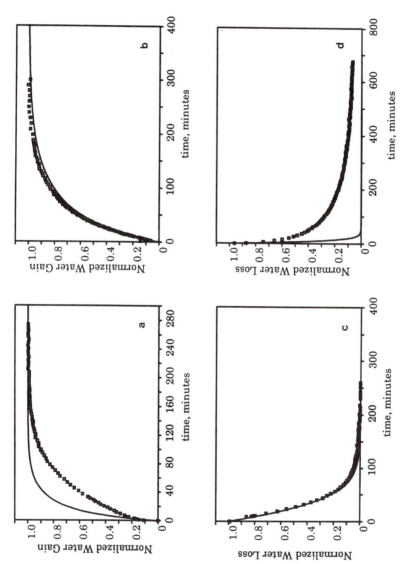

Figure 9. Water Absorption and Desorption in PrintPack EX429-325 films.
(a) 0-68% RH; (b) 68-93% RH; (c) 93-68% RH; (d) 68-0% RH at 20.5°C.

Two types of measurements were made: oxygen flux measurements during film drying and dynamic flux measurements starting with an initially dry film. These experiments will be discussed separately.

The oxygen flux through the film was followed during drying on the MOCON Ox-Tran™ 100. Immediately after removal from boiling water the film was mounted in the diffusion cell and 1 atm oxygen was flushed past the top of the film; nitrogen was flushed past the bottom of the film. After 20 to 30 minutes, the time required to flush all oxygen from the volume in the diffusion cell and connecting tubing, the MOCON Ox-Tran™ 100 sensor was valved in-line and reading were started. Time zero was when the film was first placed in the diffusion cell immediately after removal from water treatment.

Figure 10 gives oxygen flux measurements as a function of time for two different film samples boiled for different times (Note that in this and subsequent figures, the ordinate, which is oxygen flux, is written "permeability" units. While actually an oxygen flux through the film, the permeability unit is easier to interpret). We see that the recovery of oxygen barrier effectiveness is reasonably fast. For comparison the calculated oxygen permeability is added. The calculation was done assuming no water resistance in the EvOH layer with the model in section 4. The oxygen flux does not recover as rapidly as predicted from the model. This is due to the reduced drying rate brought about by low humidity in the EvOH layer discussed earlier. It is also found that the permeabilities do not reach the low value predicted from literature data. We attribute this difference to differences in materials and also variations in film thickness across the sample. In general, we found the oxygen permeabilities to be higher than those expected from measurements of EvOH film samples. However, with the very thin coextruded LEP sheet samples used we felt the film uniformity to be remarkably good.

It was interesting to find that when a film sample was boiled in water followed by "drying" in the diffusion cell at 35 °C there was no oxygen flux relaxation observed as in Figure 10. By the time the system had been flushed for 20 minutes, there was no observed maximum in the oxygen flux curve as with the 20 °C dried samples. This suggests that the most of the water diffused out in the short period of time during the flushing operation at 35 °C. Calculation of the drying rate for the film at 35 °C indicated that it should occur much more rapidly than at 20 °C. This observation re-enforces the notion that it is important to remove water from the EvOH to recover barrier; also, it confirms the measurements in Figure 10 and shows that they are not just the result of oxygen flushing from the cell and tubing following loading of the film.

In the second series of experiments, dynamic flux measurements were made on initially dry samples which were subsequently boiled for specified time periods. Curve (a) in Figure 11 is the oxygen flux as a function of time for a film sample measured at 0% RH on both sides of the film and at 35°C. Curve (b) in Figure 11 was for the same film sample which was boiled in water for 20 minutes and then quenched at 5°C for 2 minutes before being mounted on the Ox-Tran™. Curve (c) in Figure 11 is the same film sample boiled for an additional one hour and then quenched in 5°C water before mounting in the diffusion cell. There are several observations that can be made from this figure.

Following the first boiling in water the oxygen permeability of the film decreased. After the second treatment with boiling water the apparent oxygen permeability increased. Also, we note that there is a pronounced shoulder observed in Figure 11c which suggests a parallel mechanism for oxygen transport through the film. If we take all three curves in Figure 11 and find

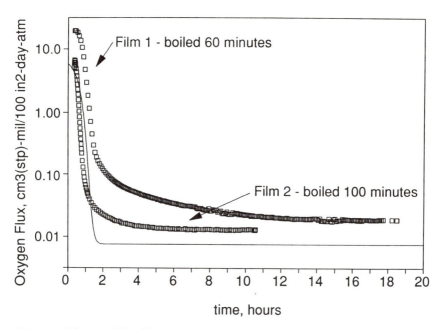

Figure 10. EX429-325 Film Oxygen Flux Following Boiling Water Treatment. Measurements made at 20°C. Other samples measured at 35°C showed no flux drop due to rapid drying at the elevated temperature.

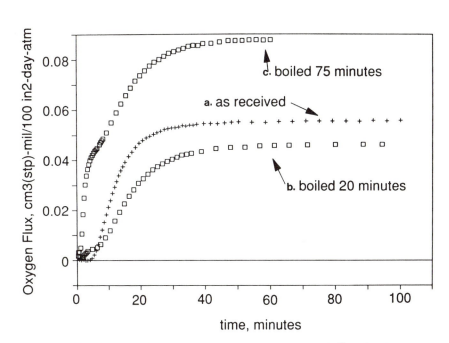

Figure 11. Dynamic Permeation Measurements at 35°C, 0% RH using Procedure Described in Ref. (24). Curves (a) film sample as received; (b) film sample from (a) boiled in water for 20 minutes and then quenched in 5°C water after boiling; (c) film sample from (b) boiled an additional 55 minutes and then quenched in 5°C water.

a common "zero" point by removing the shoulder that appeared in Figure 11b and 11c we recover the graph in Figure 12.

In Figure 12 we see that the two measurements of the samples which were boiled in water have a lower oxygen permeability than the film as received. We suspect that this is due to densification of the amorphous phase or loss of free volume in a process which is analogous to "aging" of the EvOH. Interestingly, there appears to be no change at all in the oxygen permeability of the EvOH with continued boiling suggesting that at 100°C the EvOH is relatively unaffected by hot water. This was also observed in several other film samples which received up to two hours of boiling water treatment. Since the coextruded film samples were very thin, independent water sorption measurements showed that by 20 minutes the EvOH was completely saturated at the 100°C water activity. The level of water sorption was between 25 to 30 g water/100 g EvOH-32 which is much higher than that usually found in food packages due to their much thicker outer walls.

The existence of the shoulder on the side of the oxygen flux curve was observed in several different film tests. The shoulder was shown to not be an experimental artifact by performing separate experiments with 1 atm and 0.21 atm oxygen as the driving force. When results from these experiments were calculated on a permeability basis the curves superimposed exactly.

It is not clear at this point what is the exact cause of this shoulder. It possible that during the boiling of the film mechanical damage was done which would give rise to pinholes. To test for this we took the film from Figure 11c and reduced the diffusion area with an aluminum foil mask to measure the oxygen permeability on a reduced section of the film which was approximately a quarter of the original 50 cm$^2$ area. In this measurement a shoulder was present, however, by proportion it was only about quarter as large as expected, after scaling the diffusion area, and thus suggests that the effect is not homogeneous and points toward macroscopic damage in the film.

However, it is reasonable to expect that some of the observed shoulder can be attributed to "pits" or excess free volume formed in the EvOH due to rapid cooling following water boiling. In this scenario the path of drying might be important. We can imagine that the EvOH layer swells considerably with the absorption of water and as it dries it the effective glass transition temperature will eventually drop below the storage temperature. Presumably any remaining "pockets" of water remaining when the glass transition temperature is reached would be frozen into the matrix. This is consistent with a higher water solubility at 100°C than at 20°C and is the reason we quenched the film samples immediately following the boiling process. Different film samples, which were not quenched in water following boiling, and which were placed in the diffusion cell at 35°C did not exhibit as pronounced a shoulder.

A better understanding of the interaction of water and EvOH throughout the retort and drying processes is vital to learn how to maximize long term barrier properties in the package. We are unaware of any systematic and thorough studies in this area.

CONCLUDING REMARKS

The retortable food package is an example of the general problem of measuring and calculating gas sorption and permeation through multilayer food packages. Because of the complexity of the structures and the great lack of detailed understanding of the basic transport mechanisms of water and oxygen through most packaging materials there is a great reliance on package testing. While

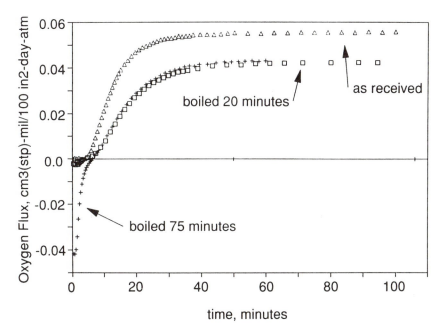

Figure 12. Oxygen Flux Curves From Figure 11 Normalized to Common Zero. EvOH oxygen permeability did not increase with treatment in boiling water.

necessary, package testing is time consuming and costly; a better appreciation of the details of gas transport would allow more efficient and intelligent testing strategies to be developed.

An example in which it is possible to make reliable estimates of performance is a plastic carbonated beverage package. Since carbon dioxide is the primary gas of interest it is possible to apply a straightforward one-dimensional model for a soda can (25). Further, the transport and sorption models for carbon dioxide and most plastics have received considerable attention in the literature.

A similar situation does not presently exist for multilayer retortable food packages. In our initial study of LEP and PEP food packages we found that there are very few experimental studies on water permeation in polymers. The studies which we did find were usually limited to low temperature, far removed from retort conditions. The effect of water activity on transport and sorption behavior has not received much attention especially for water-sensitive polymers such as EvOH. In EvOH-type systems the problem is especially acute because EvOH exhibits case II sorption at moderate humidities and its effective glass transition is depressed below storage temperatures at high humidities. Thus, modelling the transport in this type of material is very complex and does not lend itself to analysis with existing models.

Product development requires that testing be done on the final package to insure its performance is acceptable. While it is not possible to eliminate time-consuming test programs for package development, a sound model for gas transport in package structures can assist in explaining "anomalous" results and to reduce the testing that is required. In particular, a good model enables interpolation between measurements made on different package structures and also cautious extrapolation, depending on the level of sophistication. A proper balance between "fundamental" measurements such as gas solubility and diffusion coefficient over a range of conditions should be made with specific package testing. We hope that these problems will receive attention in future years because of their immense importance to the packaging industry.

ACKNOWLEDGMENTS

We want to thank Lorraine Rogers and Stan Hobbs for making layer thickness measurements on the PrintPack LEP films. Dennis Coyle made several good suggestions on how to solve the generalized package simulation model. Joe Hogan and Bill Ward have both provided encouragement and support for this work. Jill Cyr provided LEP and PEP packages for experiments and measurements on several LEP and PEP packages. Finally, we want to thank the first reviewer of this paper for the detailed comments and suggestions. This work was funded by GE Company.

LITERATURE CITED

1.  Eval Company of American, Technical Bulletin No. 110 "Gas Barrier Properties of EVAL™ Resins", p 6, Ohmah, NE.
2.  O'Kata, H., R. Moritani, and Y. Motoishi, Proceedings of COEX '87, 1987.
3.  Wachtel, J.A., B.C. Tsai, and C.J. Farrell, "Retortable Plastic Cans Keep Air Out, Flavor In", **Plastics Eng** 1985, 41, p 41.
4.  Tsai, B.C. and B. Jenkins, "Effect of Retorting on the Barrier Properties of EVOH", **Plastic Film & Sheet** 1988, 4, p 63.

5. Kyoichiro, I. and Y. Motoishi, "Retort Characteristics of EVAL Compound Packaging Materials", Kuraray Co., Ltd. Kurashiki Works, Chemical Products R&D Room, Internal Report 1984.

6. Kiang, W. W., Proceedings of Future-Pak '86, 1986, p 391, Rider Associates, Inc, Whippany, NJ.

7. Marcus, S.A. and P.T. DeLassus, "Recycle and Permeation Studies on Barrier Plastics Containers", Second International Ryder Conference on Packaging Innovations, 1984, p 137.

8. Bresnahan, W.T., "Oxygen Permeation of Retorted EvOH Multilayer Containers", Second International Conference on New Innovations in Packaging Technologies and Markets, December 3-5, 1984.

9. Iwasaki, H. and Yasuhiro Ogino, "Permeability of Laminated Ethylene-vinyl Alcohol Copolymer to Oxygen (Studies on Gas Barrier Resin 'EVAL'-No. 4", **Kobunshi Ronbunshu** 1978, 35 pp 487-492.

10. Hopfenberg, H.B., A. Apicella, and D.E. Saleeby, "Factors Affecting Water Sorption in and Solute Release From Glassy Ethylene-Vinyl Alcohol Copolymers", **J. Mem. Sci** 1981, 8, pp 273-282.

11. Gaeta, S., A. Apicella, and H.B. Hopfenberg, "Kinetics and Equilibria Associated with the Absorption and Desorption of Water and Lithium Chloride in an Ethylene-Vinyl Alcohol Copolymer", J Mem Sci 1982, 12, pp 195-205.

12. Gupta, R.P. and R.C. Laible, "Study of Hydrogen Bonding in poly(vinyl alcohol) by a Nuclear Magnetic Resonance Method", **J Poly Sci** A3, 1965, p 3951.

13. Alger, M.M. and T.J. Stanley, "Oxygen Barrier of LEP Food Containers: Effect of Humidity on the Water Vapor Transmission Rate", GE Corporate Research Memo Report MR45763 (1989).

14. Robeson, L.M., and S.T. Crisafulli, "Microcavity Formation in Engineering Polymers Exposed to Hot Water", **J Appl Pol Sci** 1983, 28, pp 2925-2936.

15. Narkis, M., L. Nicolais, A. Apicella, and J.P. Bell, "Hot Water Aging of Polycarbonate", **Pol Eng Sci** 1984, 24, pp 211-217.

16. Pryde, C.A., and M.Y. Hellman, "Solid State Hydrolysis of Bisphenol-A Polycarbonate, I. Effect of Phenolic End Groups", J App Pol Sci 1980, 25, pp 2573-2587.

17. Narkis, M., S. Chaouat-Sibony, L. Nicolais, A. Apicella, and J.P. Bell, "Water Effects on Polycarbonate", **Poly Comm** 1985, 26, pp 339-342.

18. Ito, E., and Y. Kobayashi, "Changes in Physical Properties of Polycarbonate by Absorbed Water", J App Pol Sci 1978, 22, pp 1143-1149.

19. DeLassus, P.T. and D.J. Grieser, "Factors Affecting the Moisture Barrier of Vinylidene Chloride-Vinyl Chloride Copolymer Films", J Vinyl Tech 1980, 2, pp 195-199.

20. Rudorfer, von D., O. Pesta, and H. Tschamler, Chemie Kunststoffe-Aktuell 1975, 29, pp 176-178.

21. Suh, H.K. and J.K. Lee, "The Studies of Water-Vapor Transmission Rate and Its Proofness on the Various Commerical Polymer Films", J Korean **Chem Soc** 1979, 23, pp 329-337.

22. Howsmon, G.J., and N.A. Peppas, "Mathematical Analysis of Transport Properties of Polymer Films for Food Packaging. VI Coupling of Moisture and Oxygen Transport Using Langmuir Sorption Isotherms", J Appl Poly Sci 1986, 31, pp 2071-2082.

23. Tsai, B.C., Test Report July, 1988 on Retort Performance of LEP Food Packages.

24. Alger, M.M., and T.J. Stanley, "Measurement of Transport Parameters in Polymer Films: Flux Overshoot Produced by a Step Change in Temperature", **J Mem Sci** 1989, 40, pp 87-99.

25. Alger, M.M., T.J. Stanley, and J. Day, "Gas Transport in Multilayer Packaging Structures", **Pol Eng and Sci** 1989, <u>29</u>, pp 639-644.

26. Apicella, A. and H.B. Hopfenberg, "Water-Swelling Behavior of an Ethylene-Vinyl Alcohol Copolymer in the Presence of Sorbed Sodium Chloride", **J Mem Sci** 1982, <u>27</u>, pp 1139-1148.

27. Stannett, V.T., G.R. Ranade, and W.J. Koros, "Characterization of Water Vapor Transport in Glassy Polyacrylonitrile by Combined Permeation and Sorption Techniques", **J Mem Sci** 1982, <u>10</u>, pp 219-233.

RECEIVED January 16, 1990

# Chapter 11

# Performance of High-Barrier Resins with Platelet-Type Fillers

## Effects of Platelet Orientation

T. C. Bissot

Polymer Products Department, Experimental Station, E. I. du Pont de Nemours and Company, Wilmington, DE 19880-0323

The use of platelet-type fillers, preferably fine particle size mica, in ethylene vinyl alcohol (EVOH) copolymers increases oxygen barrier performance approximately threefold. The benefit is ascribed to the increased diffusion path length (tortuous path) produced by the overlapping platelets. Typical polymer processing operations used in packaging lead to the preferred orientation with the platelets aligned parallel to the surfaces. An exception was found in a blow molded bottle test where nonparallel alignment gave minimum barrier improvement. This result is explained by the mismatched die swell between the structural and barrier layers of the multilayer parison. Changes in polymer rheology or the type of machine reduced or eliminated the problem and gave the expected barrier performance.

Plastics continue to expand into food packaging applications traditionally served by metal and glass containers. The oxygen barrier properties of the plastic food container is frequently a major consideration affecting its suitability for a specific application. Polymers used in packaging films and containers can be classified by their relative permeation to oxygen. Of the many classes of polymers used in packaging, only three can be designated as high barrier materials, i.e., those having an oxygen permeation value of less than 1 $cm^3$-mil/100 $in^2$-day-atm (Table I).

0097-6156/90/0423-0225$06.00/0
© 1990 American Chemical Society

Table I.  Melt Processible High Barrier Resins

| Resin | Oxygen Permeation Value* | |
|---|---|---|
| | Dry, 25°C | 80% RH, 25°C |
| EVOH – 30 mole % E | .01 | .11 |
| EVOH – 44 mole % E | .09 | .24 |
| PVDC – high barrier extrudable | .15 | .15 |
| AN – copolymer (BAREX) | .8 | .8 |

$*cm^3$ $O_2$-mil/100 $in^2$-day-atm

Among the high barrier resins, the ethylene vinyl alcohol (EVOH) copolymers are showing the most rapid growth. Familiar containers relying on EVOH oxygen barrier are the squeezable ketchup bottle and the single serving, shelf-stable and microwaveable entree container.

Introductory guides to the selection and use of fillers (1,2) generally start by classifying fillers according to shape classes such as spheres, cubes, blocks, flakes and fibers. The first three classes have shape or aspect ratios close to 1 and, therefore, cannot display orientation in polymer systems. Only flakes (or platelets) and fibers (needles) with significant aspect ratios can show orientation.

The most common reasons for adding fillers, especially high aspect ratio fillers, to polymers is to improve physical properties such as increased modulus (stiffness) or reduced creep.

In addition to this major use to improve mechanical properties, high aspect ratio flake-type fillers have been added to polymers for a variety of other purposes. They include improved thermal stability (3), high voltage resistance (4), electrical conductivity, radiation shielding (5) and optical and aesthetic effects (6).

High aspect ratio flakes or platelets have also been previously used to improve the gas barrier properties of low and medium barrier polymers (7,8).

The diffusion of small molecules in a polymer film can be described by Fick's first law of diffusion

$$J = -Ddc/dx \qquad (1)$$

which simply states that the flux of a gas permeating the film at a constant concentration or pressure differential will be inversely proportional to the distance the diffusing gas must travel. The addition of platelet fillers distributed in the polymer film can greatly increase this diffusion distance by creating a tortuous path for the diffusing species. A recent paper by Cussler, et al., (9) uses the more picturesque term "wiggle length" to describe segments of this pathway.

Several attempts have been made in the literature to model this system of platelets dispersed in a polymer film in order to predict the relative reduction in flux in a platelet filled system, $J_o/J_n$, where $J_o$ is the flux measured with an unfilled film and $J_n$ is the flux measured with the platelet-filled film.

An early model proposed by Prager (10) assumed random orientation of fillers of various shapes. His predictions showed that while platelets would give a greater improvement than cylinders or spheres, the relative improvement with a 20 volume % loading of randomly oriented platelets would yield only about a 40% reduction in permeability. This would be a $J_o/J_n$ of 1.67.

There have also been models based on very regular arrays of the platelets distributed in a film matrix. An early model proposed by Barrer (11,12) modeled the system as a uniform dispersion of a lattice of rectangular parallelepipeds. Using this model, Murthy (13) calculated that a large reduction in permeation could be achieved with a platelet filler volume of 20%. A sample calculation from data in the latter paper predicted that 20 volume % of a platelet filler having a long dimension of 5 microns and a thickness of 0.1 micron would yield 144 layers of platelets in a 50 micron membrane. There would be a spacing of .32 micron between layers of platelets with a 1.0 micron separation between the rectangular platelets in a layer. The calculated ratio of the diffusion coefficient of the filled composite to that of the matrix is 0.04. This represents a $J_o/J_n$ of 25.

A recent paper by Cussler, et al., (9) using a regular array model also predicted quite high barrier improvements. Their equation (Equation 2) predicted that aspect ratio and volume fraction of filler would be the major variables. They incorporated a universal correction designated, $\mu$, as a geometric factor to correct for the reality that available platelet fillers were not shaped like uniform rectangular parallelepipeds of uniform size and shape.

$$J_o/J_n = 1 + \mu\alpha^2(\phi^2/(1-\phi)) \tag{2}$$

Substituting values in the Cussler equation, which we will show later to be in a practical range, of 12 volume % loading ($\phi$) of a filler having an aspect ratio of 23 ($\alpha$) and assuming an ideal geometric factor of 1 would predict a relative barrier improvement, $J_o/J_n$ of 10.

There are very few reports in the literature documenting the improvement in gas barrier properties that can be achieved using commercially available platelet-type fillers in melt processed polymer films and none showing the improvement in that class of polymers previously identified as high barrier polymers.

Some of the best data on a practical system is that of Murthy, et al., (13) based on blends of talc in polyethylene and nylon films. Their data demonstrated relative barrier improvements ($J_o/J_n$) of 1.8-1.9 using filler loadings in a practical range of about 12 volume %.

EXPERIMENTAL

There are basically only three types of platelet-type fillers which can be considered for use in thin barrier films. These are aluminum flake, mica and talc (Table II). Other types of platelets, such as glass, stainless steel or brass flakes and certain aluminum silicate minerals, such as kaolin clay, are either too large in particle size or have too low an aspect ratio to be useful. With these three preferred fillers, comparable particle size products can be obtained by selection of appropriate grinding and size classification techniques. The 7 micron average platelet diameter is a good median particle size for blending into polymers designed for extrusion into thin films and layers in coextrusions.

Table II. Representative Platelet-Type Fillers

| Type<br>Supplier Grade | Al Flake<br>Reynolds Metal Co.<br>40XD | Mica<br>Franklin Mineral<br>#3 MICROMESH | Talc<br>Pfizer<br>MP 12-50 |
|---|---|---|---|
| Screen Analysis | 99.5% <325 | 100% <400 | 100% <400 |
| Average Part<br>Diameter $\mu$m<br>(D) | 7 | 7 | 7 |
| Platelet Thick-<br>ness $\mu$m (T) | 0.2 | 0.3 | 1 |
| Aspect Ratio<br>D/T | 35/1 | 23/1 | 7/1 |
| Density, g/cc | 2.7 | 2.77–2.88 | 2.7–2.9 |
| Price, $/lb | 4.00 | .30 | .06 |
| Price/EVOH<br>Volume<br>Equivalent | 9.00 | .70 | .14 |

Table II shows that there are major differences in aspect ratio and cost between these three fillers. The relative costs of the barrier enhancing platelets and the matrix resin are a major consideration for the practical viability of this approach. The bottom row of this table shows the price of filler on a volume basis assuming it is displacing an equivalent volume of EVOH resin. A price of $2.40/lb was assumed for the EVOH resin.

During initial studies, blends of several types of ethylene vinyl alcohol copolymers were made with all three types of platelet fillers, aluminum flake, mica and talc. Blend loadings were from 9 to 33 wt % filler. Thin films, 1 to 2 mils in thickness, were melt pressed from these composites and used to measure oxygen and water permeation rates.

Subsequent larger scale evaluations were made using both pilot and commercial scale polymer processing equipment. These included both single and multiple layer cast and blown film lines for flexible structures. Rigid containers with mica-filled EVOH were made by both melt and solid phase pressure forming of multilayer sheet and by coextrusion blow molding. Most of the rigid container testing was done on bottles made on a Bekum five-layer coextrusion blow molding machine using a one-liter Boston round bottle mold.

A Mocon (Modern Controls, Inc.) instrument modified for accurate humidity control was used to measure the oxygen permeation value (OPV) on film structures. These were generally measured at 30°C and 80% relative humidity and are reported as $cm^3$ $O_2$/100 $in^2$–day–atm.

The barrier performance of rigid containers was measured on both a second modified MOCON instrument and by a head space technique in which nitrogen purged and sealed containers were stored in an oxygen environment for periods of time. Containers were generally measured at 23°C with 100% relative humidity (RH) inside and either 50 or 75 RH outside. Container oxygen transmission rates (OTR) are expressed as $cm^3 O_2$/container-day-atm.

Platelet sizes, distribution and orientation of the EVOH blends in the film and container structures were examined by SEM. Samples were prepared by cryogenically fracturing the coextruded structures and the using an oxygen plasma etch on the exposed cross section to erode away some of the polymer. This exposed the edges of the fractured mica platelets, giving micrographs with good contrast and clearly defined platelet orientation.

RESULTS ON SINGLE LAYER FILMS

Figure 1 shows the oxygen permeation values for a series of blends of aluminum, mica and talc in an EVOH resin containing 44 mole % ethylene. The relative effectiveness of the three fillers are proportional to their aspect ratios as predicted by theory. Aluminum flake was the most effective, but the high cost and the hazard of handling dry aluminum flake rule out the commercial use-fulness of such blends. Mica was clearly superior to talc as would be anticipated on the basis of the aspect ratio data.

Balancing relative costs and effectiveness favored mica as the preferred platelet-type filler for enhancing the barrier of EVOH resins.

Larger scale testing confirmed the earlier scouting tests. Additional permeability data obtained with cast films made with a 30 mole % ethylene EVOH blended with various levels of mica are shown in Figure 2. A loading of 23.1 wt % mica was selected as a good compromise for obtaining low permeability while maintaining good polymer processing characteristics. At this loading, barrier testing on single layer films showed an improvement in the $J_o/J_n$ ratio of 3 to 5 times that of unfilled resin. Thicker films tended to show the higher level of improvement.

The initial samples submitted for customers' evaluations were simply labeled SELAR OH plus mica. This name stuck and the products are generally referred to as SELAR OH Plus. Two grades are currently being offered as commercial products (Table III). At a mica loading of 23%, the melt flow of the blend is reduced to approximately one-half that of the unfilled resin.

Table III.   SELAR OH Plus Grade Summary

| Grade Description | EVOH Melt Flow* | EVOH % E | Mica Wt % | Blend Melt Flow* |
|---|---|---|---|---|
| SELAR OH 3002P3 | 3 | 30 | 23.1 | 1.5 |
| SELAR OH 4408P3 | 16 | 44 | 23.1 | 8.0 |

*g/10 min, 2160 g, 210°C

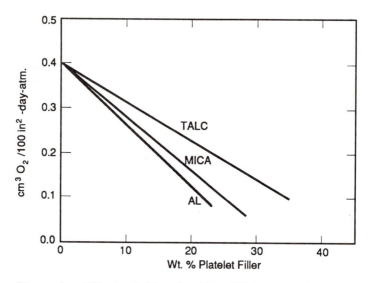

Figure 1.  Effect of Platelet-Type Fillers on the Oxygen
Permeability of EVOH Resins.

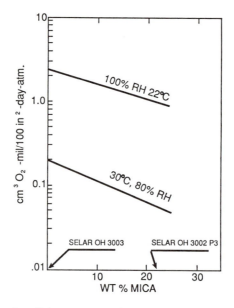

Figure 2.  Barrier Enhancement of SELAR OH Resin With Mica Filler.

As inferred from the prior discussion of models, in order to obtain maximum benefit with a high aspect ratio flake filler, it is necessary for the flakes or platelets to be aligned with their major plane parallel to the surface of the film or sheet. Fortunately, most polymer processing operations for both flexible and rigid packaging favor this preferred orientation. Flow patterns in the melt during the extrusion or coextrusion of film or sheet align the platelets parallel to the surfaces. Further improvement in the desired orientation is achieved when there is biaxial stretching of the melt following extrusion such as occurs with blown film or thermoforming operations. Figure 3, which shows the SEM of a typical cross section of a SELAR OH Plus barrier layer, illustrates this parallel alignment.

RESULTS ON COEXTRUDED STRUCTURES

Ethylene vinyl alcohol barrier resins are never used as single layer films in packaging but rather are utilized in coextruded structures combined with less expensive structural polymers.

FILMS. Table IV shows data on a five-layer flexible film structure made with mica-filled EVOH, an unfilled-EVOH control and a blend of the two. Approximately a three-fold improvement was measured on the multilayer film structure in good agreement with the previous monolayer data.

Table IV.  Oxygen Permeation of Coex Films Containing SELAR OH Resins

| Barrier Layer Resin | Barrier Layer Thickness Mil | Oxygen Permeation Value,* 30°C, 75% RH |
|---|---|---|
| SELAR OH 3003 | 0.3 | 0.109 |
| SELAR OH Blend 50% 3003 50% 3004P3 | 0.3 | 0.063 |
| SELAR OH 3004P3 | 0.3 | 0.037 |

Structure is LDPE/BYNEL E208/SELAR OH/E208/MDPE
Total thickness = 2.2 mil
*$cm^3$-mil/100 $in^2$-day-atm

THERMOFORMED CONTAINERS.  Table V shows data on a melt-formed, rigid container. This container, together with its unfilled EVOH control, was subjected to a simulated retort cycle at 121°C. The mica-filled container in this example showed an improvement of 2.5 times that of the control.

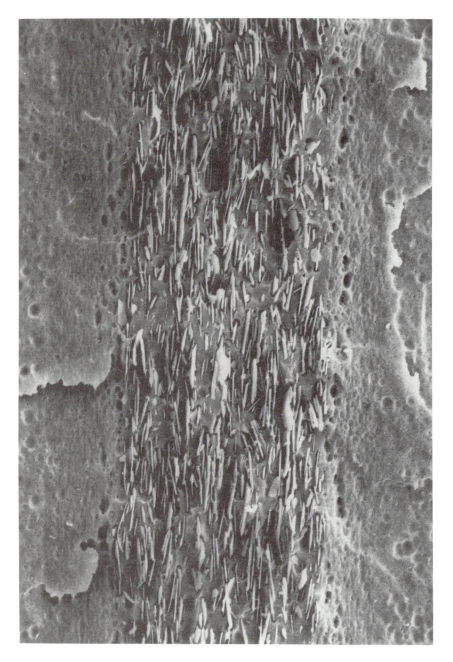

Figure 3.  SEM Micrograph of Cross-Section of Coextruded Film
Illustrating Mica Platelet Orientation in EVOH Barrier Layer.

Table V.   Melt-Formed, Single-Serving Entree Cup

| Barrier | $cm^3 O_2$/Container-Day-Atm |
|---|---|
| SELAR OH Plus Mica | .061 |
| Unfilled EVOH Resin | .152 |

*Tested after simulated 121°C sterilization cycle

BLOW-MOLDED CONTAINERS.   Initial evaluation of SELAR OH Plus in extruded blow molding containers was disappointing. Barrier improvements of only 30 to 40% were measured compared to the expected 300%. The resolution of this discrepancy provided some valuable insights into the role of platelet orientation in barrier improvement and the effects of die swell in multilayer extrusion blow molding.

The clue to the low barrier enhancement ratios came from SEM and TEM micrographs detailing platelet orientation in the sidewall barrier layer. When the sidewall barrier layer was cross sectioned in a vertical or extrusion direction, these photographs showed that the mica platelets were not aligned parallel to the bottle side-walls. Instead the platelets showed a periodic alternating pattern similar to a classic sine wave, with most platelets aligned at a 45° angle to the surface (Figure 4). When sectioned in a "hoop" or cross extrusion direction, the exposed edges of the platelets appeared parallel to the surface. This platelet orientation would be estimated to give at best a 40% improvement. This value is based on the increased diffusion path length corresponding to the hypotenuse of a 45° right triangle.

This unusual platelet orientation is believed to be caused by mismatched die swell in the different layers of the coextruded tube or parison used to form the bottle. The polypropylene (PP) used as the structural layer in these blow molding tests was a low melt flow grade recommended for blow molding applications. This resin normally gives a high degree of die swell which was readily observable as the parison exited the annular die under melt temperature and extrusion rate conditions optimized to make a bottle with uniform thickness sidewalls. Die swell at an annular die can be of two types, an increase in the diameter of tube versus that of the die opening and an increase in the thickness of the tube wall versus that of the die slit opening. It is the latter die swell phenomena that is responsible for the nonparallel mica platelet orientation. Filled polymers and especially platelet-filled polymers are well known to exhibit considerably lower die swell than unfilled resins. The individual layers in a multilayer extrusion must maintain their same individual percentage of the total structure throughout the melt flow pathway. Therefore, an internal EVOH layer must also increase in thickness as the structural PP layers undergo die swell exiting the die lips. Apparently in the case of a platelet-filled EVOH, this layer can become thicker only by folding back on itself, thus resulting in the sine wave pattern.

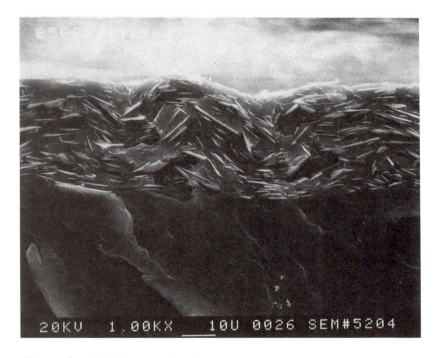

Figure 4.  SEM Micrograph of Extrusion Direction Cross-Section
of Coextruded Blow Molded Bottle Showing "Sine Wave" Pattern of
Mica Platelets.

In a shuttle machine such as was used in this program, the multilayer tubular polymer melt is extruded downward with only gravity exerting a downward pull. When the tube is extruded to the proper length, it is transferred to a mold, the top cut off, the bottom is pinched shut and air pressure is introduced into the top. This expands the hot polymer tube into the shape of the finished bottle. Typically there is a three-fold or greater expansion in the hoop direction but practically none in the vertical or extrusion direction. This causes the axis of the mica platelets in the hoop direction (perpendicular to the extruder flow direction) to align parallel to the bottle surface. However, in the vertical or extruder direction, where no stretching takes place, the axes of the mica platelets retain the alternating tipped orientation or sine wave pattern developed in the swollen zone below the die lips.

Once this unbalanced die swell mechanism was diagnosed, several solutions were suggested. One solution is to increase the die swell of the barrier resin. Since the degree of die swell is usually related to the molecular weight of the resin, a lower melt flow index (MI) EVOH resin blend with mica was evaluated. This was a 3 MI resin reduced to 1.5 MI after blending with mica. A 7 MI resin blend was used in the initial trials. Blow molded bottles made with this resin still showed the oscillating sine wave pattern of the mica platelets in extrusion direction cross sections. However, the average angle of the platelets was reduced to about 30°. Barrier testing of these bottles showed an improvement two times that of the unfilled control. This was definitely an improvement over the 30–40% previously observed.

A second solution is to reduce the die swell of the polypropylene layers. Known techniques to reduce die swell include using a lower molecular weight polymer, modifying the die configuration to extend the polymer compressed zone, adding slip agents or increasing the melt temperature. The latter was the simpliest test to run. Increasing the melt temperature of the polypropylene by 50°F gave an immediate and visible reduction in die swell. Extrusion direction cross sectioning of the sidewall showed only traces of the sine wave pattern with the average mica platelet reduced to an angle of about 15°. However, this set of bottles did not do well in barrier testing because of low bottle weight and large variability in sidewall thickness.

The third and most successful approach was accomplished on a wheel-type blow molding machine. Unlike the shuttle machine where only gravity acts on a downward extruded parison, a wheel machine draws or pulls the parison from the die. This allows extension in the extrusion direction and the opportunity to thin down the barrier layer and reorient the mica platelets.

A set of bottles made on a customer's wheel-type machine showed negligible disorientation of the mica platelets. Barrier results on these samples gave the expected three-fold improvement (Table VI). Summarizing these results on platelet orientation in blow molded bottles versus related barrier enhancement ($J_o/J_n$) (Table VII) shows that the same level of improvement is achieved as in other container processes when the platelets are aligned parallel to the container walls.

Table VI.  Co-Ex Blow Molded Bottles*

| Barrier | $cm^3 O_2$/Container-Day-Atm |
|---|---|
| SELAR OH Plus Mica | .0066 |
| Unfilled EVOH Resin | .019 |

*Experimental bottle made with ketchup-style mold on wheel machine

Table VII.  Effect of Platelet Orientation on Barrier
Enhancement - Coextruded Blow Molded Bottles

| Process Modification | Platelet Orientation to Plane of Wall | Barrier Enhancement $J_n/J$ |
|---|---|---|
| Initial Test | 45° | 1.35 |
| Low MI "OH Plus" | 30° | 1.9 |
| Above Plus Reduced PP Die-Swell | 15° | * |
| Wheel Machine | <5° | 2.9 |

*Poor wall thickness control

   While potential solutions for maximizing the barrier
performance of SELAR OH Plus in blow molded bottles have been
identified, the product has not found commercial acceptance in
containers such as squeezable ketchup bottles.  The mica imparts a
slight yellow color to the resin which the consumer finds objec-
tionable.
   The major opportunities for this new barrier resin composite
are in shallow solid phase pressure thermoformed containers and in
melt phase thermoformed containers.  These rigid containers are
typically pigmented, thus eliminating the problem of the yellow tint
(Table VIII).

Table VIII.  SELAR OH Plus - Recommended Applications

| Process | Suitability | Remarks |
|---|---|---|
| Coex Blow Molding | Barrier okay from wheel machine | Color may be unacceptable |
| Solid Phase Thermoforming | Okay on shallow draw containers | Barrier splits on deep draws |
| Melt Phase Thermoforming | Improved barrier demonstrated | Containers usually pigmented |
| Flexible Films | Improved barrier demonstrated | Poor flex crack color and transparency deficiencies |

## SUMMARY

1. A three-fold improvement in gas barrier has been demonstrated with a mica platelet filler in EVOH. This suggests that a value of about 0.3 would be appropriate as the geometric factor in the Cussler model using irregular flakes of nonuniform size typical of commercially available fillers.

2. Platelets must be aligned parallel to the surfaces of the film or membrane. The advantage of a tortuous path are rapidly lost when the platelets are tipped in the direction of gas permeation.

3. A phenomena of imbalanced die swell during coextrusion has been identified. This can produce thickening of the internal layer by a folding mechanism.

4. Practical packaging applications for platelet-filled high barrier resins have been identified.

## LITERATURE CITED

1. Katz, H. S.; Milewski, J. V., Handbook of Fillers for Plastics; Van Nostrand: New York, 1987.
2. Milewski, J. V.; Katz, H. S., Handbook of Reinforcements for Plastics, Van Nostrand: New York, 1987.
3. Transmet Corporation, Designing for Thermal Conductivity with Flake Filled Composites; TM-403: Columbus, Ohio.
4. Japanese Patent 61,204,244, Preparation of a Polyolefin Composition With High Voltage Resistance by Blending a Polyolefin With Mica and a Foaming Agent.
5. Japanese Patent 62,212,244, Lead Flake-Containing Polymer Shields for Housings for Electronic Controllers.
6. Inkpen, S. L.; et al., Electrical Induction of Patterns in Metallic Paints, Ind. Eng. Chem. Rev. 27, 58-64 (1988).
7. US Patent 3,463,350, Foodstuff Container Made from High Density Polyethylene and Mica, to Continental Can Company.
8. US Patent 4,528,235, Polymer Films Containing Platelet Particles, Sachs; et al., to Allied Corporation.

9.  Cussler, E. L.; Hughes, S. E.; Ward, W. J.; Aris, R., "Barrier
    Membranes"; J. Memb. Sci., 38, 161–174 (1988).
10. Prager, S., J. Chem. Phys., 33, 122–127 (1960).
11. Barrer, R. M.; Petropoulos, J. H., "Diffusion in Heterogeneous
    Media:  Lattices of Parallelepipeds in a Continuous Phase," Br. J.
    App. Phy., 12, 691–7 (1961).
12. Barrer, R. M., Diffusion in Polymers, J. Crank and G. S. Park,
    Eds.; Academic:  New York, 1968, pp. 165–217.
13. Murthy, N. S., et al., "Structure and Properties of Talc–Filled
    Polyethylene and Nylon 6 Films," Journal of Applied Poly. Science,
    31, 2569–2582 (1986).

RECEIVED October 17, 1989

# Chapter 12

# Permeability of Competitive Oxygen-Barrier Resins

## Orientability and Effect of Orientation

R. Shastri[1], H. C. Roehrs[1], C. N. Brown[1], and S. E. Dollinger[2]

[1]Materials Science and Development Laboratory, Central Research, The Dow Chemical Company, 1702 Building, Midland, MI 48674
[2]Films Research Laboratory, Dow Chemical U.S.A., Granville, OH 43023

In barrier packaging, it is generally accepted that orientation of the barrier resin reduces its gas permeability. Most of the published data, however, corresponds to low barrier resins. No supporting data exists yet for either high or intermediate barrier resins. To fill that void, the orientability and the sensitivity of oxygen permeability to orientation for six competitive oxygen barrier resins were evaluated. The barrier resins evaluated in this investigation include: an experimental grade Vinylidene chloride/vinyl chloride (VDC) copolymer, aromatic nylon MXD-6, amorphous nylon SELAR PA 3426, polyacrylic-imide XHTA-50A and two EVOH resins - EVAL EP-E105 and SOARNOL D. With the exception of both the EVOH grades, the remaining four barrier resins were orientable in the solid state. Generally, amorphous resins SELAR PA 3426 and polyacrylic-imide XHTA-50A were easier to orient than the semicrystalline resins, VDC copolymer and aromatic nylon MXD-6. In the case of both EVAL EP-E105 and SOARNOL D resins, all attempts to orient from the solid state were unsuccessful.

The effect of orientation on permeability is dependent on the morphological nature of the barrier resin. Semicrystalline polymers, VDC copolymer, and aromatic nylon MXD-6

0097–6156/90/0423–0239$06.00/0
© 1990 American Chemical Society

showed little if any improvements in
permeability at low orientation levels
and in the absence of additional heat
treatment.  The VDC copolymer, in fact,
showed higher permeability - 1.5 times
the permeability of the unoriented film -
with biaxial orientation.  This is
believed to be due to microvoid
development as a result of solid state
orientation of the VDC copolymer after
crystallinity is fully developed.  The
effect of orientation on amorphous
barrier resins, SELAR PA 3426 and XHTA-
50A is to decrease permeability by 5-30%
in both resins depending on the level of
orientation.

With the growing demand for coextruded products, barrier
plastics have shown significant growth in the last
several years.  Historically, the high barrier resins
market has been dominated by three leading materials --
vinylidene chloride (VDC) copolymers, ethylene vinyl
alcohol (EVOH) copolymers, and nitrile resins.  Since
1985, however, there has been a lot of interest worldwide
in the development of moderate to intermediate barrier
resins, as apparent from the introduction of a number of
such resins, notably, aromatic nylon MXD-6 from
Mitsubishi Gas Chemical Company, amorphous nylons SELAR
PA by Du Pont and NovamidX21 by Mitsubishi Chemical
Industries, polyacrylic-imide copolymer EXL (introduced
earlier as XHTA) by Rohm and Haas and copolyester B010 by
Mitsui/Owens-Illinois.

Understanding the structure-property relationships
in barrier polymers is an area of primary interest to
package designers.  In the case of most barrier resins,
the effect of environmental variables such as temperature
and relative humidity on permeability are generally well
known while a wealth of knowledge is being acquired on
the effects of geometric variables such as thickness, and
polymer structural parameters such as chemical structure,
crystallinity and molecular packing.

An additional key parameter which influences the
permeability of a polymer besides those cited above is
the state of orientation.  In the design of package
structures, orientation is of particular value as most
package forming processes inherently induce some degree
of orientation in the fabricated structure.  Though
improvements in transport characteristics with
orientation have been reported earlier in literature (1-
5), they pertain primarily to low barrier polymers such
as polyethylene (PE) and polypropylene (PP).  The reduced
permeability with orientation has been attributed to
enhanced crystallinity developed from molecular alignment
accompanying orientation.  There is little reported

evidence, however, of the effect of orientation on the permeabilities of medium and high barrier resins other than some recently published data on EVOH copolymers (6-13).
It is, therefore, important to evaluate the effects of orientation on the permeabilities of competitive medium to high oxygen barrier polymers. This study was initiated to fill that void.


Experimental

Materials. The competitive oxygen barrier resins investigated include: Dow Chemical Company's vinylidene chloride/vinyl chloride (VDC) copolymer (experimental grade XU 32009.02), Mitsubishi Gas Chemical Company's aromatic nylon MXD-6, Du Pont's amorphous nylon SELAR PA 3426, Rohm & Haas's polyacrylic-imide XHTA-50A and two EVOH resins -- Kuraray's EVAL EP-E105 (44 mole% ethylene content) and Nippon Gohsei's SOARNOL D (29 mole% ethylene content).

Film Preparation. Extrusion cast films of each barrier resin, approximately 7-10 mil in thickness, were produced by coextrusion as three-layer structures, 30 to 32 mils thick, between either HDPE or PP sacrificial skins.
Compression molded control films of VDC copolymer, 1-2 mil in thickness, were prepared by reheating extrusion cast films between the platens of a molding press.

Orientation. Solid state post orientation of individual films was accomplished with a laboratory scale T. M. Long Stretcher. Using a standard 4-inch square stretching head, the film samples were oriented up to 3.5X in either one or both directions. For biaxial orientation, simultaneous stretching mode was employed. The orientation temperature selected for each barrier resin was the lowest possible temperature above its glass transition temperature which would allow uniform stretching. For the amorphous polymers, the orientation temperatures ranged from 5° to 25°C above their glass transition temperatures, while optimal orientation temperatures for the semicrystalline resins ranged from 25° to 50°C below the corresponding crystalline melting temperatures. The orientation conditions are summarized in Table I.

Oxygen Barrier Measurements. Oxygen permeabilities of unoriented control films and oriented films were measured at 23.5°C and 65% RH on an Oxtran 1050 permeability tester.

Table I.  Orientation of Barrier Resins

| Barrier Resins | Orientation Temp. (°C) | Stretching Rate (in/sec) | Observations |
|---|---|---|---|
| VDC copolymer | 120 | 4 | non-uniform stretching |
| EVAL EP-E105 | 140-150 | 1 - 4 | very difficult; splitting |
| SOARNOL D | 140-150 | 0.2 - 1 | very difficult; splitting |
| MXD-6 | 105 | 4 | orientable below $T_{cc}$ |
| SELAR PA 3426 | 140 | 2 | easily orientable; has to be adequately dry |
| XHTA-50A | 175 | 4 | easily orientable |

Results and Discussion

Orientability. With the exception of both the EVOH grades, the remaining barrier resins were orientable. Generally, the amorphous resins SELAR PA 3426 and polyacrylic-imide XHTA-50A facilitated more uniform stretching when compared to semicrystalline resins, VDC copolymer and aromatic nylon MXD-6.

Efforts to orient VDC copolymer films resulted in uneven stretching. Adjusting orientation conditions yielded no apparent improvements in orientability. The non-uniform orientation of this film is attributed to the development of sufficient crystallinity in the extrusion cast film prior to orientation.

Orientation of aromatic nylon MXD-6 was rendered somewhat difficult by the occurrence of rapid cold crystallization above 100°C. The cold crystallization phenomenon as seen in figure 1 is similar to the well known cold crystallization behavior observed in other crystallizable polyamides and PET. To compensate for this, very short heat-up times had to be employed for successful orientation.

In the case of both EVAL EP-E105 and SOARNOL D resins, attempts to uniaxially orient to 2.5X were unsuccessful, as the films split and cracked upon drawing. Reducing the stretch rate and increasing the orientation temperature yielded no significant effect. As such, biaxial orientation of these two EVOH films was not attempted.

The difficulty experienced in orientation of EVOH films is not surprising. Ikari (6) indeed reports that drawing below the softening point of EVOH resins yields slight unevenness and cracks are likely to appear when the draw ratio is increased. He attributes this to the crystallization accompanying stretch orientation. This is especially true for low ethylene content EVOH grades.

Though we were unsuccessful in orienting EVOH resins, monoaxially as well as biaxially oriented EVOH films are commercially available (EVAL EF-XL monoaxially oriented film from Kuraray and EXCEED biaxially oriented film from Okura Industrial Company). Review of literature indicates that to improve orientation in EVOH resins requires one or more of the following: (1) rapidly quench the sheet from melt to retard crystallization prior to stretching; (2) plasticize the sheet (14-16) with water or glycols to improve chain mobility during stretching; (3) blend a small amount of nylon 6,12 (7) or thermoplastic elastomer (TPE) block copolymer (17) (as in SOARNOL STS recently introduced by Nippon Gohsei); or (4) choose higher ethylene content grades. In fact, Kuraray has two specially designed higher ethylene content grades -- EVAL EP-K (38 mole% ethylene content) to allow deep draw forming without

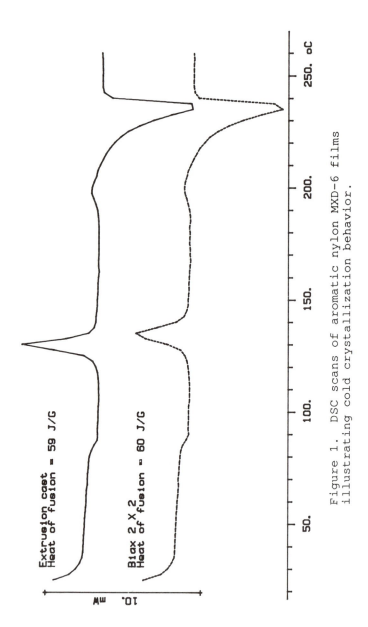

Figure 1. DSC scans of aromatic nylon MXD-6 films illustrating cold crystallization behavior.

fibrillation of the barrier layer (8) in solid pressure forming and EVAL EP-G (48 mole% ethylene content) for biaxially oriented film structures. The reduced vinyl alcohol content, in effect, lowers the amount of hydroxyl groups which usually act as a crosslink between molecules, thereby affecting the orientation behavior during drawing (18).

Solid state orientation of both amorphous resins – amorphous nylon SELAR PA 3426 and XHTA 50A yields rigid, crystal clear films. In the orientation of SELAR PA 3426 amorphous nylon, the films were dried overnight at 60°C to reduce the absorbed moisture level in the film prior to orientation. This was necessary to avoid foaming in the film when heated above its glass transition temperature.

Effect of Orientation. The measured oxygen permeability data are summarized in Tables II and III. The permeability values for the unoriented films are consistent with the expected values.

Semicrystalline polymers, VDC copolymer and aromatic nylon MXD-6 (Table II) showed little if any reduction in permeability at these moderate orientation levels. In fact, recent unpublished work has shown that aromatic nylon MXD-6 exhibits an initial increase in permeability up to 3X orientation followed by a significant reduction in permeability at higher orientation levels. The VDC copolymer also showed higher permeability with moderate biaxial orientation -- 1.5 times the permeability of the unoriented film. This is believed to be due to orientation of the polymer after crystallinity is fully developed. If the orientation of VDC copolymers is induced prior to full development of crystallinity in the material, one would not expect to see an increased oxygen permeability. In commercial practice, therefore, forming of VDC copolymer structures is normally done on rapidly quenched polymer to orient it while still in the amorphous state at temperatures near or above the $T_m$ of VDC copolymer.

Two plausible explanations for the observed higher permeability of oriented VDC copolymers appear likely. They are, formation of microvoids during solid state orientation or a change in the nature or size of crystallites during orientation. DSC characterization of the same films showed a noticeable difference in their crystallinities -- 24% for the biaxially oriented film vs. 33% for the extrusion cast film. Also the melting transition appears to be somewhat broadened for the biaxially oriented film (Figure 2) suggesting an increase in the distribution of crystallite size. However, wide angle x-ray diffraction studies reveal no indication of any structural differences between the extrusion cast and the biaxially oriented film. This seems to support the

Table II. Effect of Orientation on Oxygen Barrier
Characteristics of Semicrystalline Barrier Resins

|  | $O_2$ permeability @23.5°C & 65% RH (cc-mil/100 in$^2$ 24 hr. atm) |
|---|---|
| A. VDC Copolymer (experimental grade XU 32009.02) | |
| Compression molded film | $0.20 \pm 0.02$ |
| Extrusion cast film | $0.20 \pm 0.01$ |
| Biaxially oriented - 2.5 x 2.5 | $0.30 \pm 0.01$ |
| B. Aromatic Nylon MXD-6 | |
| Extrusion cast | $0.37 \pm 0.09$ |
| Biaxially oriented - 2 x 2 | 0.39 |

Table III. Effect of Orientation on Oxygen Barrier
Characteristics of Amorphous Barrier Resins

|  | $O_2$ permeability @23.5°C & 65% RH (cc-mil/100 in$^2$ 24 hr. atm) |
|---|---|
| A. Amorphous Nylon SELAR PA 3426 | |
| Extrusion cast | $1.40 \pm 0.31$ |
| Uniaxially oriented - 2.5x | $1.14 \pm 0.07$ |
| Biaxilly oriented - 2.5 x 2.5 | $1.01 \pm 0.01$ |
| B. Polyacrylic-imide XHTA-50A | |
| Extrusion cast | $3.12 \pm 0.17$ |
| Uniaxially oriented - 2x | $2.95 \pm 0.04$ |
| 2.5x | 2.84 |
| Biaxially oriented - 2 x 2 | $2.76 \pm 0.03$ |

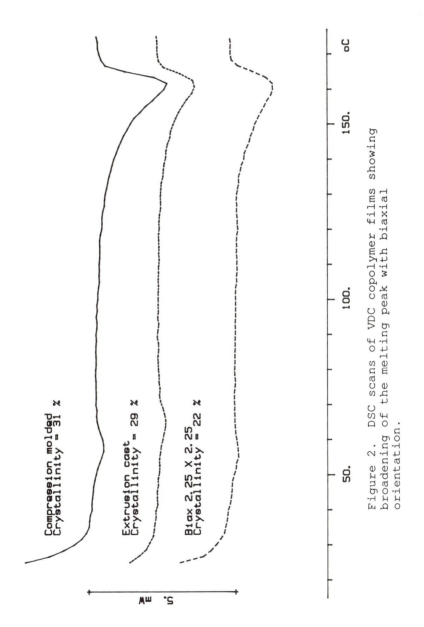

Figure 2.   DSC scans of VDC copolymer films showing broadening of the melting peak with biaxial orientation.

contention of microvoids development from solid state
orientation of the VDC copolymer. Indeed, this
hypothesis is corroborated by the observations of non-
uniform stretching of VDC copolymer films. The
characteristic necking behavior is associated with
yielded and unyielded zones, which is expected to lead
to formation of microcrazes responsible for the observed
higher permeability.

The effect of orientation on amorphous barrier
resins SELAR PA 3426 and XHTA-50A are shown in Table III.
The measured oxygen permeability of 1.4 cc-mil/100 in$^2$ 24
hr. atm for unoriented extrusion cast SELAR PA 3426 film
is reduced by 19% upon uniaxial orientation, while the
permeability of biaxially oriented films are 28% lower
than that of extrusion cast film. Similarly, the
decrease in oxygen permeability with 2X and 2.5X uniaxial
orientation in XHTA-50A are 5% and 9% respectively, from
the measured value of 3.1 cc-mil/100 in$^2 \cdot$ 24 hr. atm for
unoriented film. The corresponding reduction in
permeability with 2X x 2X biaxial orientation is 12%.

The effect of orientation on oxygen permeability of
the medium and high barrier resins is seen to be
dependent upon the morphological nature of the barrier
resin prior to orientation. A plot of the oxygen
transmission rates as a function of the overall draw
ratio (figure 3) illustrates this clearly. While the
semicrystalline polymers, VDC copolymer, and aromatic
nylon MXD-6, show little change in the permeability with
moderate amounts of orientation in the solid state,
orientation of the amorphous polymers SELAR PA 3426 and
XHTA-50A causes reduction in the permeability by 5-30% in
both resins, depending upon the overall level of
orientation.

The permeability reductions seen in amorphous
barrier resins agree very well with the expected changes
as per predictive relationships correlating the polymer
structural index, "Permachor" values to its gas
permeability (19-21). However, for the semicrystalline
polymers, the changes observed in the permeabilities are
significantly lower than the reductions of over 30%
predicted from empirical relations. According to Salame
(19-21), such large reductions are expected from the
increase in tortuosity caused by the lining up of
crystallites. Yet, experimentally, such greater degrees
of reduction in permeability are typically observed only
at orientation levels greater than 5X. Paulos and Thomas
(3) indeed found reduction in the oxygen permeability in
HDPE of less than 10% at 2-2.5X orientations. Swaroop
and Gordon (22) also report relatively insignificant
change in oxygen permeability of PET at draw ratios less
than three. Our results for MXD-6 and VDC copolymer are
consistent with these findings.

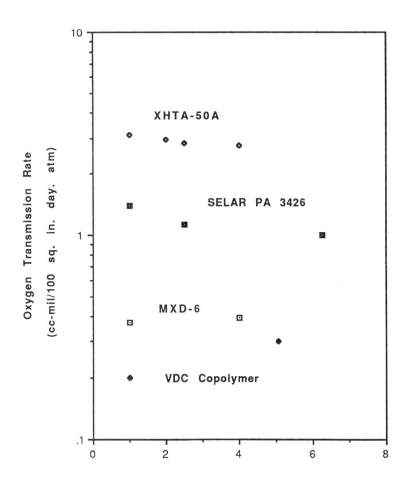

**Overall Draw Ratio**

Figure 3. Oxygen transmission rates as a function of the overall draw ratio, illustrating the dependence on the morphological nature of the barrier resin prior to orientation.

Conclusion

The results here suggest that amorphous polymers are more
readily orientable than semicrystalline polymers. The
difficulty in orientability of semi-crystalline resins
can be attributed to the fact that crystallinity is fully
developed in the films prior to the orientation step. .In
commercial practice, therefore, these polymers are
rapidly quenched to permit orientation while still in the
amorphous state.
    The effect of orientation on the oxygen permeability
is dependent upon the morphological nature of the barrier
resin prior to orientation. For the semicrystalline
polymers, VDC copolymer, and aromatic nylon MXD-6, little
change was observed in the permeability with orientation
from the solid-state. However, solid-state orientation
of the amorphous polymers SELAR PA 3426 and XHTA-50A was
shown to decrease permeability by 5-30% in both resins,
depending upon the level of orientation.

Literature Cited

1.  De Candia, F.; Vittoria, V.; Peterlin, A. Journal
    Polym. Sci., Polym. Phys. 1985, 23, pp. 1217-1234.
2.  Holden, P.S.; Orchard, G. A. J; Ward, I. M. Journal
    Polym. Sci., Polym. Phys. 1985, 23, pp. 709-731.
3.  Paulos, J. P.; Thomas, E. L. Journal Appl. Polym.
    Sci. 1980, 25, pp. 15-23.
4.  De Candia, F.; Vittoria, V.; Rizzo, G.; Titomanlio,
    G. Journal Macromol. Sci., Phys. 1986, B25 (3), pp.
    365-378.
5.  Devries A. J. Polym. Eng. Sci., 1983, 23 (5), pp.
    241-246.
6.  Ikari, K. Proceedings of the International
    Coextrusion Conference COEX Europe '86, 1986,
    Cologne, pp. 21-56.
7.  Okata, H.; Okaya, T.; Kawai, S. Proceedings of the
    International Coextrusion Conference COEX America'86,
    1986, pp. 63-87.
8.  Foster, R. H. Polym. News, 1986, 11, pp. 264-271.
9.  Culter, J. D. Journal Plast. Film & Sheeting, 1985 1,
    pp. 215-225.
10. Mitsutani, A.; Morimoto, O. Research and Development
    Report No. 30, Nippon Chemtec Consulting Inc., JAPAN,
    1985.
11. Schroeder, G. O. FUTURE-PAK '84, Second
    International Ryder Conference on Packaging
    Innovations, 1984, pp. 333-355.
12. Ikari, K. Proceedings of the Second International
    Conference on Coextrusion Markets and Technology COEX
    '82, 1982, pp. 1-42.

13. EVAL Technical Bulletin, Kuraray Co. Ltd., No. KIC-102.
14. Japan Kokai 50-144776, "Saponified Ethylene-Vinyl Acetate Copolymer Films," assigned to Toyobo, 1975.
15. Japan Kokai 50-144777, "Saponified Ethylene-Vinyl Acetate Copolymer Films," assigned to Toyobo, 1975.
16. Chiba, T., et al. US 3,419,654, 1968.
17. Moriyama, T.; Asano, K.; Iwanami, T. Proceedings of the International Coextrusion Conference COEX Europe '86, 1986, pp. 69-103
18. Yoshida, H.; , Tomizawa, K.; Kobayashi, Y. Journal Appl. Polym. Sci., 1979, 24, pp. 2277-2287.
19. Salame, M. Polym. Eng. Sci., 1986, 26 (22), pp. 1543-1546.
20. Salame, M. Journal Plast. Film & Sheeting, 1986, 2, pp. 321-334.
21. Salame, M. FUTURE-PAK '84, Second International Ryder Conference on Packaging Innovations, 1984, pp. 119-136.
22. Swaroop N.; Gordon, G. A. Polym. Eng. Sci., 1980, 20 (1), pp. 78-81.

RECEIVED November 14, 1989

# Chapter 13

# Polymer Blends

## Morphology and Solvent Barriers

P. M. Subramanian

Polymer Products Department, E. I. du Pont de Nemours and Company, Experimental Station, P.O. Box 80323, Wilmington, DE 19880–0323

Articles such as containers, films, etc., prepared from polyolefins suffer from significant permeation of organic solvents and flavors, making polyolefins less desirable for a variety of packaging and industrial applications (e.g., automotive fuel tanks). Enhancement of the permeability barrier properties of polyolefins by blending with a polymer having high barrier properties has limited applicability in most cases, because such barrier polymers are relatively expensive and the conventional blend technology offers only limited barrier enhancements at low concentration of these additives because of the homogeneous nature of their dispersion. Described here are polymer blends having "laminar" morphologies of the dispersed barrier polymer phase in the matrix of the polyolefin resulting in significant enhancement of the hydrocarbon barrier properties of the polyolefin composition. Thus, using 3-10% of a polyamide as the barrier polymer, permeability loss can be reduced by a factor greater than 200 times over that from unmodified polyethylene. This significant enhancement of hydrocarbon barrier properties approximates that obtained by a complex multilayer coextruded system, but achieved here by using a simple process with only a single extruder system. The "laminar" or platelet morphologies in these structures can be established by microscopic examination of the cross section of the fabricated article. These can also be indirectly established by nondestructive ultrasonic examination of the molded article. Using this technology, small and large containers with good physical properties and permeability barrier to hydrocarbon fuels as well as fuels containing alcohols have been prepared and characterized.

Polyolefins such as polyethylene (PE) and polypropylene (PP) are ideal candidates for the fabrication of a wide variety of films, containers, trays, etc., for various packaging and industrial applications. While these hydrocarbon polymers

0097–6156/90/0423–0252$06.00/0
© 1990 American Chemical Society

have outstanding permeability barrier properties against permeation of water, their high permeability to many organic solvents and gases makes these undesirable for applications where high barriers are required. Improved barrier properties are required for packaging applications such as containers and films in the food, cosmetics and other industries. A variety of techniques have been developed to impart enhanced barrier properties; e.g., coextrusion with a thin layer of the barrier polymer, surface modification by treatment with highly oxidizing gases, and often secondary coatings using expensive techniques. Barrier enhancements by using conventional polymer blending technologies have been less successful because of the need for high concentrations of the barrier polymers to achieve significant reduction in permeability. Such high concentrations result in poor economics and generally lower impact toughness. Demanding applications such as automobile fuel tanks require that they pass stringent toughness requirements, while being economical.

Earlier work done in these laboratories has shown that polymer blends wherein the barrier polymers are dispersed as thin platelets parallel to the surface of the fabricated article, have significantly improved permeability barrier properties than the conventional "homogeneous", uniform, dispersions (1,2). The high barriers demonstrated were then achieved by blending 15-20% polyamides with a linear polyethylene, which achieved performance comparable to that obtained by adding as much as 50% nylon by conventional blending. While this performance has been satisfactory for a variety of applications, some of the more demanding uses require that the barrier polymer be used more efficiently. Described here are such blend compositions which show substantial resistance to permeation of hydrocarbon solvents and their mixtures.

*Experimental*

*Materials.* High density polyethylene (HDPE) having different molecular weights, and specific gravity of 0.951 (Marlex 5202, HXM 50100, made by Phillips 66 Co), were used for extrusion applications. Polyamides used were a semicrystalline copolyamide of adipic acid, hexamethylene diamine and caprolactam, and a copolyamide containing isophthalic acid as well. An anhydride modified polyethylene (3-5) as an interlaminar adhesive/compatibilizer was also used. The combinations are generally included in "Selar" barrier materials supplied by E. I. du Pont de Nemours & Co.

*Processing.* The pellets of the "Selar" (polyamide/compatibilizer) were blended with linear, high density polyethylene (HDPE) in the ratios described in the various tables. The blend of the pellets was fed to an extruder equipped with a screw and mixing hardware chosen for low shear/mixing (as described in U.S. 4,444,817) and containers made by extrusion blowmolding using blowmolding machines manufactured by Johnson Controls (USA) or Battenfeld Fischer (Germany) and other equipment manufacturers. The containers described in Table I were cylindrical ("Boston Round") with a capacity of 1 l. and had a finished neck for tight sealing.

The large containers (e.g., automotive fuel tanks) were made using low concentrations, about 4% of the modified polyamide, in large "accumulator type" blowmolding machines (e.g. manufactured by Sterling Extruders, [USA]; Kautex, [Germany]), equipped with low mixing extruder screws and other hardware. The polyethylene used for these (e.g., "Marlex" HXM 50100) differed from the above in having much higher molecular weight and melt viscosity.

The polyethylene and the blowmolding equipment used are well known in the industry. While the brand names may be different, in general, these are comparable in their physical properties and machine capabilities.

*Permeability Testing.* Permeabilities of these containers were determined gravimetrically by measuring weight loss at different intervals. Elevated temperatures were often used to carry out accelerated testing. The containers were filled with various organic solvents (toluene, xylene, etc) to various levels and the weight loss measured. These measurements are described in ASTM D2684-73 and is the preferred technique for evaluating irregular–shaped articles. The associated weight losses are generally reported as gram-mil/100 sq inch/day (24 hrs) (gmd).

*Microstructure/Morphology.* Microtomed sections of the fabricated part were examined under an optical microscope at various magnifications. For contrast, a polarizer was used. In the "laminar" type of articles, the nylon platelets were large enough to be visible with the naked eye, especially when the nylon was dyed for contrast. The microstructure can also be indirectly established by ultrasound techniques. For these measurements, the specimen, in a water bath, was exposed to a pulse of 10 MHz ultrasound. The echos from the front and back sides of the walls of the specimen were recorded on an oscilloscope. These echos are separated in time by an amount proportional to the thickness of the specimen. The echos from the discontinuities within the sample, such as layers or voids appear in time, between the front and rear surfaces as a function of their location. These signals form a "fingerprint" pattern depending on the morphology of the nylon particles and can be correlated with the actual permeability (6-7) obtained by weight loss measurements.

### Results and Discussion

Homogeneous polymer blends from incompatible polymers such as polyethylene and nylon generally result in poor dispersions, with the dispersed particles being very large with poor adhesion to the matrix. This is shown in Figure 1 in which a cross section of an article made from a blend of polyethylene and nylon (80:20) was examined under a microscope. The nylon particles are large and reveal incompatibility. The homogeneity of these blends can be improved substantially by incorporating a polymeric surfactant or compatibilizer. Such blends, especially in the case of polyolefins and nylon, are well known in the literature as "homogeneous" blends wherein the dispersed phase is homogeneously distributed in the matrix polymer phase (8-9). Such technology is also generally applied for toughening of nylon (10-11) where a functionalized elastomer is dispersed homogeneously as small micron-sized particles in the nylon matrix. These examples show the good dispersion of nylon in the polyolefin phase, or the reverse, depending upon the concentrations. The good interfacial adhesion in such cases is indicated by the fine dispersion of the polyamide shown in Figure 2 in which a section of the wall of a container from a composition containing 15% nylon, dispersed homogeneously in polyethylene in the presence of a carboxylated polyethylene, was examined under a microscope. The small, spherical submicron nylon particles are uniformly distributed within the polyethylene matrix.

10 μm

Figure 1.   Polyethylene–nylon blend (85 : 15).   View of cross-section of molded article.   (Reprinted with permission from Ref. 1.   Copyright 1984 Technical Association of Pulp and Paper Industry.)

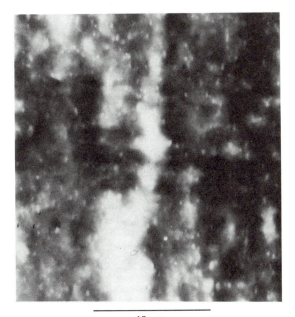

10 μm

Figure 2.   Polyethylene–nylon–anhydride modified polyethylene blend (85 : 15 : 3). View of cross-section of molded article.   (Reprinted with permission from Ref. 1. Copyright 1984 Technical Association of Pulp and Paper Industry.)

For permeability properties, the continuous phase has the predominant effect. Homogeneous blends of a permeable thermoplastic polymer and a barrier polymer, usually are very permeable until there is a substantial volume fraction of the latter. These permeability reductions arise because of the longer path-length that the permeant molecule has to travel, compared with the unmodified matrix polymer and can be calculated using expressions derived by Nielsen and others (12). The permeation of hydrocarbons from the containers made from such a polyethylene-nylon blend is shown in Figure 3 where the solvent (unleaded gasoline) was placed in a blowmolded container of about 1 liter capacity and the loss measured by determining the weight difference at room temperature (23 ° C).

Blowmolded containers from the polyethylene- nylon blend, where the nylon is distributed as thin platelets or lamellae as described in references (1) and (2) show substantial resistance to permeation, as shown in Figure 4. While these barrier properties are outstanding as given, current work demonstrates that by better control of the rheological properties of the polymers as well as the melt processing parameters, substantially higher efficiencies can be obtained (Table I) (13). Solvent permeabilities for different solvents in 1 1 containers weighing 55 g. and 70 g. weight (greater wall thickness), respectively are given for different solvents as a function of the concentration of the polyamide. As the concentration of nylon is increased, the hydrocarbon permeabilities become substantially lower than that in containers made from unmodified polyethylene. Since the resistance of nylon to permeation of polar solvents is poor, the enhancement of barrier properties are also diminished. This can be seen in the case of tetrahydrofuran (THF) permeability. Nonetheless, the performance is significantly better than polyethylene alone. Once the critical laminar structure is established, further enhancement takes place at a slower rate with increasing nylon content.

*Permeation Rates in "Commercial" Containers.* The properties mentioned above were obtained in containers blow molded under experimental conditions. Similar performance has been demonstrated on a routine commercial basis. Table II describes the permeabilities in these containers. The results are comparable to that of containers made under laboratory conditions.

*Large Containers.* In general, most of the solvent containers and articles used in the packaging industry are small (e.g. 0.5 1–20 l capacity). These are often made by blowmolding using medium size equipment. Large containers such as automotive fuel tanks are made using very complex, large equipment that are often of the "accumulator-head type". The raw materials for such, as well as the design and configuration of the extruders, dies, molds, etc. and the processing conditions are critical because of the need to attain the stringent impact toughness requirements as well as barrier properties. "Laminar" blend technology allows one to accomplish this by choosing the appropriate low levels of the nylon containing "Selar" compositions. Figure 5 shows the permeability of xylene in a PE/nylon laminar walled container as a function of the nylon concentration.

While the permeability properties of individual containers are important for a variety of applications and are meaningful, the permeability of fuel tanks in an automobile is not tested alone but tested as a total system by resorting to a "shed test" in which the permeability of the whole car is determined. The results, therefore, include the permeability losses not only from the fuel tank, but also the losses from other parts of the system — e.g., fuel lines, carburetors, etc. Figure 6 shows the permeability properties of various "Selar" compositions

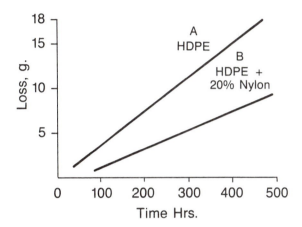

Figure 3. Hydrocarbon (unleaded gasoline) permeability (weight loss, g) in blow molded containers from polyethylene (HDPE) and polyethylene/nylon blends. (Reprinted with permission from Ref. 1. Copyright 1984 Technical Association of Pulp and Paper Industry.)

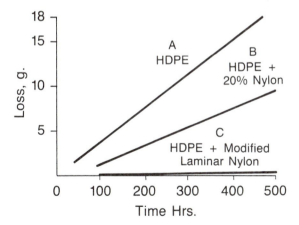

Figure 4. Hydrocarbon permeability in containers from polyethylene, polyethylene–nylon homogeneous blend and polyethylene–nylon laminar blend. (Reprinted with permission from Ref. 1. Copyright 1984 Technical Association of Pulp and Paper Industry.)

TABLE I

SOLVENT PERMEATION IN HDPE/NYLON
LAMINAR WALLED BOTTLES
(WEIGHT LOSS, GMD*)

| | $A^{++}$ SERIES | | | | $B^{++}$ SERIES | | | |
|---|---|---|---|---|---|---|---|---|
| "SELAR"[+] CONC % | 0 | 4 | 8 | 12 | 0 | 4 | 8 | 12 |
| XYLENE, GMD | 557 | 7.2 | 1.03 | 0.73 | 544 | 2.9 | 0.83 | 0.79 |
| TOLUENE " | 689 | 16.7 | 4.50 | 2.1 | 638 | – | 3.8 | – |
| HEXANE " | 52.9 | 0.28 | 0.013 | 0.007 | 47.2 | 0.18 | 0.009 | 0.005 |
| TETRA- " HYDROFURAN | 36.9 | 6.0 | 2.15 | 1.05 | 35.4 | 5.33 | 1.5 | 0.67 |

*DETERMINED AT 60°C IN EXPERIMENTAL CONTAINERS, 1 L. CAPACITY.
GMD  -  GRAMS-MIL/DAY (ASTM D-2694)

[+]"SELAR" RB - ADHESION MODIFIED NYLON SOLD BY
　　　　　　　E. I. DU PONT DE NEMOURS
　　　　　　　WILMINGTON, DE

[++]A SERIES - LIGHT WALLED, 55 G WEIGHT CONTAINERS
  B SERIES - THICKER WALLED, 70 G WEIGHT CONTAINERS

TABLE II

PERMEABILITY (WEIGHT LOSS)
IN INDUSTRIAL, LAMINAR WALLED CONTAINERS
FROM POLYETHYLENE AND NYLON ("SELAR" RB)

| SOLVENT | WEIGHT LOSS % | RELATIVE IMPROVEMENT (HDPE CONTAINERS = 1) |
|---|---|---|
| HALOGENATED HYDROCARBONS | | |
| TRICHLOROETHANE | 0.04 | 170 |
| 0-DICHLOROBENZENE | 0.03 | 90 |
| AROMATICS | | |
| XYLENE | 0.12 | 350 |
| TOLUENE | 0.25 | 200 |
| OXYGENATED SOLVENTS | | |
| METHYL ETHYL KETONE | 0.45 | 20 |
| ETHYL ACETATE | 0.50 | 35 |
| METHYL ALCOHOL | 0.55 | 5 |

DATA OBTAINED WITH 1 LITER BOTTLES AT 10% "SELAR" RB (70 GR. BOTTLE)
MADE UNDER "INDUSTRIAL" CONDITIONS.

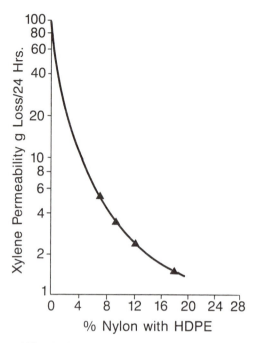

Figure 5.   Permeability (weight loss, g) as a function of nylon concentration, in laminar-walled containers. (Reprinted with permission from Ref. 1. Copyright 1984 Technical Association of Pulp and Paper Industry.)

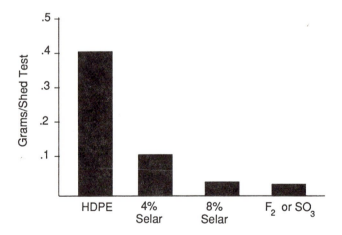

Figure 6.   Permeation (weight loss, g) of a fuel tank with different barrier enhancement treatments, to unleaded gasoline, by the shed test. (HDPE,HDPE + Selar at 4% and 8% surface modified HDPE containers (modified with $F_2$ or $SO_3$). (Reprinted with permission from Ref. 7. Copyright 1989 Society of Automotive Engineers.)

as well as fuel-tanks made by surface modification by exposure to fluorine ($F_2$) or sulfur trioxide ($SO_3$). "Selar" nylon at 4% concentration in a polyethylene blend shows substantial reduction in permeability over polyethylene. At such small concentrations, the excellent toughness of the polyethylene is not compromised.

*Permeability to Hydrocarbon Fuels Containing Methanol.* Mixed solvents behave unusually in permeation in that their rate of permeation and selectivity are not the average of their components. For example, permeation of hydrocarbon fuels containing low molecular weight alcohols even at low concentrations are significantly higher than that of pure hydrocarbons, when the barrier treatments are made using surface treatments with oxidizing gases, or the "laminar blends" using 66/6 copolyamides or polycaprolactam. This permeability is further aggravated when the mixed fuel is contaminated with moisture. However, amorphous copolyamides ("Selar" II) containing aromatic acids, distributed as thin platelets, are found to be more effective in reducing the permeability loss of such hydrocarbon-alcohol (xylene:methanol 85:15) mixtures *(14)* (Figure 7).

*Long-Term Permeability in Polyethylene-Nylon Laminar Walled Containers.* Polyolefin containers often are subject to loss of properties with aging. These result from plasticization, crystallization, and stress cracking. Figure 8 shows the permeability properties of 66/6 copolyamide based "laminar" blends. The permeability performance was found to be unchanged during over 5 years of storage in the two separate samples shown here.

*Impact Toughness of Large Containers.* Polyethylene-nylon blends generally show lower impact toughness properties than pure polyethylene when the nylon is dispersed homogeneously. However, large containers with laminar distribution of the low levels of nylon and fabricated by blow molding under industrial conditions (using properly designed molds and processing conditions) have passed the low temperature (-40 °C) drop impact tests (>6 meters height) readily. Instrumented impact toughness and fracture toughness properties also do not show significant changes from polyethylene when the modifying nylons are used at these small concentrations.

*Morphology of the Polymer Blend.* The high efficiency of the barrier polymer results from the platelet type dispersion of the nylon, under the controlled processing conditions which requires that the ingredients be combined and melt processed under low shear processing conditions. Melt compounding and dispersive mixing tend to reduce the effectiveness of the nylon platelets. Unlike the nearly spherical small particles that are obtained (and desirable) for homogeneous blends, the nylon in these laminar blends are dispersed as large thin platelets parallel to the walls of the article as shown in Figure 9 where sections of the container are viewed under a microscope. The heavy black lines represent nylon platelets. Many of these dispersed nylon platelets are large and often can be seen without the aid of the microscope, especially when they are dyed selectively (e.g. with bromothymol blue or Congo red).

Nondestructive techniques using ultrasound and computer analyses have also been developed in order to establish the morphology of the nylon dispersion rapidly. These tests are useful for fast nondestructive testing of the fabricated articles such as bottles, films, containers, and fuel tanks, etc. Figure 10 shows the response of ultrasonic analysis of these containers. The signal obtained in the "laminar-walled" container shows distinctive peaks corresponding to the

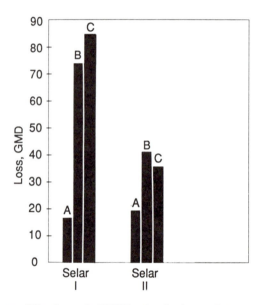

Figure 7. Permeability losses in HDPE–nylon laminar-walled containers using different nylon compositions (Selar I and Selar II) : [A] xylene [B] xylene: methanol 85 : 15  [C]  xylene : methanol : water  (96.5 : 3.5: 0.5). Permeability denoted as (grams–mils/day, GMD). (Reprinted with permission from Ref. 7. Copyright 1989 Society of Automotive Engineers.)

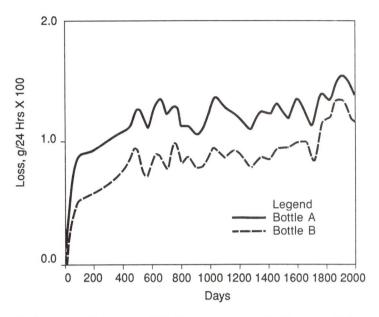

Figure 8.   Long term xylene permeability in two separate 1 gallon laminar-walled containers made from HDPE Selar. Permeation (weight loss) denoted as g/24 hours × 100.

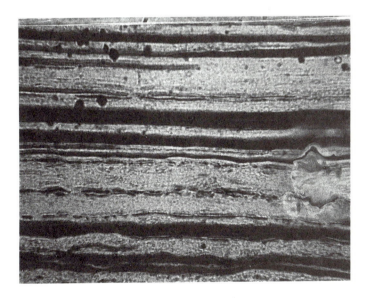

Figure 9.   Cross-sectional view of a section of the wall of a HDPE–nylon laminar-walled container (80 × magnification).

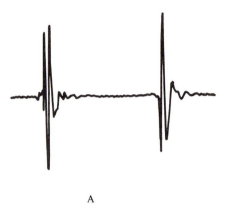

A

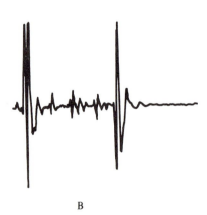

B

Figure 10.   Ultrasound response of container walls.   (A) High-density polyethylene
(unmodified)   (B) HDPE/Selar RB (5%) laminar-walled containers.

laminar dispersion of the polyamide and can be indirectly correlated with the actual permeability loss. This technique offers a fast and reliable route for fast characterization and quality control of the fabricated parts.

These analytical techniques as well as the availability of appropriate melt-processing hardware, have made the "laminar" dispersion system amenable to fabrication of high-barrier containers and other articles, routinely.

## Summary and Conclusions

Polymer-polymer blends afford a unique technology for modification of properties. The resulting properties are extremely dependent on the morphology of the polymer phases. The "laminar" dispersion of small amounts of a polymer with good permeability barriers, in a matrix of a polymer which is generally very permeable, offers significant barrier enhancements, unlike uniform, homogeneous dispersions of the same polymers. Such barrier enhancements have been demonstrated in the case of polyethylene and small amounts of a nylon. Their morphologies have been identified and are described. Such technology has been applied to the blow molding of small and large containers with enhanced barrier properties. These morphologies and the use of "tailored" nylons give a route to fabrication of small and large containers with barrier to permeability of hydrocarbons and other types of solvents and their mixtures, without adversely affecting their physical and mechanical properties.

## Literature Cited

1. P. M. Subramanian, *Conference Proceedings - Technical Association of Pulp and Paper Industry*; Laminations and Coating Conference, *341* (1984).
2. P. M. Subramanian, V. Mehra, *Society of Plastics Engineers (SPE) ANTEC Proceedings, 301* (1986).
3. E. A. Flexman, Jr. and F. C. Starr, U.S. Patent 4,026,967 (1977).
4. T. J. Grail and R. A. Steinkamf, U.S. Patent 3,953,655 (1976).
5. P. M. Subramanian, U.S. Patent 4,444,817 (1984).
6. T. W. Harding, (E. I. du Pont de Nemours) private communication.
7. R. L. Bell, V. Mehra, *SAE (Society of Automotive Engineers) Technical Paper #890442*, International Congress, Feb. 27 - March 3 (1989).
8. A. D. Armstrong, U.S. Patent 3,373,222 (1968).
9. R. B. Mesrobian, P. E. Sellers, D. Ademaitis, U.S. Patent 3,373,224 (1968).
10. B. N. Epstein, U.S. Patent 4,174,358 (1979).
11. S. Y. Hobbs, R. C. Bopp, and V. H. Watkins, *Polymer Engineering and Science, 23*, 380 (1983).
12. L. E. Nielsen, *J. Appl. Polym. Sci., 23*, 1907 (1979).
13. J. M. Torradas, private communications.
14. V. Mehra, private communications.

RECEIVED January 30, 1990

# Chapter 14

# Improvement in Barrier Properties of Polymers via Sulfonation and Reductive Metallization

W. E. Walles[1]

Central Research, The Dow Chemical Company, 1702 Building, Midland, MI 48674

This paper considers various methods to create a thin diffusion barrier at the surface of a plastic article such as a packaging film or container after it has been given its intended shape. Such an approach permits one to choose the optimum polymer type based on strength, processability, cost, etc. without being constrained by barrier properties. The advantages of surface sulfonated containers are presented for barrier applications such as polyethylene drums and tanks for solvents and automotive gasoline. The improvement of barrier properties for critical gases such as oxygen are also treated. The barrier treatment can be applied during blow molding as an integrated process or as a separate step afterwards. For very demanding future applications, such as all-plastic thermos bottles and flat foam panels with vacuum insulation, the sulfonation step can be followed by reductive metallization. This results in a further dramatic reduction in permeation, especially when combined with a high barrier polymer overcoat to cover minor imperfections in the metallized surface.

Chemical modification of the surface of engineering polymers is an attractive approach to the problem of economically producing a barrier or a permselective structure with tailored properties. The following discussion will focus on applications of this technology

[1]Current address: Coalition Technologies, Ltd., 6648 River Road, Freeland, MI 48623

0097–6156/90/0423–0266$06.00/0
© 1990 American Chemical Society

involving sulfonation(1-8). Related applications involving
asymmetric membranes for gas separations have also been
demonstrated, but will not be considered in depth here.
This latter membrane application involves selectively
altering the permeability of one gas relative to another
without causing unacceptable reductions in the permeability
of the desired penetrant. In the case of polyethersulfone,
it has been shown to be possible to essentially double the
selectivity of an asymmetric membrane for the $CO_2/CH_4$ and
$O_2/N_2$ systems while reducing the permeability of the
desired $CO_2$ and $O_2$ only by about a factor of two (9).

The barrier work involves reducing the permeability of
various polymers to both liquids and gases. In the case of
liquids, the objective is to develop medium level diffusion
barriers to volatile organics such as hydrocarbon fuels
which have a tendency to swell and plasticize many
containers. For gases, the objectives are more demanding
and require the development of high, or effectively
complete barrier for applications such as preventing long
term diffusion of air components into a vacuum retained by
plastics in insulating applications.

Typically, chemically modified surface layers involve
thicknesses ranging from less than 1 micrometer(μ) up to
20μ, so the overall *mechanical* properties of the treated
objects are hardly affected by the process. Even if one
limits the discussion to materials containing C-H bonds,
practically all engineering plastics are covered except
pure fluorocarbons and some silicones. Clearly, various
gases can react with the carbon-hydrogen bonds on the
surface of a plastic article and can reduce the diffusion
coefficient of penetrants in the material. The choice of
sulfonation as the preferred treatment, therefore, is not
based solely on the ability to modify transport properties.

Rationale for the Use of Surface Sulfonation

Our exploratory experiments involving the use of
fluorine treatments resulted in containers with a good
barrier to gasoline; however, upon repeated flexing, a loss
of barrier properties can result. The barrier layers in
these cases were up to 0.1 μ thick. Additional studies
using 25μ films of polystyrene, polyethylene and
polypropylene exposed to fluorine for a longer time led to
total pulverization of the films. The loss in mechanical
properties appears to be associated with the localization
of energy associated with the C-F bond formation, which in
turn can break neighboring C-C bonds of the polymer
backbone. Under certain rigorously controlled treatment
protocols, the use of fluorine has proven workable(10), but
it was felt that a less aggressive agent allowed added
process flexibility for our work.

Surface chlorination can produce good barrier layers
with less tendency to provide flexure failures.
Unfortunately, chlorine reacts too slowly to be practical
as a treatment agent in the absence of promotion by UV
light.  It is clearly not convenient to use UV in the
interior of containers, so alternative activators were
sought.  Chemical activation of the chlorination reaction
was shown to be possible using a mixture of 10% $SO_3$/90% $Cl_2$
as such or further diluted with air or nitrogen at 25°C.
The process led to significantly uptakes of both Cl and $SO_3$
(4) and produced good barriers to gasolines, particularly
to the ethanol- and methanol-containing types.
    Treatment of the inside of high density polyethylene
(HDPE) automotive gas tanks with about 20% $SO_3$ in air
followed by air purging with subsequent neutralization with
$NH_3$ gas resulted in an excellent gasoline barrier (See Fig.
1).  Concentrations of about 75 and 200 micrograms $SO_3$ per
$cm^2$ of surface reduces permeation losses by 90 and 99%,
respectively in ambient temperature permeation tests.   As
indicated in Fig.2, the barrier layer was found to have
$-SO_3^-NH_4^+$ groups  to a depth of 20-25 micrometers.  The
treated layer has good resistance to flexure failures.
When a 25 micrometer polyethylene film was similarly
sulfonated, it was found to have roughly 200% elongation at
break and twice the breaking strength of the untreated
polyethylene.
    Of the first series of one thousand station wagons
equipped with these tanks, several were retrieved from
junked cars with gasoline exposure times corresponding to
more than 90,000 miles each.  In all of these cases, the
sulfur surface concentration, barrier properties and
mechanical properties of the tank walls were fully intact.

Nature of the Barrier Layer

    Based on infrared spectroscopic analysis, it is clear
that following $SO_3$ exposure,  substantially all of the
sulfur present is in the form of:

$$\overset{|}{\underset{|}{-C}}-SO_3H$$

With typical treatment procedures, it appears that the
sulfonated polymer units are interspersed between 0-500
units of unsulfonated polymer units as illustrated by the
following formula

$$-\underset{SO_3H}{CR}-C(R_1)(R_2)-[C(H)(R)-C(R_1)(R_2)]_x-\underset{SO_3H}{CR}-C(R_1)(R_2)-$$

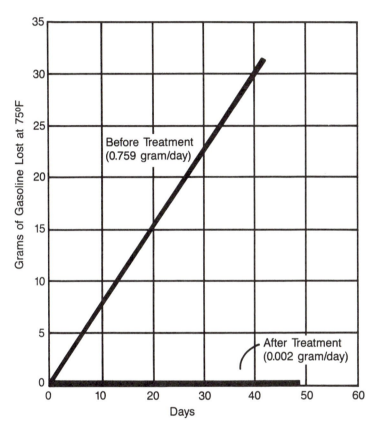

Figure 1: Effects of surface sulfonation of the amount of gasoline lost from one pint bottles with 25 mil wall thickness at 75°F as a function of time.

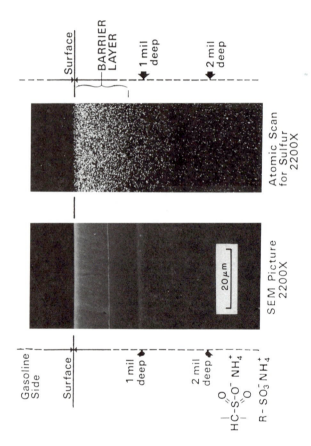

Figure 2: Photograph of the cross section of a surface sulfonated polyethylene container wall showing the scanning electron microscope and sulfur atomic scan profiles.

where x = 0-500 and R, $R_1$ & $R_2$ are hydrogens, chlorines, alkyl groups, etc., depending upon the specific polymer being treated. The sulfonate group can be used to graft ethylene oxide and other epoxides on the surface if desired, to promote the hydrophillic nature of the barrier layer (11). Essentially all commercially available films containing either a CH or an NH bond have been found to be treatable via the sulfonation process (11, 22). For the NH containing materials, the analogous structure, $NSO_3H$, to that shown for the CH bond reaction occurs.

   The barrier layer formed by the sulfonation treatments is shiny yellowish to light brown for treatment of simple polyolefins in the presence of inert gases due to the presence of some unsaturated groups in addition to the primary sulfonate group. A detailed identification of the structures produced was published recently(2). In the presence of oxygen, simultaneous oxidation and sulfonation can occur to produce a dark brown or even black color. Even at relatively low concentration, the presence of complex chromophoric groups resulting from the further reaction of such hydroxy, keto and carboxylic acid groups are apparent, so the use of color as a indicator of the extent of sulfonation is only approximate. When it is desirable to reduce or remove the color, bleaching with an agent such as sodium hypochlorite or hydrogen peroxide is effective(3).

Sulfonation Examples

   The time  required to produce a desired barrier improvement, varies with the type of polymer, its degree of crystallinity and the particular penetrant being considered. For example, a gasoline container or gas tank fabricated of polyethylene generally requires a degree of sulfonation equivalent to about 0.1 mg to 1 mg $SO_3$ per $cm^2$. For methylene chloride or other halogenated lower hydrocarbons, a higher degree of sulfonation, equivalent to about 1.5 mg to 20 mg $SO_3$ per $cm^2$ is desirable. On the other hand, if the container is fabricated from polypropylene, a degree of sulfonation as low as 0.015 mg $SO_3$ per $cm^2$ can be effective; however, higher levels are generally used(3).

   To prevent the conversion of the sulfur trioxide to sulfuric acid which inhibits the sulfonation process, it is important to exclude water vapor from the carrier gas used to bring the sulfur trioxide into contact with the part to be treated. A conventional drying tube is adequate for this purpose. As would be expected, for a given type of substrate, the degree of treatment varies with the concentration of sulfur trioxide and the temperature used. For instance, if a substrate such as polyethylene is exposed at 25°C to 2 volume percent sulfur trioxide in air for 15 to 20 minutes to give a desired degree of treatment,

only 2-3 minutes are required with 18 percent sulfur
trioxide at 25°C. Moreover, at 35°C, the time required
when using 18 volume percent sulfur trioxide is shortened
to only 1-2 minutes(3). Clearly, the exposure time and
concentration of sulfur trioxide are inversely related at
any temperature, and the exposure time at a given
concentration is an exponentially decreasing function of
temperature for each material. Due to the complexity of
the modification process, the detailed processing
relationships must be determined empirically for each
material.

Under optimum conditions at sufficiently dilute
conditions, studies with a quartz spring microbalance
indicate that the reactive gas penetration is a diffusively
controlled process. Both weight gain and depth of
penetration tend to follow a square root of time
relationship. Under high concentrations, the generation of
heat affects the kinetics and can result in an uptake rate
that is almost directly proportional to time instead of
square root of time. At high sulfur trioxide contents,
or in the presence of fuming $H_2SO_4$ or $SO_3HCl$, a sharp front
of extremely highly sulfonated polymer is formed, thereby
producing the linear uptake behavior. Upon wet
neutralization with aqueous NaOH, this sharp layer swells
and can slough off. Due to its smaller molecular size as
compared to fuming $H_2SO_4$ or $SO_3HCl$, sulfur trioxide is able
to penetrate more deeply into a treated article during a
typical exposure period. More details of the procedure
have been discussed elsewhere(3,5,6).

Neutralization procedures also provide opportunities
to tailor the barrier layer. Up to two-thirds of the $-SO_3H$
groups are typically not reached by an aqueous NaOH
neutralization, while the tiny $NH_3$ gas can diffuse to and
neutralize essentially all acid groups in a convenient
period. The barrier properties of the $-SO_3^-M^+$ layer are
strongly dependent upon the nature of the counterion, $M^+$.
In addition to the use of $NH_3$, various metal ions can be
introduced from water solution via ion exchange. Of some
27 common metal ions studied, the best barrier results were
obtained with Li, Na, Cu, Mg, Sr, V, Mn, Co and Ni. As
shown in Figure 3 at a surface concentration of about 70
microgram of $SO_3/cm^2$, which for a 25 micrometer film is
about 1% equivalent bulk sulfur, very different barrier
properties result. Under dry conditions $Na^+$ and $Li^+$ are
six times and twelve times more effective, respectively,
in creating an oxygen barrier as compared to $NH_4^+$. Because
of its hydrophillic nature, the efficacy of the barrier
layer can be influenced significantly by the presence of
humidity.

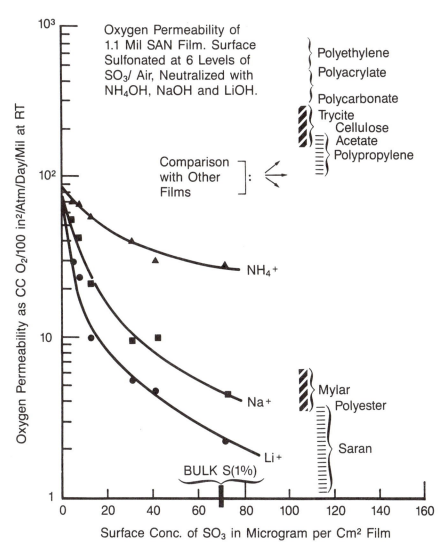

Figure 3: Demonstration of the dramatic reductions in oxygen permeability at 25°C following surface sulfonation to different degrees. The effects of introducing different ions in place of the $NH_4^+$ is also illustrated. The oxygen permeabilities of various resins are shown for comparison.

Reductive Metallization

A natural extension of the above ideas involves the chemical reduction of heavy metal ions, present as counterions within the treated surface region. In the case of copper and silver, highly reflective metal layers can be obtained with a metallic-type total barrier in some cases. The electrical surface conductivity, which is ionic before the reduction, is converted to electronic after the reduction to free metal, thereby permitting rapid electroplating if desired. Moreover, for certain applications, this approach can be used to produce an extremely high barrier to the passage of gases. For example novel thermal insulation panels for refrigerators with plastic-walled panels enclosing a vacuumized space with very little gas but with load-bearing fine powders have been developed (13-22).

In addition to adding structural integrity, correctly chosen powders added to the evacuated cavity in a vacuum container help to eliminate gas from the system or generate a product or products which can be adsorbed by another gas adsorbing material such as activated carbon. For example, carbon dioxide is readily removed in this manner with mixtures of carbon and lithium alkoxides such as lithium isopropoxide. The lithium isopropoxide reacts with carbon dioxide to form lithium carbonate and to liberate diisopropyl ether which is strongly adsorbed by activated carbon (21). Although carbon dioxide is also adsorbed by the activated carbon, it slowly desorbs, reenters the gas phase and is subsequently converted to diisopropyl ether where it is permanently removed from the gas phase. Similar approaches are possible with other gases, and in some cases, simple activated charcoal is sufficient.

For demanding insulation applications, vacuum within the plastic container must be maintained for 30 years or more. To achieve such a super barrier, an ultrathin, very regular metallic layer can be created by combining the previously discussed sulfonation treatments with reductive metallization. The metal layer is extensive enough so that at least 95-99% of the surface is covered by metal. An additional problem in using molded plastics to enclose vacuum and to maintain it via an ultrathin metallic barrier layer arises from irregular metallization. This latter problem, caused by residual mold stresses, can be overcome by first sulfonating the surface lightly to produce a water wettable surface, followed by the application of a thin latex coating which dries to yield a low stress surface. This surface, in turn can be sulfo-metallized as discussed below. For even higher integrity, the metal layer can be further coated with a layer of a barrier copolymer such as vinylidene chloride/vinyl chloride. The barrier polymer is most conveniently applied in latex form to the exposed surface of the metal layer to form a coating of a few micrometers thick. Suitable barrier polymers include latexes that will give a continuous film at temperatures

below the heat distortion temperature of the the organic
polymer used to make the plastic part itself.
    A typical example of the process to produce a
metallized layer involves the development of a sulfonated
layer of micrometer thickness using the techniques
discussed above.   A degree of surface sulfonation of only
0.2 micrograms of sulfur trioxide equivalents per square
centimeter is achieved by contacting the surface with dry
air containing 2 percent sulfur trioxide  at 25°C for
approximately one second.   A metallizing bath based on one
part of each of the following is conveniently used:
0.6% Ag(NH$_3$)$_2$NO$_3$ in H$_2$O, 0.3% NaOH in  H$_2$O with 0.15%

glucose and 0.15% fructose in  H$_2$O (21).  The sulfonated
surfaces are dipped into the bath and metallization is
completed within 1 minute.   The thickness of the metal
layers in this case are approximately 0.03 micrometers as
determined by microscopy.   In this case, the location of
each metal atom is originally fixed by the diffusion of SO$_3$
into the polymer, thereby establishing the position of the
species at the time of reduction:

$$-\overset{\overset{\textstyle O}{\|}}{\underset{\underset{\textstyle O}{\|}}{C}}-S-O^-NH_4{}^+\ Ag°$$

A closely related second method to achieve an even denser
metallic layer involves the introduction of tin ions
instead of silver followed by a spray of water-based
silver-ammonia complex,  freshly mixed with reducer.  The
tin catalyzes the deposition of silver, resulting in the
formation of colloidal metallic silver.  Upon drying, a
very regular, shiny metallic silver layer with a thickness
ranging from 10 to 600 atoms is produced as indicated
formally below:

$$-\overset{\overset{\textstyle O}{\|}}{\underset{\underset{\textstyle O}{\|}}{C}}-S-O^-NH_4{}^+\ Sn°\ Ag°_{10-600}$$

    Based on experience, under the above conditions,
production of at least 80-120 silver atoms per sulfonated
center is adequate to provide a total metallic-type barrier
to the passage of air. These extremely thin silver layers,
when part of a vacuum-treated thermal insulation panel,
conduct very little heat around the rims of the panel as is
needed to achieve the desired high thermal insulation. For
example, the thermal conductivities of an appropriately
formed panel is typically as low as 0.007 watts m /m$^2$°C as
compared to values for polystyrene and polyurethane foams
of 0.03-0.04 watts m /m$^2$°C and 0.016-0.025 watts m /m$^2$°C,
respectively (3).
    To illustrate the value of the treatment, a vacuum
container with a silver layer approximately 0.01 mil is
compared with various cases in Table I.   The 90/8/2 vinyl

chloride/acrylonitrile/sulfoethyl methacrylate barrier
terpolymer coating is applied to the completed metallized
container. The latex had a particle size of about 0.22
micrometer and was applied by dipping in a 50% solids
latex. The excess latex is allowed to run off the surface
and the wall is dried at 60°C for 15 minutes. The
container is a 1 quart vacuum bottle having a surface
facing the evacuated enclosed space of 1500 cm$^2$ with an
evacuated volume of 500 cm$^3$, and a standard styrene/
acrylonitrile boundary wall thickness, excluding the
thickness of the metal and barrier plastic of 80 mils. The
space enclosed by the boundary wall is evacuated to a
pressure of 0.01 mm Hg, and 80g of activated charcoal of
0.2 micron size is added to the enclosed space under
vacuum. Prior to addition to the enclosed space, the
charcoal is activated for 48 hours at 10$^{-7}$ mm Hg. The
permeation rates determined using a mass spectrometer were
used to estimate the projected container life based on the
fact that the activated charcoal which fills the evacuated
space can adsorb a total of about 24-80 cc(STP) of air. The
projected life, therefore, was estimated by dividing the
gas adsorbing capacity by the air permeation rate.

Table I: Comparison of metallized and non metallized
        containers (14)

| Sample No. | Coating | Air Permeation Rate (cc(STP) air/day) | Projected Container Life |
|---|---|---|---|
| 1 | Metal & Barrier Plastic | 0.056 | 1.2 -4 years |
| 2 | Metal alone | 3.2 | 7.5-25 days |
| 3 | Barrier Plastic alone | 1.1 | 22.5 -75 days |
| 4 | None | 9 | 3-9 days |

    Further characterization of the synergistic effects of
overcoating metal layers with the barrier polymer was done
using simple flat polystyrene films of 5 mil thicknesses
that had been surface sulfonated to provide a low degree of
sulfonation of 1.5 micrograms of sulfur trioxide per cm$^2$.
Two strips of the surface sulfonated film were coated with
different thicknesses of the same barrier polymer as that
used above. An additional sulfonated strip was not coated,
but was metallized as described above, and a final strip
was both metallized and coated with a layer of 0.22 mils of
barrier polymer. The results of oxygen transmission
measured by mass spectrometer at 25°C are reported in

Table II and clearly show the dramatic improvement in performance when the barrier polymer and metallization are used together (14). Higher degrees of sulfonation as compared to 1.5 micrograms $SO_3/cm^2$ used here, e.g., 70 micrograms $SO_3/cm^2$ discussed in connection with Fig. 3, clearly would produce much higher barriers than that shown for the uncoated lightly sulfonated sample 4. Nevertheless, even the best of these samples would be over an order of magnitude poorer than the high barrier achieved in sample 1.

Table II: Illustration of the synergistic effects of metallization and a thin barrier polymer overcoat (14)

| Sample No. | Coating | Coating Thickness $\mu g/cm^2$ or (mil) | Oxygen Transmission [cc(STP)/100 in$^2$ day atm] |
|---|---|---|---|
| 1 | Metal & Barrier Plastic | 220 (0.01) 1200 (0.22) | 0.013 |
| 2 | Barrier Plastic | 710 (0.13) | 0.43 |
| 3 | Barrier Plastic | 1420 (0.26) | 0.20 |
| 4 | Surface Sulfonated (1.5 $\mu g/cm^2$) | - | 25.0 |
| 5 | Untreated Control | - | 25.2 |

Waste Treatment Considerations

Additional practical considerations that are attractive about the sulfonation process is its safety and ease of treatment of waste components. Excess sulfur trioxide is easily removed with a gas scrubber to yield a simple to handle sulfuric acid stream, and with the convenient ammonia neutralization process described above, the principal waste stream is comprised of water with a small amount of ammonium sulfate. Methods of avoiding waste by recycling of $SO_3$ itself have also been described (7,8). Moreover, with recent developments for in-situ generation of $SO_3$ and $NH_3$, the amount of waste generated can be reduced greatly (1).

Conclusion

     Of several potentially applicable approaches to
reducing the permeability of a broad spectrum of
inexpensive engineering resins used in packaging, processes
based on surface sulfonation are unparalleled in the
diversity of modifications that can be achieved.  The
approach is relatively simple to perform during molding or
in post molding operations and leaves the intrinsic
mechanical properties of the substrate intact.  The surface
treated layers have good flexibility even at relatively
high treatment levels, since chain backbone scission and
crosslinking is rare.
     The simple sulfonated barrier layer is very effective
against nonpolar components such as fuels and even
chlorinated hydrocarbons.  Under dry conditions, the
barrier layer is also highly effective for suppressing the
permeation of fixed gases such as oxygen or nitrogen.  For
higher barrier requirements, the sulfonation treatment can
be effectively combined with reductive metallization
techniques to produce very high barriers.  When combined
with barrier polymer overcoats and gas adsorbing powdered
fillers, it is possible to produce super barriers capable
of being used in vacuum panels and bottles for thermal
insulation.  For household refrigerators, outside
dimensions are fixed because of standardized kitchen
dimensions in homes, aircraft and boats. Replacing
presently used foam insulation by vacumized plastic panels
should permit thinner insulation and therefore, more room
inside the refrigerator.  More energy efficient
refrigerators based on these highly insulating panels are
also obvious possibilities.

Literature Cited

1. Walles, W. E.   U. S. Patent No. 4,775,587, 1988.
2. Ihata, J., J. Polym. Sci., Part A, 1988, 26, 167.
3. Walles, W. E. U. S. Patent No. 3,740,258,  1973.
4. Walles, W. E. U. S. Patent No. 4,220,739,  1980.
5. Walles, W. E. U. S. Patent No. 3,613,957,  1971.
6. Walles, W. E., In-Mold Sulfonations System, U. S. Patent
   applied for.
7. Walles, W. E., Process for the generation of sulfur
   trioxide reagent and sulfonation of the surface of
   polymeric resins,U. S. Patent applied for.
8. Walles, W. E., Apparatus for the generation of sulfur
   trioxide reagent and sulfonation of the surface of
   polymeric resins,U. S. Patent applied for.
9. Chiao, C. C. U. S. Patent No. 4,717,395, 1988.
10. Gentilcore, J. F., Trialo, M.A. and Waytek, A. J.,
    Plast. Engr., 1978, 34, 40.
11. Walles, W. E. U. S. Patent No. 3,770,706,  1973.
12. Walles, W. E. U. S. Patent No. 3,625,751, 1971.

13. Walles, W. E. U. S. Patent No. 3,824,762, 1974.
14. Walles, W. E. U. S. Patent No. 3,828,960, 1974.
15. Walles, W. E. U. S. Patent No. 3,856,172, 1974.
16. Walles, W. E. U. S. Patent No. 3,921,844, 1975.
17. Walles, W. E. U. S. Patent No. 3,993,811, 1976.
18. Walles, W. E. U. S. Patent No. 4,000,246, 1976.
19. Walles, W. E. U.S. Patent No. 3,996,725, 1976.
20. Walles, W. E. U. S. Patent No. 4,457,977, 1984.
21. Cheng, C, and Walles, W. E. U. S. Patent No. 4,745,015, 1988.
22. Walles, W. E. U. S. Patent No. 3,959,561, 1975.

RECEIVED January 25, 1990

# Chapter 15

# Fluorinated High-Density Polyethylene Barrier Containers

## Performance Characteristics

**J. P. Hobbs, M. Anand, and B. A. Campion**

**Air Products and Chemicals, Inc., 733 Broad Street, Emmaus, PA 18049**

Surface fluorination of high density polyethylene
(HDPE) is very effective in improving the barrier
properties toward hydrocarbon solvents. Containers
prepared by in situ fluorination of HDPE were found to
have a two-order-of-magnitude decreased toluene
permeability, compared to untreated containers, both
at room temperature and 50°C. The fluorination
process is also effective in eliminating container
distortion or paneling which typically occurs in HDPE
containers filled with hydrocarbon solvents. The
paneling of untreated HDPE containers is due to a
simultaneous deterioration in mechanical properties of
the HDPE and the development of a weak vacuum in the
containers, both resulting from the toluene sorption
by the polymer. Fluorination of HDPE reduces solvent
sorption into the container walls, thus preserving the
polymer mechanical properties.

High density polyethylene (HDPE) is extensively used in fabricating
both industrial and household containers for containing a wide
range of chemicals. When such containers are used to package
hydrocarbon solvent-based systems, it is widely recognized that a
barrier treatment (1-3) is desirable to reduce product loss due to
permeation or absorption. This loss of product is typically
accompanied by container distortion often referred to as paneling
or buckling. This paneling may be manifested as a loss of
roundness, or in extreme cases can cause the container to fall
over, or collapse under top loading.
    The structural characteristics of surface fluorinated HDPE
have been studied by several authors (4-6). Other authors (7-14)
have examined the effects of the fluorinated HDPE barrier on the
absorption and permeation of organic compounds. Likewise, there is
a large body of information on the sorption, diffusion, and
permeation of organic compounds in HDPE (15-18) and on the effects

0097–6156/90/0423–0280$06.00/0
© 1990 American Chemical Society

of solvent exposure on the morphological structure and physical properties of HDPE (19-21). The fundamental aspects of paneling in HDPE containers caused by the exposure to solvents has not been addressed in the literature.

The objective of this paper is to elucidate some of the processes that result in paneling, and the role of the AIROPAK Process in-line fluorinated barrier layer in preventing paneling in HDPE containers.

## Experimental

Sample Preparation. The experimental containers were 16 oz. cylindrical bottles prepared by extrusion blow molding of Rexene B54-25H grade HDPE resin from Soltex Corp. (0.954 g/cc density). The melt temperature of the extrudate was 210°C and the bottles were formed by inflation of the parison to 0.72 MPa in a 10°C water-chilled mold. The containers, weighing approximately 41.5 grams, had a nominal wall thickness of 0.1 cm and a surface area of 408 cm². One-half of the containers were prepared untreated using nitrogen as the inflation gas. The other half were in situ fluorinated by the AIROPAK Process using a nitrogen-diluted fluorine as the inflation gas (Figure 1), as described by Dixon (7) and Kallish, et al (22).

Test Procedures. Blow molded fluorinated and untreated bottles were filled with 370 grams of reagent grade toluene and heat sealed with a low density polyethylene-coated aluminum seal, and capped with a polypropylene closure. Ten sample containers of each type were tested at 50 ± 2°C (50°C) and at room temperature (RT) for permeation and the onset of paneling. Empty containers, both treated and untreated, were also stored with the filled containers to serve as controls.

Over the six-week period of permeation testing, sample containers from each type and temperature set were randomly selected and used for tensile testing. The tensile specimens were prepared and tested following the procedures of ASTM Test Method D1708-79 with the following modifications: (i) the bottles were drained, and five specimens were cut parallel to the long axis of the cylinder (machine direction), (ii) the specimens were tested immediately after blotting to remove any surface solvent, (iii) a test speed of 5.08 cm/min. (2 in./min.) was used, and (iv) the tensile modulus of elasticity was calculated from the initial slope of the load-extension curve.

Upon emptying the containers, three additional specimens were cut, blotted on filter paper to remove surface solvent, weighed and dried at 60°C in an air circulating oven. The solvent loss of these specimens in (g. toluene/g. dry polymer) was defined as the dynamic toluene sorption for the tested container. Equilibrium toluene sorption was determined at RT and 50°C by immersion to constant weight of replicated cutouts from control bottles.

Additionally, toluene-filled bottles, both untreated, and in situ fluorinated, were fitted with pressure gauges and monitored for pressure changes in the containers. Also, a set of untreated and in situ fluorinated containers was filled with toluene, but not sealed, and monitored for paneling at RT.

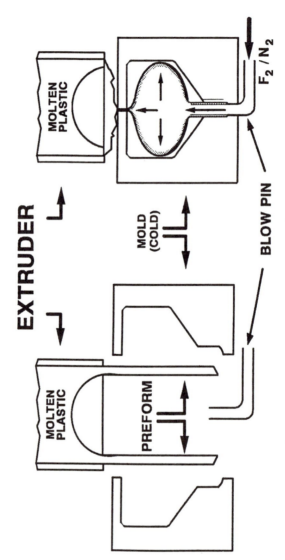

Figure 1.   In situ fluorination of blow molded container by the AIROPAK Process.

## Results and Discussion

Permeation Properties. The data shown in Figure 2 are the toluene permeation rates of the fluorinated and untreated containers; g. toluene/container per day are plotted vs. the time of toluene exposure on a logarithmic scale. These cumulative permeation rates were calculated based on the cumulative weight loss over the time of toluene exposure, as opposed to the differential permeation rates based on the differential weight loss over each time interval. The room temperature permeation rates for the in-situ fluorinated containers were less than 0.01 g./day; and, hence, have been rounded up to 0.01 g./day for illustrative purposes. In Figure 2, the flat portion of the curves for the untreated containers yielded the steady state permeation rates. From these values, the permeability coefficients ($\bar{P}$) for the untreated containers were calculated using Equation 1.

$$\bar{P} = \frac{(\Delta W)(L)(22400)}{(MW)(A)(\Delta P)} \quad (=) \quad \frac{cc(STP)cm}{cm^2 \; sec.(cm \; Hg)} \quad (1)$$

where $\Delta W$ is the steady state permeation rate in grams per second, MW is the gram molecular weight of the penetrant (92.15 g./g. mol), L is the wall thickness (0.07 cm), A is the surface area of permeation (408 cm$^2$), and $\Delta P$ is the pressure gradient of the penetrant across the wall. Under the test conditions employed, the value of $\Delta P$ is equal to the vapor pressure of toluene at the test temperature. The calculated values of $\bar{P}$ are shown in Table 1.

It may be noted in Figure 2 that the in-situ fluorinated containers exposed to toluene at 50°C appear to be just approaching steady state after 1000 hours of solvent exposure. Thus, the cumulative permeation rate of Figure 2 will result in a slight underestimation (~10%) of the steady state permeability coefficient. This underestimation was corrected for in Table 1 by using the differential weight loss rate in Equation 1.

It may be correspondingly noted that the in-situ fluorinated containers exposed to toluene at room temperature would still be far from steady state after 1000 hours of solvent exposure. The underestimation of the room temperature toluene permeability coefficient for the in-situ fluorinated containers was minimized in Table 1 by using the differential weight loss rate in Equation 1.

Table I shows that the application of an in situ fluorinated barrier has resulted in an approximately two-order-of-magnitude reduction in the steady state toluene permeation rate relative to the untreated containers.

Paneling Characteristics. Indicated by vertical arrows in Figure 2 are the times at which the onset of paneling, or buckling, of the untreated containers was observed. The paneling of the untreated containers progressed from a visually and tactilely detectable loss of roundness observed between 8 and 16 hours of solvent exposure to a severe distortion of the sidewalls within 24 hours of filling.

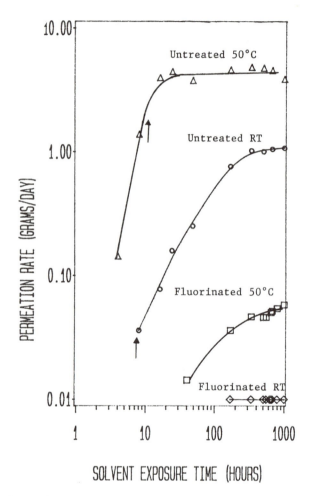

Figure 2. Toluene permeation rate for HDPE containers versus solvent exposure time. Paneling time indicated by vertical arrows.

TABLE I. STEADY STATE PERMEATION

| | Wt. Loss Rate (g. Toluene/Day) | Toluene Permeability+ Coefficient P $\frac{ccl(STP)cm}{cm^2 \ sec.(cm \ Hg)}$ |
|---|---|---|
| Untreated HDPE | | |
| RT | 1.3 | $2.2 \times 10^{-7}$ |
| 50°C | 7.2 | $3.7 \times 10^{-7}$ |
| Fluorinated HDPE | | |
| RT | <0.01 | $<2.0 \times 10^{-9}$ |
| 50°C | 0.08 | $4.5 \times 10^{-9}$ |

$+\overline{P}$ for the fluorinated HDPE is the composite permeability coefficient.

This onset of paneling preceded the attainment of steady state permeation from the untreated containers and in the case of the RT bottles, occurred at measured transient permeation rates comparable to those found at steady state for the treated containers at 50°C. No evidence of paneling was detected in the fluorinated containers at RT or 50°C with solvent exposure times of more than 1000 hours.

The tensile strength data from the container cutouts are reported in Figures 3 and 4. The tensile modulus data are reported in Figures 5 and 6; the tensile property in megaPascals is plotted vs. the storage time at RT or 50°C. No differences in tensile properties were detected between the fluorinated and untreated containers not exposed to toluene.

The data in Figures 3 through 6 show that the onset of paneling in the untreated containers accompanies a 30% to 50% loss in tensile modulus and a 15% to 20% loss in tensile strength. It is evident that the presence of the in situ fluorinated barrier layer has served to preserve the mechanical properties of the underlying HDPE.

The dynamic toluene sorption data, in (g. toluene/g. dry polymer), are shown in Figure 7. The equilibrium sorption values of toluene were measured to be about 0.075 g. solvent/g. dry polymer at RT and about 0.095 g. solvent/g. dry polymer at 50°C. This trend to increasing solvent sorption by HDPE with increased temperature has been noted by other including Corbin et al (14) for toluene and Liu et al (16) for n-hexane. The increased sorption at the higher temperature may be due to increased swelling of the polymer at the higher temperature. The lower values for the

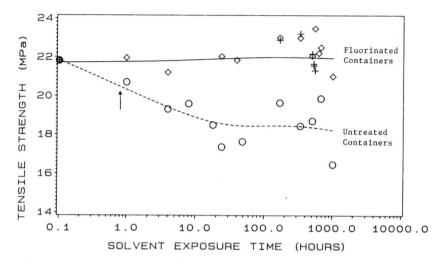

Figure 3.   Tensile strength of HDPE containers exposed to toluene
at room temperature versus solvent exposure time.   + symbols represent containers not exposed to solvent.

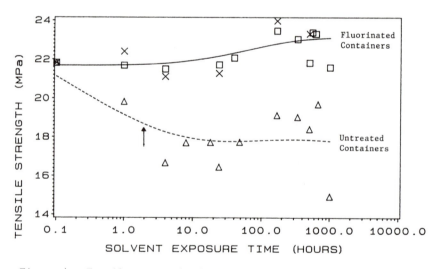

Figure 4.   Tensile strength of HDPE containers exposed to toluene
at 50°C versus solvent exposure time.   X symbols represent containers not exposed to solvent.

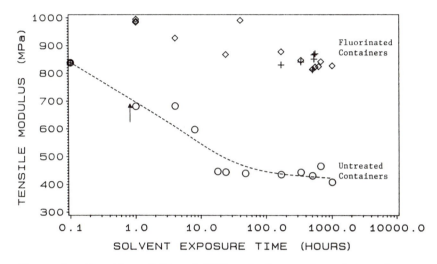

Figure 5. Tensile modulus of HDPE containers exposed to toluene at room temperature versus solvent exposure time. + symbols represent containers not exposed to solvent.

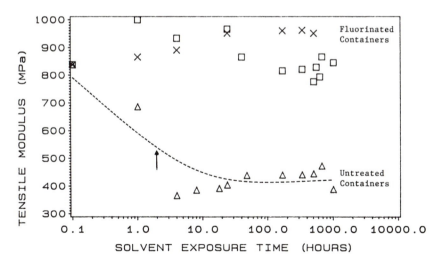

Figure 6. Tensile modulus of HDPE containers exposed to toluene at 50°C versus solvent exposure time. X symbols represent containers not exposed to solvent.

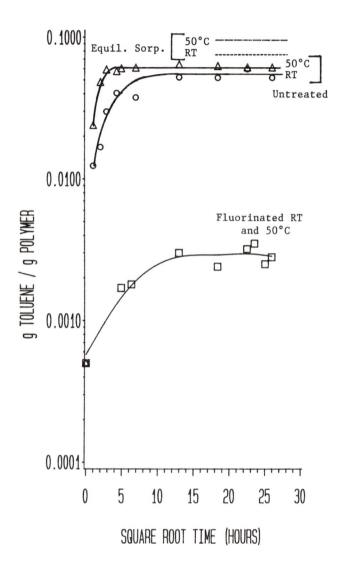

Figure 7.  Toluene sorption in HDPE containers versus solvent exposure time.

dynamic sorption measurements, relative to the equilibrium values, result from the volatilization of the solvent from the outer surface of the container.

As indicated in Figure 7, there is a 20- to 30-fold reduction in the steady state dynamic toluene sorption in the in situ fluorinated barrier container compared with the untreated. No significant difference in solvent uptake could be detected between the solvent-filled, treated containers stored at RT and 50°C. This is attributed to the low absolute value of the dynamic solvent sorption in the in situ fluorinated containers and the relatively small difference in equilibrium toluene sorption between RT and 50°C.

The tensile modulus data of Figures 5 and 6 are replotted against the dynamic toluene sorption in Figures 8 and 9, respectively. In the case of the untreated HDPE containers, the loss of mechanical properties is a result of the significant solvent sorption by the polymer. The in situ fluorinated barrier reduces the solvent sorption in the HDPE and thus allows the material to retain its mechanical properties.

It was determined that the decrease in the mechanical properties observed in the untreated containers was not a sufficient condition for inducing paneling. Simultaneously, a small vacuum inside the container was required to cause paneling. In Figure 10, the differential pressure measured between the inside and outside is plotted against time for toluene-filled containers at RT. The untreated containers quickly developed and maintained a vacuum at the limit of instrumentation (-30 mm Hg).

Figure 10 also shows that the fluorinated containers filled with toluene, upon sealing, became slightly pressurized relative to the laboratory. The pressure then fluctuated between -15 and +30 mm Hg relative to the laboratory over an extended period of time. This variability was mimicked in a closed steel cylinder used as a control, and is attributed primarily to temperature variations and changes in atmospheric pressure in the laboratory. It is important to note that the permeation rate of atmospheric gases under these storage conditions is significantly lower than that for toluene. (Air permeability coefficients for 0.914 to 0.964 density polyethylene (23) range between 0.1 and 2.0 x $10^{-10}$ cc(STP)cm/cm$^2$ sec.(cm Hg) compared to the toluene P values of Table I.) Thus, back permeation of air into the containers would not be a significant factor in equilibrating the internal pressure.

Some of the pressure fluctuations observed in the fluorinated containers produced internal vacuum levels equivalent to those found in paneled untreated containers, yet no paneling was observed in any of the fluorinated containers. We additionally found that simply filling the untreated containers with toluene without the seal resulted in an equivalent loss of the polymer mechanical properties, but did not cause paneling. This leads to the conclusion that both a negative pressure differential and a reduction in mechanical properties were necessary to induce paneling in these HDPE containers.

It was observed in experiments with untreated HDPE that panelling occurred earlier (by ¯4 hours) in the containers at room temperature vs. 50°C (Fig. 2). This may appear surprising

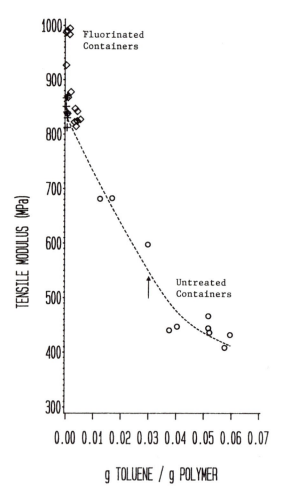

Figure 8.    Tensile modulus of HDPE containers exposed to toluene
at room temperature versus solvent sorption.    + symbols represent
containers not exposed to solvent.

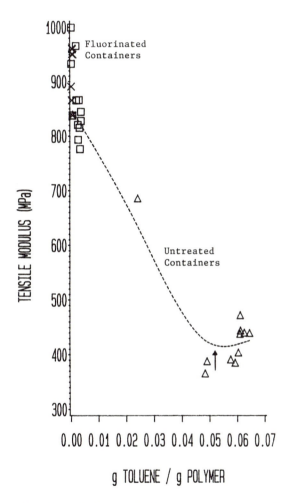

Figure 9.   Tensile modulus of HDPE containers exposed to toluene at 50°C versus solvent sorption.   X symbols represent containers not exposed to solvent.

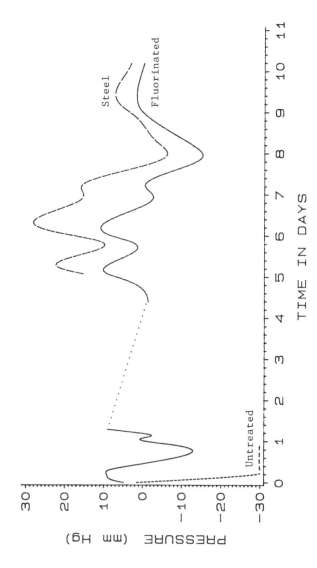

Figure 10.  Internal pressure versus time for untreated and fluorinated HDPE containers and a steel cylinder filled with toluene at room temperature.

considering that the lower sorption and permeation rates at the lower temperature. This phenomenon is believed to be due to an increase in the internal pressure of the container at 50°C because of heating of the sealed container from room temperature (container fill temperature) to 50°C (test temperature). Thus, this increased internal pressure must be compensated for by solvent loss before panelling can occur. Hence, the lower temperature untreated containers are found to panel earlier.

An estimation of the internal pressure of RT untreated containers after eight hours of toluene exposure can be made based on the toluene volumes lost due to permeation and sorption. The result is a vacuum of 7.7 mm Hg, where the 1.4 g. of solvent sorbed into the HDPE walls is the dominant factor. This indicates that during the early period of solvent exposure simple sorption of the solvent into the polymer can generate conditions leading to paneling in untreated HDPE containers.


Conclusions

The ability to package hydrocarbon solvents in HDPE containers is greatly improved by in situ fluorination using the AIROPAK Process. The barrier properties to toluene are improved 100-fold after fluorination. The propensity of an untreated HDPE container to panel in the presence of toluene is eliminated after fluorination.

It was found that the paneling of untreated HDPE containers was induced by the combination of a decrease in the mechanical properties of the container wall and the concomitant development of a vacuum inside the container. Both of these effects were due to toluene sorption into the container wall. The application of an in situ fluorinated barrier to the interior of the container resulted in the preservation of the mechanical properties and pressure inside the container.

It was additionally calculated that the back permeation of atmospheric gases into the container was at an insufficient rate to offset vacuum development due to solvent sorption and permeation.

Literature Cited

1. Buck, D. M.; Marsh, P. D.; Kallish K. J. Regional Technical Conference, Society of Plastics Engineers, 1985.
2. Tarancon, G. U.S. Patent No. 4 467 075, 1984.
3. Shefford, R. A. U.K. Patent No. GB 2 069 870B, 1983.
4. Clark, D. T.; Feast, W. J.; Musgrave, W. K. R.; Ritchie, I. J. of Polymer Science, 1975, 13, 857-90.
5. Adcock, J. L.; Inoue, S.; Lagow, R. J. J. of the American Chemical Society, 1978, 100:6, 1948-50.
6. Lagow, R. J.; Margrave, J. L. J. Polymer Sci. Polymer Letters Edition, 1974, 12, 177-84.
7. Dixon, D. D. U.S. Patent No. 3 862 284, 1975.
8. Gentilcore, J. F.; Trialo, M. A.; Woytek, A. J. Plast. Eng., 1978, 34, 40.

9. Hayes, L. J.; Dixon, D. D. J. of Applied Polymer Science, 1978, 22, 1007-13.
10. Hayes, L. J.; Dixon, D. D. J. of Applied Polymer Science, 1979, 23, 1907-13.
11. Koros, W. J.; Stannett, V. T.; Hopfenberg, H. B. Polymer Engineering and Science, 1982, 22, 738-46.
12. Bliefert, C. Metalloberflache, 1987, 41, 317-74.
13. Farrell, C. J. Ind. Eng. Chem. Res., 1988, 27, 1946-51.
14. Corbin, G. A.; Cohen, R. E.; Baddour, R. F. J. of Applied Polymer Science, 1985, 30, 1407-18.
15. Kwapong, O. Y.; Hotchkiss, J. H. J. of Food Science, 1987, 52, 761-63.
16. Liu, C. P. A.; Neogi, P. Journal of Membrane Science, 1988, 35, 207-15.
17. Ng, H. C.; Leung, W. P.; Choy, C. L. J. of Polymer Science, 1985, 23, 973-89.
18. Koszinowski, J. J. of Applied Polymer Science, 1986, 32, 4765-86.
19. Blackadder, D. A.; Keniry, J. S. J. of Applied Polymer Science, 1972, 16, 1261-80.
20. Morgan, R. J.; Nielsen, L. E. J. of Polymer Science, 1972, 10, 1575-85.
21. Starkweather, H. W., Jr. Macromolecules, 1985, 18, 958-60.
22. Kallish, K. J.; Bauman, B. D.; Goebel, K. A. Proc. Annual Technical Conference (ANTEC) Society of Plastics Engineers, 1985.
23. Brandrup, J. Polymer Handbook, John Wiley & Sons, Inc., 1975.

RECEIVED December 5, 1989

Chapter 16

# Packaging of Juices Using Polymeric Barrier Containers

J. Miltz[1], C. H. Mannheim[1], and B. R. Harte[2]

[1]Department of Food Engineering and Biotechnology, Technion—Israel Institute of Technology, Haifa, 32000, Israel
[2]School of Packaging, Michigan State University, East Lansing, MI 48824

The interaction between the polyethylene (PE) contact surface in high gas barrier and "bag in the box" packages with citrus products and model solutions was studied. The presence of corona treated polyethylene films in model solutions and juices accelerated ascorbic acid degradation and browning. In all cases d-limonene concentration in juices in contact with PE surfaces was reduced. In both orange and grapefruit juices, aseptically filled into cartons, the extent of browning and loss of ascorbic acid was greater than in same juices stored in glass jars. Sensory evaluations showed a significant difference between juices stored in carton packs and in glass jars, at ambient temperature after 10-12 weeks. The cartons were found not to be completely gas tight. In metallized films for "bag in the box" packages, discontinuities in the metallization were found which increased with handling and caused an increase in gas permeability. Storage life data for orange juice in "bag in the box" packages are presented. The effect of d-limonene absorption by sealant polymers on the latter's properties was evaluated. It was found that the mechanical properties and oxygen permeability of the sealent films were altered due to d-limonene absorption.

In the last few years, a rapid expansion in the use of high gas barrier plastic structures for food packaging has taken place. These structures include multilayer films, trays, bottles and cups, "bag in the box" and "bag in the barrel" type packages and laminated cartons.

0097–6156/90/0423–0295$06.75/0
© 1990 American Chemical Society

Trays are used for different shelf stable products, including meals. The bottles are used primarily for ketchup (but other products are on their way); the cups are used for sauces, vegetables and soups while the "bag in the box" and the laminated cartons are used primarily for aseptically packed juices and concentrates. Aseptic packaging has been used in Europe for many years. In the USA, this kind of packaging has become one of the fastest growing areas since the Food and Drug Administration (FDA) approved the use of hydrogen peroxide for sterilization of the polyetheylene (PE) contact surface (1).

The gas barrier properties of the above mentioned packages are achieved either by: (i) high gas barrier polymers like ethylene vinyl alcohol copolymers (EVOH) and vinylidene chloride copolymers (PVDC) that are used in the form of a layer or coating in the multilayer structure or (ii) a metallized film (primarily polyetheylene terephthalate - PET, but also polypropylene -PP and others) or (iii) aluminum foil. All of these structures normally contain polyetheylene (PE) as the food contacting and sealing component. Polyethylene is a very widely used packaging material and is considered to be quite inert to food. Several studies, however, have shown (2-6) that depending on the structure and primarily on the nature of its surface, PE may affect the shelf life of different foods. Polyethylene absorbs different food components and, thus, may change the composition and character of the product. Moreover, the food components that are absorbed by the package may affect the properties of the food contacting layer (7) and as a result, the properties of the package as a whole.

Whereas the structure of PE depends primarily on the polymerization conditions, the nature of its surface, in a package, depends primarily on the processing conditions. Excessive temperatures during extrusion, coextrusion or coating may cause oxidation of the surface which in turn may accelerate degradation of food components and reduce shelf life (4-6) compared to the same product packaged in glass containers.

## Factors Affecting Shelf Life of Aseptically Packed Citrus Products

The shelf life of pasteurized hot or aseptically filled juices and concentrates is limited due to chemical reactions which are influenced by the storage temperature as well as the presence of oxygen and light. In plastic laminates, as opposed to glass, transfer of gases through the packaging materials and/or seals, as well as absorption of food components into the contacting layer and migration of low molecular compounds from the packaging material into the food may also affect shelf-life (6, 8-10).

In storage studies of orange juice in Tetra Brick and glass bottles, Durr et al. (11) reconfirmed that storage temperature was the main parameter affecting shelf-life of orange juice. In addition they found that orange oil, modeled as d-limonene, was absorbed by the polyethylene contact layer of the Tetra Pack. These authors claimed that this could be considered an advantage since limonene is a precursor to off-flavor components. Marshall et al. (12) also reported absorption of d-limonene from orange juice into polyolefins. They found that the loss of d-limonene into the contact layer was directly related to the thickness of the polyolefin layer and not its oxygen permeability. Marshal et al. (12) also quantified the absorption of several terpenes and other flavorings from orange juice into low density polyethylene (LDPE). They found that the predominant compounds absorbed were the terpenes and sesquiterpenes. More than 60% of the d-limonene in the juice was absorbed by LDPE but only 45% by Surlyn. Other terpenes such as pinene and myrcene were absorbed to greater extents. Infrared spectra of LDPE showed that the terpenes were absorbed rather than adsorbed. Since many desirable flavor components in citrus juices are oil soluble, their loss by absorption above can adversely affect flavor of juices. Scanning Electron Microscope (SEM) pictures of LDPE after exposure to orange juice, showed significant swelling of the polymer (12) inferring severe localized internal stresses in the polymer. These stresses indicated a possible cause for delamination problems occasionally encountered in LDPE laminated board structures. The adhesion between LDPE and foil was much less than between Surlyn and foil. The presence of d-limonene in the polymer aided in the absorption of other compounds, such as carotenoid pigments, thus reducing color of juice. Buchner (13) also mentions reduction in aroma bearing orange oil due to absorption into the polyethylene contact layer of the aseptic packages.

Gherardi (2) and Granzer (14) found that juices and nectars deteriorated faster in carton packs than in glass bottles. In products with low fruit content, such as fruit drinks, the difference in deterioration between packages was small. Gherardi (2) found high oxygen and low carbon dioxide concentrations in the head space of carton packs as compared to low oxygen and high carbon dioxide in glass bottles. Therefore, Gherardi (2) concluded that the main cause for differences in quality between products in carton packs and glass was due to higher oxygen transfer rates into the former package. Granzer (14) also claimed that the oxygen permeability of carton packs caused the faster deterioration of juices packed in them. Granzer (14) stated, without showing proof, that the source of the oxygen permeation into the carton pack was the 22cm long side seam. Since then, Tetra Pack improved the side seam by adding a strip of PET. Koszinowski and Piringer (3) investigated the origin of off-flavors in various packages. They established gas chromatographic and mass spectrographic procedures coupled with sensory threshold odor analysis for the determination of off-flavor emanation from packaging materials. They found reaction products from the polymeric materials as well as a combination of oxidation and pyrolytic elimination reactions in polyethylene or paperboard to be responsible for off-odors.

Products packed in "bag in the box" packages, especially
in the larger sizes, also have a reduced shelf life as compared
to those packed in metallic packages, mainly due to damage
caused to the films during transportation (6,15). The reduced
shelf life in these packaging systems is probably due to flex
cracking which causes pinholes and even "windows" in the
metallized layer of the laminate thus reducing the barrier
properties (6).

In this paper a review of our previous work on the effect
of PE contacting surface in laminated cartons and of different
high barrier films (in "Bag in the Box" type packages) on the
shelf-life of citrus juices and the effect of d-limonene
absorption on the properties of sealant films will be shown.
In addition, some new unpublished results on this subject will
be reported.

## MATERIALS AND METHODS

### Materials
250 ml and 1 liter laminated cartons made of PE/Kraft
Paper/Aluminum foil/PE made by Tetra Pack and PKL respectively
were used in the present work.

The Hypa S and Pure Pack packages that were used for
comparison are basically made of the same materials but
different configurations.

The bag in the "Bag in the Box" was made from the
following laminates:
1. PE/Met PET/PE from various manufacturers.
2. PE/EVOH/PE
3. PE/Met PET/Met PET/PE
where Met. stands for metallized.
Structures 2 and 3 are considered as having high barrier
properties. The second structure has EVOH as a barrier and the
third is a face-to-face metallized polyester. The latter
structure is very stiff due to the face-to-face lamination of
the metallized polyester.

The sealant films in the small cartons were as follows:
Low density polyethylene film (LDPE) ($5.1 \times 10^{-2}$ mm thick)
obtained from Dow Chemical Co., Surlyn S-1601 (sodium type; 5.1
x $10^{-2}$ mm thick) and Surlyn S-1625 (zinc type; $7.6 \times 10^{-2}$ mm
thick) obtained from DuPont. The samples were cut into strips,
2.54 cm x 12.7 cm, and immersed into juice (7 samples per 250 ml
bottle) which had been filled into amber glass bottles and
closed with a screw cap.

The ratio of film area to volume for a 250 ml pack
(area/volume) was 0.9 $cm^2/cm^3$. Film strips were also heat
sealed together using an impulse heat sealer and immersed into
the juice (7 samples per bottle). In the 1 liter cartons only
extrusion coated LDPE was used, as received from PKL,
Combiblock, Germany.

## Methods

### Ascorbic Acid Degradation and Browning

For ascorbic acid (AA) degradation and browning studies, the 1 liter cartons were used. Orange and grapefruit juices were packed aseptically on a commercial line into cartons and glass jars. The carton structure was a laminate of LDPE/carton/aluminum/LDPE made by PKL-Combibloc, Germany. The carton dimensions were 10 x 6 x 17 cm. AA degradation and browning were measured as described by Mannheim et al. (4).

### Interaction Between Carton Packs and Juices

**Carton Strips.** The interaction between carton packs and juices were evaluated by immersing strips of carton laminates in juices stored in hermetically sealed glass jars. Surface to volume ratios of 4:1 and 6:1, as compared with the actual ratio of $0.67, cm^2/cm^3$ in 1 liter packs were used.

**Polyethylene films.** In order to elucidate the effect of the polyethylene contact surface, and separate possible contribution by the carton itself, untreated and corona treated (oxidized) polyethylene strips, made of LDPE of similar properties to that used in PKL cartons, were immersed in model solutions. The model solution consisted of 10% sucrose, 0.5% citric acid, 0.1% ascorbic acid, 0.05% emulsified commercial orange oil and 0.1% potassium sorbate. Also strips of LDPE, from the inner liner of the "bag in the box" at a surface to volume ratio double the actual ratio in a 6 gallon bag, were inserted into glass jars, which were hot filled with orange juice and stored at $35°C$.

### Microscopy
The integrity of the metallic layer in metallized PET films was studied with a Wild M21 Polarizing microscope. The magnification used was 40. The printouts were made by a D5V Universal Condenser Enlarger with a magnification of 5.3. Thus, the total magnification was 212.

### Absorption of d-limonene by Carton.
Aseptically packed orange juice in laminated carton packages (Tetra-Pack), was obtained from a local company. The packages contained 250 ml of juice. A control was maintained by immediately transferring juice from the cartons to brown glass bottles at the plant site. Quantification of d-limonene was done according to the Scott and Veldhuis titration method (16).

To determine percent recovery, a known amount of d-limonene were weighed into a distillation flask and dissolved with 25 ml isopropanol, 25 ml of water was then added. The contents were then distilled to determine the amount of d-limonene recovered.

To determine the amount of d-limonene in the carton
material, the whole carton (1 x 0.5 cm pieces) was cut up after
the juice was removed and the carton rinsed with distilled
water. The carton pieces were then placed into a distillation
flask. Seventy ml isopropanol were added so that the sample
was completely immersed. The sample/isopropanol mixture was
allowed to equilibrate for approximately 24 hr. Several
extraction times were used to determine recoveries of d-limonene
from carton stock. Seventy ml of water were then added and the
sample distilled as before.

## Effect of d-limonene Absorption on Mechanical Properties of Polymer Films

The orange juice used in this study was 100% pure
(reconstituted) obtained from a local company. The
antioxidants, Sustane W and Sustane 20 (UOP Inc., 0.02% w/w
total), and the antibacterial agent, sodium azide (Sigma
Chemical Co.) (0.02%, w/w) were added to the juice in order to
prevent oxidative and microbial changes during storage.
    To determine the effect of antioxidant and antimicrobial
agents on juice stability, pH (Orion Research Co. Analog pH
Meter, Model 301), juice color (Hunter D25 Color Difference
Meter) and microbial total counts were monitored as indicators
of juice quality.

### Stress Strain Properites
Stress-strain properties were determined as a function of
absorbant concentration using a Universal Testing Instrument
(Instron Corp., Canton, MA). The procedure used was adopted
from ASTM Standards D882-83 (1984). Ten specimens were tested
to obtain an average value. The amount of d-limonene absorbed
was determined according to the Scott and Veldhuis (16)
procedure.

### Influence  of d-limonene Absorption on Barrier Properties of Polymer Films
Sample specimens of the same LDPE and Surlyn S-1652 were
immersed into orange juice until equilibrium (d-limonene) was
established. The oxygen permeability of the control
(nonimmersed) and immersed samples were measured using an Oxtran
100 Oxygen Permeability Tester (MOCON-Modern  Controls,
Rochester, MN). Permeability measurements were performed at 95%
RH and 23°C.
### Gelbo Flexing
The different films from which the bag of the "Bag in the Box"
package is made of were flexed twenty times in a Gelbo Flexing
instrument at standard conditions (17). This test gives an
indication about the damage that may be caused to the film
during handling and transportation. The properties of the
different films were evaluated before and after flexing.
### Sensory Evaluation
The triangular taste comparison tests were carried out according
to Larmond (18).

RESULTS AND DISCUSSION

Interaction between carton packs and juices

Ascoric Acid (AA) degradation and browning of orange juices are presented in Figures 1 and 2 respectively. In all cases the rates of the deteriorative reactions were higher in cartons as compared to glass. Differences in browning and ascorbic acid degradation between juices in glass and cartons were initially small and increased with storage time. Initially, juices in glass and cartons had similar dissolved and headspace oxygen values which affected the above reactions in the same fashion. Later on, the effect of the laminated package became more pronounced due to its oxygen transmissibility as well as due to the product-package interaction  Similar results (not shown) were obtained with grapefruit juice. The rate of AA degradation may differ in different carton packs as shown in Fig. 3. The Hypa S package had the highest loss of AA.

A rapid loss of d-limonene, from 70 to 40 ppm for orange juice was observed in the 1 liter carton packs  during the first days of ambient (25°C) storage (Fig. 4). Similar results were obtained by others (11,12) who claimed absorption by the film as the cause for this phenomenon.  Again, the loss may vary between packages as shown in Fig. 5.   The loss of d-limonene can be compensated for by increasing the initial d-limonene concentration.  However, as d-limonene is a precursor to oxidative reactions of flavour compounds and since some flavour compounds are absorbed selectively at a higher rate, this is not desirable.
     Triangle taste comparisons showed (4-6) that, for orange juice after about 2.5 months storage at 25°C there was a significant (p<0.05) difference between juices in the 1 liter cartons and glass containers (Fig. 6). In grapefruit juices a significant taste difference (p<0.05) was found after about 3 months (Fig. 6). The longer period to obtain a taste difference in grapefruit juice is due to the stronger flavor of this juice which tends to mask small flavor changes. Color differences were noted easier in grapefruit juices than in orange juices. In orange juice, some carotenoids may be oxidized and have a bleaching effect which compensates for the change in appearance due to browning, thus explaining the color difference between the two juices (19).

The effect of carton in contact with orange juice on ascorbic acid retention and on browning is shown in figures 7 and 8. It can be seen that the carton accelerated the loss of AA and browning, compared to similar juices stored in glass. After 14 days there was also a significant taste difference (p<0.05) between samples in contact with carton strips and those without the strips. These results seem to indicate that the polymeric surface had an accelerating effect on some of the reactions affecting shelf-life of citrus juices. In these

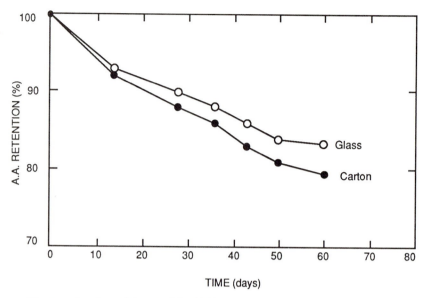

Figure 1. Ascorbic acid (AA) retention in orange juice aseptically packed into glass jars and carton packs, stored at 25°C.

Reproduced with permission from ref. 4. Copyright 1987 Institute of Food Technologists.

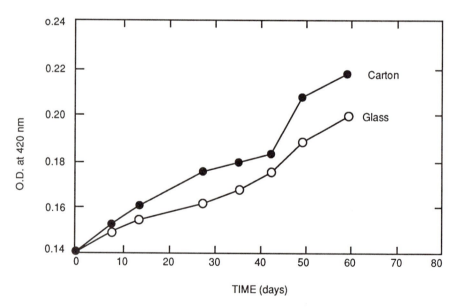

Figure 2.   Browning of aseptically filled orange juice in glass jars and carton packs, stored at 25°C.

Reproduced with permission from ref. 4. Copyright 1987 Institute of Food Technologists.

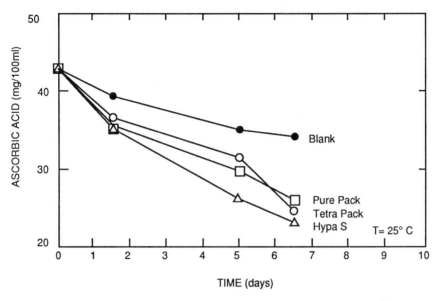

Figure 3. Ascorbic acid retention in orange juice at 25°C in various packaging materials. (BLANK=GLASS)

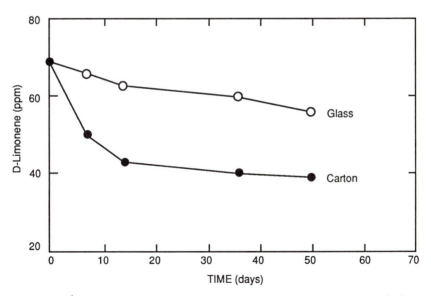

Figure 4. d-limonene concentration in orange juice packed into glass jars and carton packs, stored at 25°C.

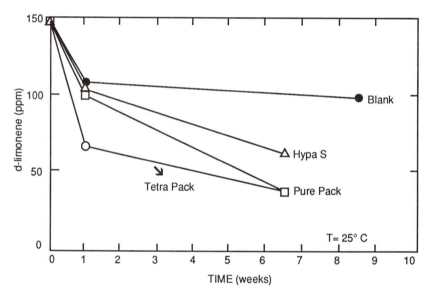

Figure 5. Concentration of d-limonene in orange juice stored at 25°C in various packaging materials. (BLANK = GLASS).

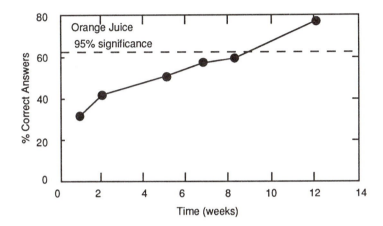

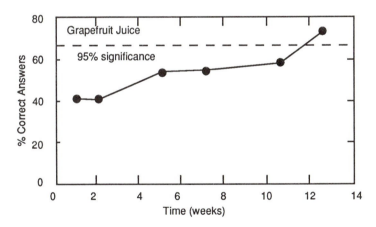

Figure   6. Triangle    Taste    comparisons    between    juices
aseptically filled into glass jars and carton   packs,    stored
at 25°C.

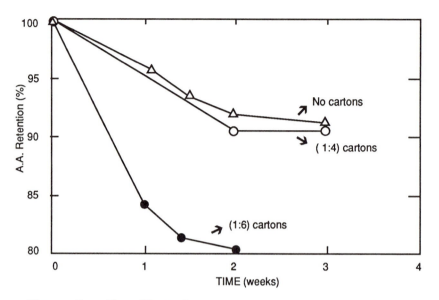

Figure 7.    The effect of orange juice with carton strips on
ascorbic acid   retention, stored at 25°C.

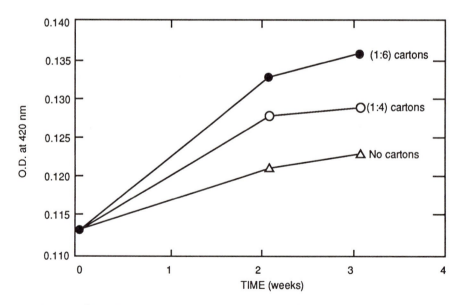

Figure 8.    Browning of orange juice as affected   by   contact
with carton   strips, stored at 25°C.   (BLANK = GLASS).

experiments, the ratio of contacting surface to juice volume was
4 to 6 times higher than that in 1 liter containers. This means
that in 1 liter containers the effect will be somewhat delayed
as compared to the above test, but in smaller packages (i.e.
250ml) the shelf life would be shorter due to a higher surface
to volume ratio.

In order to elucidate the effect of the polyethylene
contact surface, and separate any possible contribution of the
carton itself, untreated and corona treated (oxidized)
polyethylene strips, made of LDPE of similar properties to that
used in PKL cartons, were immersed in model solutions. The
model solution consisted of 10% sucrose, 0.5% citric acid, 0.1%
ascorbic acid, 0.05% emulsified commercial orange oil and 0.1%
potassium sorbate. The effect of the untreated and corona-
treated LDPE strips on ascorbic acid degradation is shown in
Figure 9. From these results it becomes clear that the
polethylene contact surface accelerated the rate of ascorbic
acid degradation with the oxidized film having a greater effect.
In the model solutions a rapid reduction (40-60% within about
six days) in d-limonene content took place, due to oxidation, in
samples containing LDPE strips as compared to a loss of only 10%
in the blanks (Figure 10). Absorption of d-limonene by the
LDPE accounted for this increased loss. In a triangle taste
comparison of water with and without LDPE strips, stored in
sealed glass jars at 35°C for 48hr, all 12 tasters were able to
distinguish between the samples, indicating the presence of a
strong off-flavor in the water which had been in contact with
LDPE strips.

Strips of LDPE, from the inner liner of the "bag in the
box" at a surface to volume ratio double the actual ratio in a 6
gallon bag, were inserted into glass jars, which were hot filled
with orange juice and stored at 35°C. In contrast to the
degradative effect of the PE contact layer of the cartons, no
detrimental effect on browning, ascorbic acid, and taste, as
compared to juice in glass, was found in this case. However,
also in this case d-limonene content was reduced to about half
the initial content in a short time.
These differences may be accounted for by the different
manufacturing techniques of the polyethylene contact surfaces.
A high extrusion temperature, followed by corona treatment which
causes oxidation, was used for the PE layer of the cartons,
whereas the PE liner for the "bag in the box" is a LDPE extruded
at a lower temperature and without corona treatment. This
study, therefore, indicates that an oxidized polymer surface may
accelerate reactions resulting in a reduced shelf-life. These
results were confirmed by Durr (10) who presented the effect of
high temperature extrusion lamination of an Ionomer on
development of off-flavors. In recent years carton packs are
coated with an additional layer of low temperature extruded PE
in order to reduce the above effects.

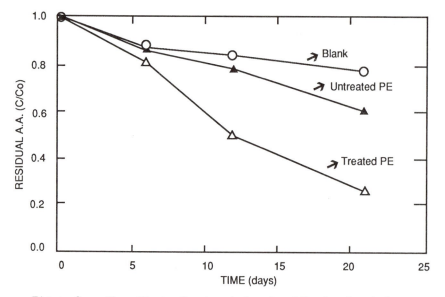

Figure 9.   The effect of untreated and oxidized polyethylene on ascorbic acid degradation in a model solution, stored at 35°C.

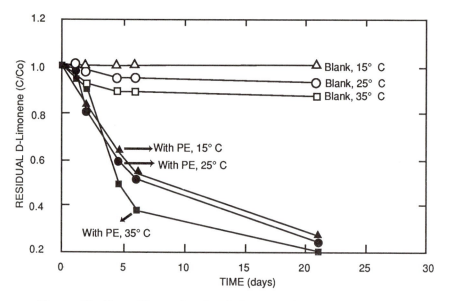

Figure 10. The effect of polyethylene strips on residual d-limonene in a model solution, stored at different temperatures. (BLANK = GLASS).

Reproduced with permission from ref. 4. Copyright 1987 Institute of Food Technologists.

Interaction in "bag in the box" packages

The oxygen permeabilities of the films used in the bag of the "bag in the box" are given in Table 1.

Table 1. Permeability Values for Different Films

| Film<br>Manufacturer | Structure | Thickness<br>(micron) | ($O_2$ Permeation rate<br>$cc/m^2$.24hr.At) |
|---|---|---|---|
| A | PE/Met PET/Met PET/PE | 76 | 0.2 |
| B | PE/Met PET/PE | 105-110 | 1.9 |
| C | PE/Met PET/PE | 115 | 1.0 |
| D | PE/Met PET/PE | 76 | 0.3 |
| E | PE/Barrier/PE | 82 | 0.8 |
| E | PE/Barrier/PE | 75 | <0.05 |

Orange juices aseptically packed in the "bag in the box" made from above films, were stored at 25°C. Results of periodic tests made on these juices, as compared to juices packed aseptically in glass jars, are presented in Figures 11 - 13. The juice in the structure PE/Met PET/Met PET/PE deteriorated fastest as compared to other structures. The juice in the different structures deteriorated at a higher rate than in glass. d-limonene in all films was reduced by 50% during the first 10 days at 25°C and after 25 days at 15°C (not shown in the figures) and then remained constant.

Flex Cracking of Metallized Layer

In figures 14 and 15 the integrity of the metallic layer in the metallized PET films is shown before handling and after 20 flexes in a Gelbo instrument representing transport conditions. It can be seen that some discontinuities in the metallic layer exist in the form of scratches even before flexing. These discontinuities become more severe during transportation causing an increase in oxygen permeation through the films (from 0.5 to 5.5 $cc/m^2$/ day for the shown film). These results conform with and explain our previous results (4-6) as well as those from other investigators (11,12,14). The results show the combined effect of oxygen transmissibility and the interaction with polyethylene on a reduced shelf-life of citrus products, in aseptic packages made of various laminates, as compared to same products in glass.

Effect of d-limonene on physical properties of packaging material

To evaluate the effect of d-limonene absorption on the properties of the package, the distribution of d-limonene betwen juice and 250 ml carton in aseptically packed orange juice

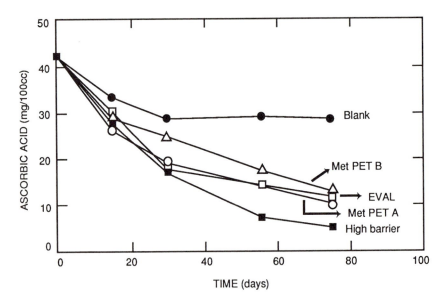

Figure 11.   Change   of   the   ascorbic  acid  concentration  in aseptic orange juice in "bag  in  the  box"  after  transport simulation, stored at 25°C.   (BLANK = GLASS).

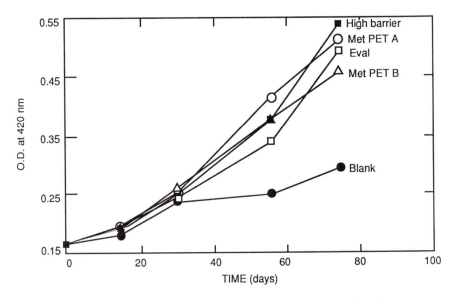

Figure 12.   Change of browning in aseptic orange juice in bag in the box after transport simulation, stored at 25°C. (BLANK = GLASS).

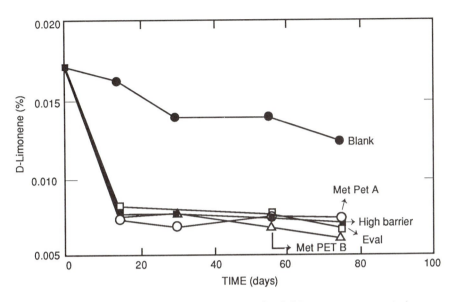

Figure 13. Change in percent of d-limonene content in aseptic orange juice in "bag in the box" after transport simulation, stored at 25°C. (BLANK = GLASS).

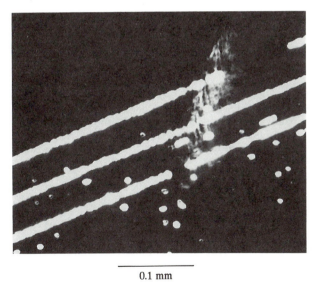

0.1 mm

Figure 14. Flaws in original metallized PET films from which the bag in the "bag in the box" package is made.

stored at 24°C/49% RH and 35°C/29% RH was studied. The results
are shown in Table 2.

Table 2: Distribution of d-limonene Between Juice and
Package Material in Aseptically Packed Orange Juice
During Storage at 24°C and 35°C[*]

| | | Storage Time (days) | | | | | | | | | |
|---|---|---|---|---|---|---|---|---|---|---|---|
| | | 0 | | 3 | | 6 | | 11 | | 18 | | 25 | |
| Sample | Initial | 24°C | 35°C | 24°C | 35°C | 24°C | 35°C | 24°C | 35°C | 24°C | 35°C |
| Juice in glass bottle (mg/250 ml) | 25.0 | 24.7 | 23.4 | 24.7 | 23.0 | 24.0 | 22.6 | 23.8 | 21.1 | 23.7 | 20.3 |
| Juice in carton package (mg/250 ml) | 25.0 | 19.9 | 17.8 | 18.6 | 16.9 | 17.2 | 16.5 | 16.9 | 14.0 | 14.0 | 12.1 |
| Carton package (mg/ package) | 2.5 | 7.5 | 7.6 | 9.9 | 8.6 | 10.7 | 9.2 | 11.6 | 10.1 | 11.6 | 10.6 |

* Average of two determinations.

The carton originally contained 2.5 mg d-limonene/ package. The
carton material had already absorbed some d-limonene when the
samples were put in storage (Day 0) because product was obtained
1 day after packing. It can be seen that while in the glass
bottles the residual d-limonene was 95% and 88% at 24°C and 35°C
respectively after 25 days of storage, the corresponding values
in the carton packages were 56% and 48% only, as a result of
absorption by the package.

Absorption of d-limonene by the different sealant films
during storage at 24°C, 49% RH is shown in Table 3. Within 3
days storage, all of the films had rapidly absorbed d-limonene.
After 3 days, the rate of absorption in LDPE and Surlyn (sodium
type) decreased, while Surlyn (zinc type) had reached
saturation. 12 and 18 days were required to reach saturation
for Surlyn (sodium type) and LDPE respectively. The amount of
d-limonene absorbed at equilibrium was 6.4 mg/100 $cm^2$ for
Surlyn (zinc type). The carboxy and zinc groups in the Surlyn
probably alter the lipophilic character of the polymer, but do
not prevent absorption of flavor components.

Table 3: Distribution of d-limonene Between Orange Juice and
Sealant Films During Storage at 24°C, 49% RH

| Content of d-limonene (mg) Storage Time (days) | Juice | LDPE | Juice | S-1601 | Juice | S-1652 |
|---|---|---|---|---|---|---|
| 0 | 46.3 | 0 | 46.3 | 0 | 35.5 | 0 |
| 3 | 36.1 | 6.8 | 29.3 | 10.8 | 26.6 | 7.0 |
| 6 | 33.9 | 8.3 | 26.6 | 12.0 | 23.4 | 7.3 |
| 12 | 28.5 | 10.2 | 23.5 | 14.7 | 23.2 | 7.5 |
| 18 | 26.6 | 11.8 | 22.5 | 14.7 | 22.4 | 7.5 |
| 27 | 26.5 | 11.9 | 22.3 | 14.5 | 21.6 | 7.5 |

The values are represented as mg/250ml in juice and mg/225.8 $cm^2$ in films.
The results are the means of triplicates.
S-1601: Surlyn, sodium type
S-1652: Surlyn, zinc type

The effect of the amount of d-limonene absorbed by the sealant films on their mechanical properties is shown in figures 16-19. In Fig. 16, the effect of d-limonene on the modulus of elasticity are shown. As can be seen, d-limonene has a plasticizing effect on the sealant films causing a decrease in their stiffness. The effect on the modulus of elasticity for the sodium type Surlyn was the largest and for the zinc type the smallest while the effect on the modulus of PE was in between. The effect of d-limonene on the tensile strength is shown in Fig. 17. The plasticizing effect of d-limonene, as represented by the strength of the sealants is seen again. The sodium type Surlyn seems to be affected most while the tensile strength of PE remained almost unaffected by d-limonene absorption.

In Fig. 18 the effect of d-limonene absorption on the elongation at break is shown. Whereas almost no effect on this property was found for the Surlyn type films, the elongation at break of LDPE in the machine direction increased while that in the cross direction decreased in the presence of d-limonene. This suggests that the properties of this film were not balanced in both directions.

d-limonene absorption also reduced the seal strength of the different sealants as seen in Fig. 19.

The oxygen permeability constants for LDPE, Surlyn (sodium type) and Surlyn (zinc type) (non immersed) were 88.9, 111.8, and 116.8 (cc.mm/$m^2$.day.atm), respectively. After d-limonene was absorbed by the films, the permeability constants increased to 406.4, 279.4 and 185.4 respectively. Mohney et al. (20) and Baner (21) also found that absorption of organic flavor constituents by films increased the permeability of the films to the flavor constituents.

The preceding data, therefore, clearly show that package/product interaction in high gas barrier packages may

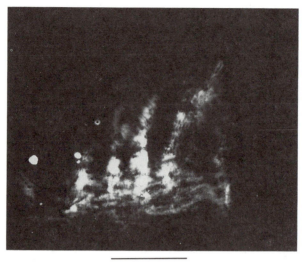

0.1 mm

Figure 15. Flaws in metallized PET film after 20 flexes in a
Gelbo instrument.

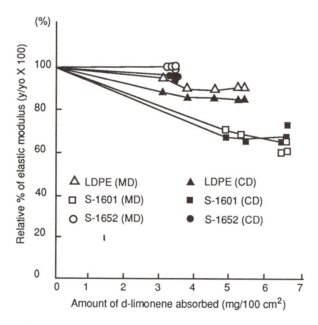

Figure 16. Relationship between d-limonene content and
modulus of elasticity for test films.

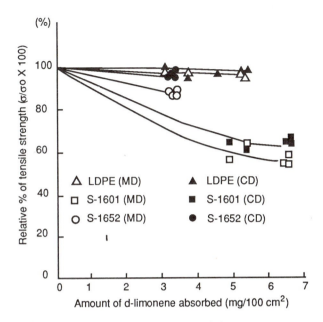

Figure 17. Relationship between d-limonene content and tensile strength for the test films.

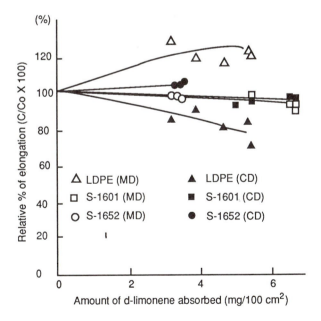

Figure 18. Relationship between d-limonene content and percent elongation at break for the test films.

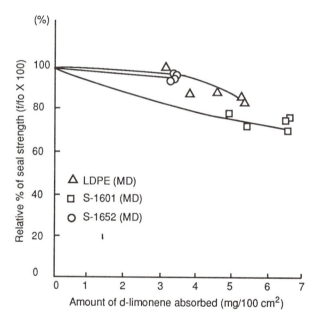

Figure 19.    Relationship between d-limonene content and seal
strength for the test films.

cause a reduction in the shelf life of citrus juices as well as alter the properties of the packaging material. A longer shelf life of these products could be expected for lower PE coating temperatures and better seals.

LITERATURE CITED

1.  U.S. Food and Drug Administration. Fed. Register. 1981, 46, 6.
2.  Gherardi, S.T. Proc. Cong. of Fruit Juice Prod. Munich, 1982, 143.
3.  Koszinvoski, J., Piringer, O. Deutsche Lebensmittel Rundschau., 1983, 79, (6) 179.
4.  Mannheim, C.H., Miltz, J., and Letzter, A. J. Food Sci. 1987, 52, 737.
5.  Miltz, J. and Mannheim, C.H. In Food Product-Package Compatibility; Gray, J.I., Harte B.R. and Miltz, J. Ed., Technomic, 1987, 245.
6.  Mannheim, C.H., Miltz, J. and Passy, N. In ACS Symp. Series 365; Hotchkiss,J. Ed., 1988, 68.
7.  Hirose, K., Harte, B.R., Giacin, J.R., Miltz, J. and Stine ,C. In ACS Symp. Series 365; Hotchkiss, J. Ed., 1988, 28.
8.  Gilbert, S.G. Food Technol. 1985, 39 (12), 54.
9.  Miltz, J., ELisha, C. and Mannheim, C.H. J. Food Process. Preserv. 1980, 4, 281.
10. Fernandes, M.H., Gilbert, S.G., Paik, S.W. and Steir, F. J. Food Sci. 1986, 51, 722.
11. Durr, P., Schobinger, U.and Waldvogel, R. Lebensmittel Verpackung. 1981, 20, 91.
12. Marschall, M.R., Adams, J.P. and Williams, J.W. In Aseptipak Schotland Business Research Inc. Princeton. 1985, p 299.
13. Buchner, N. Neue Verpackung. 1985, 4, 26.
14. Granzer, R. Proc. Cong. of Fruit Juice Prod., Munich, 1982. 161.
15. Carlson,U.R. Food Technol. 1984, 38 (12), 47.
16. Scott, W.C. and Veldhuis, M.K. J.AOAC, 1966, 49 (3), 628.
17. ASTM F-392.
18. Larmond, E. Methods for Sensory Evaluation of Food; Publication 1284, Canada Dept. of Agriculture. 1970.
19. Passy, N. and Mannheim, C.H. J. Food Eng. 1983, 2, 19.
20. Mohney, S.M., Hernandez, R.J., Giacin, J.R., Harte, B.R. and Miltz, J. J. Food Sci. 1988, 53 (1), 253.
21. Baner, A.L. 13th Annual IAPRI Symposium, 1986, Oslo, Norway

RECEIVED November 14, 1989

Chapter 17

# Loss of Flavor Compounds from Aseptically Processed Food Products Packaged in Aseptic Containers

A. P. Hansen and D. K. Arora

Department of Food Science, North Carolina State University, Box 7624, Raleigh, NC 27695-7624

The loss of natural flavor compounds in aseptically processed and packaged milk and cream and the aroma changes in the headspace of UHT milk were monitored during storage. The concentration of methanal, propanal, butanal and nonanal decreased over six months for two processes (138 °C for 20.4 s and 149 °C for 3.4 s). Total aldehyde concentration decreased over a six month storage period. Headspace analysis of UHT processed milk revealed a loss of higher molecular weight flavor compounds at 12 week storage. The loss of flavor compounds may be due to the interaction with the packaging material (low density polyethylene) and with the proteins. The loss of flavor compounds in UHT processed and packaged milk and cream varied from 50-97% over a 12 month storage period.

Food packaging is undergoing enormous changes due to the shift in consumer interest towards convenience foods. Among food packaging materials, the use of plastics is expected to increase by 55% by 1990 (1). The market share of flexible and rigid plastic packaging is expected to increase from $19 billion to $44 billion by the year 2000 (2).
　　Since the approval of hydrogen peroxide for sterilizaion of polyethylene food-contact surfaces in 1981, there has been an enormous proliferation in aseptic processing and packaging of food products. Besides convenience, aseptic packaging has resulted in improved

0097–6156/90/0423–0318$06.00/0
© 1990 American Chemical Society

product quality and in lower packaging/operating cost.
Milk-based products, fruit juices and juice drinks are
the major aseptic products in the market (3-5). More than
50 U.S. companies are currently marketing fruit juices
and drinks in aseptic packages (6). Tomato-based
products, soups, wine and mineral water are expected to
be introduced soon (6). The sale of fruit juice and juice
drinks is estimated to be $60 million by the year 2000
(6). The U.S. market may have a volume of over 15 billion
aseptic packages by the year 2000 (6).
     Most aseptic foods are packaged in a variety of
polymeric materials. The plastic polymers used in aseptic
packages are either in a pure or a coextruded form to
optimize barrier properties. These plastics are not as
inert as metal and glass containers resulting in a
shorter shelf life for laminated packages when compared
to metal and glass containers.
     A variety of interactions such as permeability,
migration, light penetration and sorption between
plastics and food products have been described (7). The
sorption of flavor compounds by polymeric materials has
not been extensively discussed. Sorption refers to the
scalping of flavors and aroma components of food products
by plastics. Flavor scalping may influence the sensory
quality and the acceptability of food products by the
consumer.
     Most of the aseptic packaging is done in containers
where a low density polyethylene (LDPE) layer comes in
contact with the food product. It has been shown that
LDPE film does not provide a good barrier to the loss of
flavor compounds. Morris Salame (8) reported 71% and 66%
absorption of non-polar and polar flavor compounds,
respectively, by polyethylene (PE). The sorption of
flavor compounds into PE film increased with carbon chain
length (9). Greater sorption of flavor compounds was
observed in less crystalline PE film, which suggested
that high crystalline films may reduce flavor compound
sorption into the PE. Kwapong and Hotchkiss (10) reported
that d-limonene was absorbed to a higher degree than
benzaldehyde, citral and ethylbutyrate into LDPE and
ionomer resin (Surlyn). Meyers and Halek (11) reported
30-40% loss in hydrocarbon terpenes and 10% loss in
oxygenated terpenes when exposed to LDPE. Mannheim et al.
(12) reported a 25% reduction in d-limonene content over
14 days in orange juice samples containing LDPE strips.
Marcy et al. (13) reported lower flavor scores in
concentrated orange juice and concentrated orange drink
stored in aseptic flexible bags.
     Recently, Charara (14) reviewed sorption of flavor
compounds into packaging materials. This comprehensive
review showed that polymeric materials in contact with
food products caused a loss of flavor compounds in food
products and model food systems. Low levels of fruit pulp
content in juice resulted in higher absorption of flavor

compounds (terpene hydrocarbons) and that the degree of
absorptivity depended upon the nature of the flavor
compounds.

In dairy products the lipid component is an
important contributor of dairy flavor. Compounds such as
mono- and di-carbonyls and volatile fatty acids impart
flavor to milk though present only in low concentrations.
Saturated and unsaturated aldehydes and various ketones
affect flavor though present in concentrations of ppm or
ppb (15).

Most of the flavored aseptic foods in the market are
fortified with essential flavors to compensate for any
flavor compounds lost during processing and storage. This
empirical technique of fortification may work for some
food products; in others this process may lead to further
deterioration as the sorption process is driven by a
concentration gradient. Loss of one component in a
complex flavor mixture may result in a loss of flavor
intensity or a change in flavor notes.

Different polymeric materials with high aroma and
flavor barrier properties are currently being tested.
Recently, it was reported that ethylene vinyl alcohol
(EVOH) coextruded containers resulted in no loss of
essential oil and 24% loss of vitamin C in orange juice
(16). Aaron Brody of Schotland Business Research, Inc.
(17) reported that EVOH improved gas and aroma barrier
properties and will be used extensively in barrier films.
To minimize flavor compounds sorption into packaging
materials, impregnation of the polymeric layer with
suitable flavor compounds has been suggested (7,17).

The reports on sorption of flavor compounds by
packaging materials suggest that more elaborate and
detailed research on the interactions between flavor
compounds and packaging materials is needed to produce
aseptic products having a more stable flavor profile, a
longer shelf life and increased consumer acceptance.

In this paper we will discuss the effect of UHT
processing and storage on the loss of natural flavors in
milk and cream and on the aroma changes in the headspace
of aseptically processed and packaged milk during
storage.

## Materials and Methods

Fresh milk and cream were obtained from the N.C. State
University dairy farm and standardized to 3.25% fat and
10% fat, respectively. Milk was processed at 149°C for
3.4 s for short residence time (SRT) and 138°C for 20.4 s
for long residence time (LRT). Cream was processed at
138°C for 20.4 s, at 143°C for 7.8 s, at 149°C for 3.4 s
and at 149°C for 20.4 s. All samples were preheated to
78°C. The milk was cooled in a No-Bac Aro-Vac
(Cherry-Burell) to a filling temperature of 7°C. Aseptic
packaging was accomplished by a Tetra Brik Model AB3-250

filler. Milk and cream samples were stored at 24°C and
40°C and analyzed monthly for up to one year.
The milk and cream were centrifuged, and the fat
fraction was removed. The fat was extracted with carbonyl
free hexane, and the fat extract was immediately reacted
with 2,4-dinitrophenylhydrazine. The carbonyl compounds
formed respective 2,4-dinitrophenylhydrazones (2,4-DNPH).
The hydrazones were separated into monocarbonyls and then
further purified using thin layer chromatography (18).
Alkanal 2,4-DNPH were analyzed by gas chromatography
using a 1.83 m glass column (i.d. 4 mm) packed with 2%
OV-1 on acid washed Chromosorb W (180-200 mesh) (18,19).
    Headspace analysis was conducted by gas
chromatography according to the method of Keller and
Kleyn (20) with the modification described by Nash (21).
Headspace analysis of milk samples was conducted
biweekly.

## Results and Discussion

Loss of Flavor Compounds in UHT Processed Milk.   The
flavor compound profiles of UHT processed milk (138°C for
20.4 s, LRT) are shown in Figure 1 and Figure 2 for   0
and 6 mo storage at 24°C, respectively. Peaks 2, 3 and 5
corresponding to propanal, butanal and nonanal,
respectively, disappeared over 6 mo. Similarly, UHT
processed milk (149°C for 3.4 s, SRT) stored for 6 mo at
24° C showed significant decreases in the peak areas
under methanal, propanal and butanal while the nonanal
peak disappeared (Figure 3 and Figure 4). At zero time
SRT samples retained a higher concentration of flavor
compounds than LRT samples.
    The individual concentrations of flavor compounds in
UHT processed milk at LRT and SRT stored at 24°C and 40°C
is shown in Figure 5 - Figure 8. In LRT processed milk,
higher molecular weight aldehydes and total aldehydes
decreased in concentration. In all other instances there
was a net loss in aldehydes. In SRT processed milk, all
of the individual aldehydes decreased as well as the
total aldehydes at both 24°C and 40°C storage. The loss of
aldehydes may be attributed to binding by the packaging
materials and by the proteins in milk which have
undergone greater chemical and physical alteration caused
by the extended processing hold time.
    The headspace analysis of aromatic compounds of UHT
processed milk for time zero is shown in Figure 9. The
flavor compound profile after storage for 12 wk is shown
in Figure 10. The higher molecular weight flavor
compounds have almost disappeared. These results agree
with Shimoda et al. (9) who reported greater sorption of
flavor compounds into PE as carbon chain length
increased. Figure 11 shows the flavor compound profile of
UHT processed milk at 24 wk. At this storage period the
low molecular weight flavor compounds were also reduced.

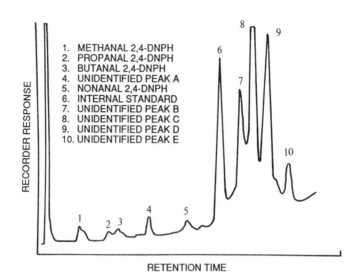

Figure 1. Time zero gas chromatogram of alkanal 2,4-DNPH from UHT milk processed at 138°C for 20.4 s (LRT). (Reproduced with permission from Ref. 18. Copyright 1982 American Dairy Science Association.)

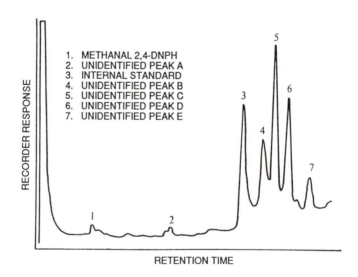

Figure 2. Gas chromatograph of alkanal 2,4-DNPH from UHT milk processed at 138°C for 20.4 s and stored six mo at 24°C (LRT). (Reproduced with permission from Ref. 18. Copyright 1982 American Dairy Science Association.)

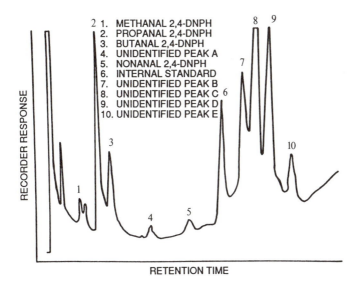

Figure 3.   Time   zero   gas   chromatogram   of   alkanal
2,4-DNPH from UHT milk processed at 149°C for 3.4 s
(SRT). (Reproduced with permission from Ref. 18.
Copyright 1982 American Dairy Science Association.)

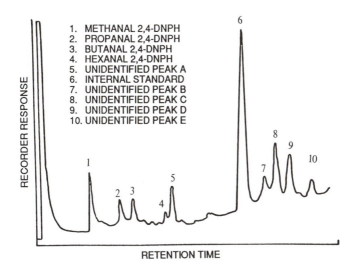

Figure 4.   Gas chromatogram of alkanal 2,4-DNPH from
UHT milk processed at 149°C for 3.4 s and stored six
mo at 24°C (SRT). (Reproduced with permission from
Ref. 18. Copyright 1982 American Dairy Science
Association.)

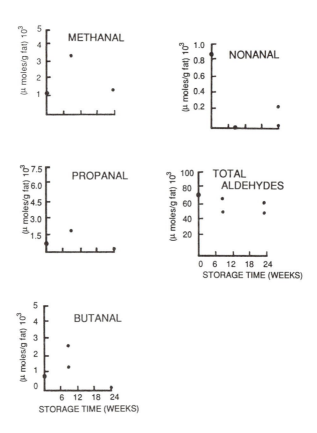

Figure 5. Changes in alkanal concentrations in UHT milk processed at 138° C for 20.4 s and stored at 24° C (LRT).

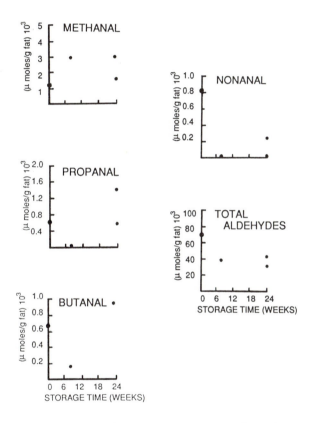

Figure 6. Changes in alkanal concentrations in UHT milk processed at 138° C for 20.4 s and stored at 40° C (LRT).

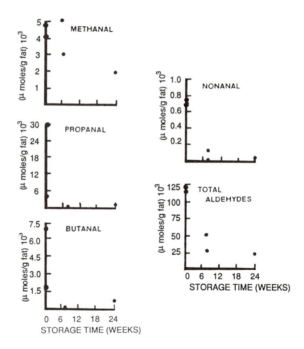

Figure 7.  Changes in alkanal concentrations in UHT milk processed at 149°C for 3.4 s and stored at 24°C (SRT).

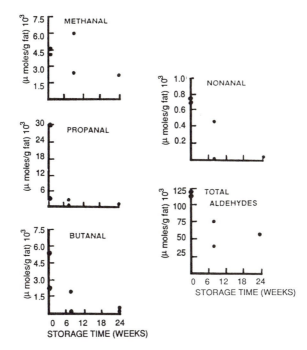

Figure 8.   Changes in alkanal concentrations in UHT milk processed at 149°C for 3.4 s and stored at 40°C (SRT).

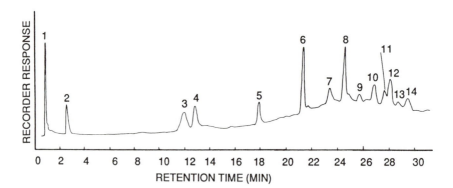

Figure 9.   Time zero gas chromatogram of headspace
volatiles from direct-processed UHT milk.

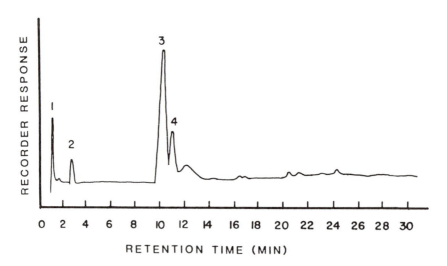

Figure 10.   Gas chromatogram of headspace volatiles
from direct-processed UHT milk stored 12 wk.

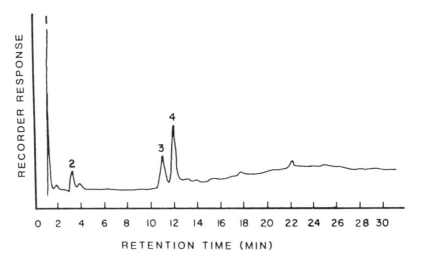

Figure 11.  Gas chromatogram of headspace volatiles from direct-processed UHT milk stored 24 wk.

Table 1.  Concentration of Flavor Compounds in UHT Cream
Over Twelve Months Storage

| Flavor Compounds | Months | | | | |
|---|---|---|---|---|---|
| ($\mu$moles/g fat) | 0 | 3 | 6 | 9 | 12 |
| Butanal ($\times 10^4$) | 0.65 | 0.66 | 0.65 | 0.52 | 0.33 |
| Hexanal ($\times 10^4$) | 0.21 | 0.21 | 0.21 | 0.23 | 0.26 |
| Heptanal ($\times 10^4$) | 0.36 | 0.25 | 0.18 | 0.14 | 0.09 |
| Octanal ($\times 10^4$) | 1.14 | 1.13 | 0.98 | 0.68 | 0.27 |
| Nonanal ($\times 10^4$) | 2.00 | 1.56 | 1.10 | 0.71 | 0.26 |
| Decanal ($\times 10^4$) | 2.18 | 1.55 | 1.07 | 0.80 | 0.46 |
| Total Carbonyls ($\times 10^2$) | 8.06 | 5.28 | 3.21 | 1.90 | 0.24 |

The loss of flavor compounds may have been due to binding by the packaging materials or due to interacting with the chemical constituents of foods.

**Loss of Flavor Compounds in UHT Processed Cream.** Table 1 shows the concentrations of flavor compounds in UHT processed cream packaged in Tetra Brik packages over 12 mo storage. The concentration of flavor compounds represents an average of four processes. Butanal concentration decreased 50%, aldehydes ($C_7$-$C_{10}$) decreased 75-85% and total carbonyls (aldehydes, ketones, enals, dienals, ketoacids, ketoglycerides, and dicarbonyls) decreased 97%.

**Loss of Flavor Compounds in Model Systems.** Preliminary work conducted in our laboratory showed sorption of aldehydes by LDPE used in aseptic packages (Hansen, A. P.; Arora, D. K., North Carolina State University, Raleigh, unpublished data). The sorption rate was related to concentration of aldehydes and to time, as both parameters increased so did the amount of flavor compounds sorbed into the film.

A second mechanism, the possible interaction of aldehydes with amino acids or peptides forming Schiff base reaction products, was also studied. Initial results (Hansen, A. P.; Heinis J. J., North Carolina State University, Raleigh, unpublished data) showed a decrease in the aldehyde concentration when model systems containing various aldehydes and amino acids were heat processed. Losses ranging from 8 to 50% of various aldehydes have been observed in the pasteurization of aldehydes and amino acids mixtures in simulated milk ultrafiltrate.

## Summary

Flavor is one of the most important attributes of a food product because it often determines whether the food is accepted or rejected. Reduction of flavor compounds sorbed into packaging material may produce more flavor stable aseptic food products. UHT processed milk and cream stored at 24°C and 40°C showed decrease in both aldehyde and total carbonyl concentrations. Further research is needed to determine the degree of binding of flavor compounds to chemical constituents of food products and to polymers used in aseptic packaging.

## Acknowledgments

The authors would like to thank M. S. Armagost for his assistance in preparation of this manuscript.

Literature Cited

1.  Hotchkiss, J.H.   In Food and Packaging Interactions;
    Hotchkiss,   J.H.,   Ed.;   American   Chem.   Soc.:
    Washington DC, 1988; pp. 1-10.
2.  Anonymous.  1987. Food Eng. 1987, 59, 54.
3.  Anonymous.  Food Processing 1983, 44, 30-32.
4.  LaBell, F.  Food Processing 1986, 47, 104-105.
5.  Swientek, R.J.  Food Processing 1984, 45, 28-30.
6.  Sacharow, S.  Prepared Foods 1986, 155, 31-32.
7.  Landois-Garza, J.; Hotchkiss, J.H.  Food Eng 1987,
    59, 39,42.
8.  Salame, M.  1988. Paper presented at Bev-Pak '88.
9.  Shimoda, M.; Ikegami, T.; Osajima, Y.  J. Sci. Food
    Agric. 1988, 42, 157-163.
10. Kwapong, O.Y.; Hotchkiss, J.H.  J. Food Sci. 1987,
    52, 761-763, 785.
11. Meyers, M.A.; Halek, G.W.  Proc. Inst. Food Tech.
    46th Annual Meeting, June 15-18, 1986, p 131 (Abst.
    161).
12. Mannheim, C.H.; Miltz, J.; Letzter, A.  J. Food Sci.
    1987, 52, 737-740.
13. Marcy, J.E.; Hansen, A.P.; Graumilch, T.R.  J. Food
    Sci. 1989, 54, 227-230.
14. Charara, Z.N.  M.S. Thesis, University of Florida,
    1987.
15. Kinsella, J.E.  Chemistry and Industry 1969. No. 2,
    p. 36.
16. Anonymous.  Packaging Digest Magazine 1988, 25,
    86,89,91-92,94.
17. Brody, A.L.  Food Engineering 1987, 59, 49-50,52.
18. Earley, R.R.; Hansen, A.P.  J. Dairy Sci. 1982, 65,
    11-16.
19. Hutchens, R.K.  M.S. Thesis, N.C. State University,
    Raleigh, 1979.
20. Keller, W.J.; Kleyn, D.H.  J. Dairy Sci. 1972, 55,
    564.
21. Nash, J.B.  M.S. Thesis, N.C. State University.,
    Raleigh, 1984.

RECEIVED November 14, 1989

# Chapter 18

# Diffusion and Sorption of Linear Esters in Selected Polymer Films

G. Strandburg[1], P. T. DeLassus[2], and B. A. Howell[1]

[1]Department of Chemistry, Central Michigan University, Mt. Pleasant, MI 48859
[2]Barrier Resins and Fabrication Laboratory, The Dow Chemical Company, 1603 Building, Midland, MI 48674

Polymers are being used in more sophisticated food-packaging applications. These applications must do more than keep the food safe from oxygen damage. They must preserve the attractive flavor profile of the food. In short, the package must assist in flavor management.

Permeation of flavor (and aroma) compounds in polymers can be part of flavor management in two general ways. First, unintended species can enter the food by permeation from the environment or by migration from the package itself.

Second, the flavor and aroma molecules of the food may leave by permeation through the package or by simple sorption by the package. The flavor profile may be made unacceptable by broad diminution or by selective removal of a few components.

This paper will describe experiments that concern this second area of flavor management. The experiments were directed toward the interactions between linear esters and some of the films that are important for food-packaging applications. The analysis of the data will focus on the important solution process with special attention for the thermodynamics.

Framework for Experiments

Aromas. Nine linear esters were included in this study. They are listed in Table I. Esters comprise an important flavor group. Not all of these esters are found in food. These specific esters provide a systematic approach to a complex problem. They were purchased from Aldrich Chemical Company as part of the "Flavors and Fragrances Kit #1: Esters."

0097–6156/90/0423–0333$06.00/0

Table I
Linear Esters for Permeation

| Ester Name | Number of Carbons | Boiling Point °C | % Purity | CAS Number |
|---|---|---|---|---|
| Methylbutyrate | 5 | 102 | 98+ | 623-42-7 |
| Ethylpropionate | 5 | 99 | 97+ | 105-37-3 |
| Ethylbutyrate | 6 | 120 | 98+ | 105-54-4 |
| Propylbutyrate | 7 | 142 | 98+ | 105-66-8 |
| Ethylvalerate | 7 | 144 | 98+ | 539-82-2 |
| Ethylhexanoate | 8 | 168 | 98+ | 123-66-0 |
| Ethylheptanoate | 9 | 188 | 98+ | 106-30-9 |
| Ethyloctanoate | 10 | 206 | 98+ | 106-32-1 |
| Hexylbutyate | 10 | 208 | 98+ | 2639-63-6 |

Films. Three films were included in this study. Low density polyethylene (LDPE) was included as a representative polyolefin. It is not considered to be a barrier polymer. It has permeabilities to selected aroma compounds slightly higher than the permeabilities of polypropylene and high density polyethylene (1). A vinylidene chloride copolymer (co-VDC) film was included as an example of a barrier that is useful in both dry and humid conditions. The film was made from a Dow resin which has been designed for rigid packaging applications. A hydrolyzed ethylene-vinylacetate copolymer (EVOH) film was included as an example of a barrier film that is humidity sensitive. The polymer was a blend of resins with total composition of 38 mole% ethylene.

These films were tested as monolayers about $2.5 \times 10^{-5}$m(=1 mil) thick. They would typically be used in multilayer structures; however, the individual permeabilities can be combined to predict the behavior of multilayers.

Instrument. The analytical instrument for these experiments is shown schematically in Figure 1. This instrument was designed and built at The Dow Chemical Company for measuring the permeation of flavor and aroma compounds through polymer films. The gas handling section of the instrument contains the plumbing, containers of aroma solutions, and the experimental film. This enclosure is insulated, and the temperature can be controlled, ± 1°C, from subambient to about 150°C. The detector is a Hewlett-Packard 5970 mass spectrometer. More details are available (2).

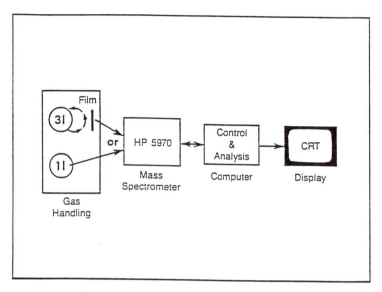

Figure 1.  Schematic of Instrument


This instrument is used in two general ways.  First, with each new compound, a mass spectrum is generated to identify the most populous and unique ion fragments.  Second, for actual permeation experiments, the instrument is programmed to monitor selected ion fragments, typically three, for each of the aromas involved.  When a single permeant is used, the three most populous ion fragments may be chosen.  However, when mixed permeants are used, significant degeneracies are avoided.  All the experiments of this study were done with a single permeant.  The response of the instrument was determined to be linear in the operating range and beyond, both at higher and lower concentrations.

In all experiments, the following sequence was used.  First, a calibration experiment was completed by preparing a very dilute solution of the permeant in nitrogen in the one liter flask and introducing some of the gaseous solution directly to the mass spectrometer via a specially designed interface.  A known concentration and a measured response yield a calibration constant.

Then a permeation experiment was completed by preparing a dilute solution of the permeant in nitrogen in the three liter flask and routing this gaseous solution past the purged, experimental film. The detector response was recorded as a function of time for subsequent analysis.

The computer was used to program and control the mass
spectrometer and to collect and analyze the data.  The computer was
equipped with a CRT and a printer.
The advantages of this instrument are a) it can be used with
very dilute solutions of permeants, b) multiple permeants may be
used in a single experiment, c) humidity may be used by adding water
or a saturated salt solution to the three liter flask, d) the entire
progress of the experiment can be monitored as a function of time,
and e) a wide range of temperatures can be used.

Variables.  A range of temperatures was used with the barrier films
-- co-VDC and EVOH.  Warm temperatures were used because these
barrier polymers had low permeabilities and low diffusion
coefficients at cooler temperatures.  Experiments at lower
temperatures would have taken prohibitively long and may have been
below the detection limit of the instrument.  Data at several
temperatures allow a more sophisticated analysis of the permeation
process.  The LDPE film was tested at a single, low temperature.
A single concentration of permeant was used for all the
experiments.  Approximately two microliters of flavor/aroma compound
were used in the three liter flask.  This corresponds to about 150
ppm(molar) and varies from ester to ester depending upon the
molecular weight.  Previous experiments have shown that, in this
range of concentrations, the permeation process is not a strong
function of the permeant concentration (3,4). Table II lists the
partial pressure ($P_x$) at the experimental temperatures for the five
carbon ester and the nine carbon ester.  Also included are the vapor
pressures ($P_o$) at these temperatures, and the ratio $P_x/P_o$.

Table II. Partial Pressure at the Experimental Temperature for Five-and
Nine-Carbon Esters

| Ester | T(C) | $P_o$ (Pa)[1] | $P_x$ (Pa)[2] | $P_x/P_o$ |
|-------|------|------------|------------|----------|
| C-5 | 105 | $1.1 \times 10^5$ | 19 | $1.7 \times 10^{-4}$ |
| C-5 | 95 | $8.1 \times 10^4$ | 18 | $2.2 \times 10^{-4}$ |
| C-5 | 85 | $5.9 \times 10^4$ | 18 | $3.0 \times 10^{-4}$ |
| C-5 | 25 | $4.5 \times 10^3$ | 15 | $3.3 \times 10^{-3}$ |
| C-9 | 105 | $7.1 \times 10^3$ | 12 | $1.7 \times 10^{-3}$ |
| C-9 | 95 | $4.5 \times 10^3$ | 12 | $2.7 \times 10^{-3}$ |
| C-9 | 85 | $2.8 \times 10^3$ | 11 | $3.9 \times 10^{-3}$ |
| C-9 | 25 | $8.3 \times 10^1$ | 9 | $1.1 \times 10^{-1}$ |

1) From the Antoine equation [5]
2) For a 2 μL sample

Humidity was tested only qualitatively in a single experiment. In an experiment with the EVOH film and ethylvalerate at 110°C, 0.25 milliliters of water was injected into the three liter flask after the permeation experiment had reached steady state. Within seconds, the permeation rate increased to above the detection limit of the mass spectrometer. This was consistent with prior experience wherein the permeability and the diffusivity in an EVOH film increase by a factor of about 1000 in the presence of moisture (6). No more experiments were done with humidity for this study.

Framework for Analysis.

For a thorough analysis of aroma transport in food packaging, the permeability (P) and its component parts - the solubility coefficient (S) and the diffusion coefficient or diffusivity (D) are needed. These three parameters are related as shown in Equation 1.

$$P = D \times S \tag{1}$$

The permeability is useful for describing the transport rate at steady state. The solubility coefficient is useful for describing the amount of aroma that will be absorbed by the package wall. The diffusion coefficient is useful for describing how quickly the permeant aroma molecules move in the film and how much time is required to reach steady state.

Equation 2 describes steady state permeation where $\Delta M_x$ is the

$$\frac{\Delta M_x}{\Delta t} = \frac{P \ A \ \Delta p_x}{L} \tag{2}$$

quantity of permeant x that goes through a film of area A and thickness L in a time interval $\Delta t$. The driving force for the permeation is given as the pressure difference of the permeant across the barrier, $\Delta p_x$. In an experiment, $\Delta M_x/\Delta t$ is measured at steady state while A and L are known and $\Delta p_x$ is either measured or calculated separately.

The diffusion coefficient can be determined from the transient portion of a complete permeation experiment. Figure 2 shows how the transport rate or detector response varies with time during a complete experiment. At the beginning of an experiment, t = 0, a clean film is exposed to the permeant on the upstream side.

Initially, the permeation rate is effectively zero. Then "break through" occurs, and the transport rate rises to steady state. The diffusion coefficient can be calculated in either of two ways. Equation 3 uses $t_{1/2}$, the time to reach a transport rate that is one-half of the steady state rate (7).

$$D = \frac{L^2}{7.2 \ t_{1/2}} \tag{3}$$

Equation 4 uses the slope in the transient part of the curve and the

$$D = \frac{0.176 \; L^2 \; (\text{slope})}{Rss} \qquad (4)$$

steady state response Rss (8). After P has been determined with Equation 2 and D has been determined with Equation 3 (or 4), and then S can be determined with Equation 1.

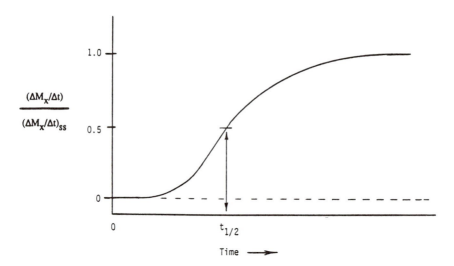

Figure 2. Relative Transport Rate as a Function of Time

Many sets of units are used to report the permeability in the literature. This paper will use SI units in the following way. For Equation 2 to be valid, the permeability must have dimensions of quantity of gas times thickness divided by area-time-pressure. If the kilogram is used to describe the quantity of gas and the pascal is used as the unit of pressure, then units of permeability are $kg \cdot m/m^2 \cdot s \cdot Pa$. This unit is very large and a cumbersome exponent often results. A more convenient unit, the Modified Zobel Unit, (1 MZU = $10^{-20}$ $kg \cdot m/m^2 \cdot s \cdot Pa$), was developed for flavor permeation. Zobel proposed a similar unit earlier (9). The units of the diffusion coefficient are $m^2/s$. The units of the solubility coefficient are $kg/m^3 Pa$.

Equations 1, 2, and 3 are working definitions of P, D, and S. No corrections are made for the crystallinity in any of the experimental samples of this study.

For most simple cases, P, D, and S are simple functions of temperature as given in Equations 5, 6, and 7.

$$P (T) = P_o \exp (-E_p/RT) \qquad (5)$$

$$D (T) = D_o \exp (-E_D/RT) \qquad (6)$$

$$S (T) = S_o \exp (-\Delta H_s/RT) \qquad (7)$$

where, $P_o$, $D_o$, and $S_o$ are constants, T is the absolute temperature, and R is the gas constant. $E_p$ is the activation energy for permeation. $E_D$ is the activation energy for diffusion, and $\Delta H_s$ is the heat of solution. If log P data are plotted on the vertical axis of a graph and $T^{-1}$ is plotted on the horizontal axis, a straight line will result. The slope will be $-0.43 E_p/R$. The diffusion coefficient and solubility coefficient behave similarly. Equations 1, 5, 6, and 7 can be manipulated to yield equation 8 which relates the activation energies and the heat of solution.

$$E_p = E_D + \Delta H_s \qquad (8)$$

Equations 1-7 were developed for gas permeation through rubbery polymers. They must be used with caution with glassy polymers and/or organic vapors that interact strongly with the barrier.

The heat of solution can be separated into two basic parts - namely, the heat of condensation ($\Delta H_c$) and the heat of mixing ($\Delta H_m$). Equation 9 expresses the relationship

$$\Delta H_s = \Delta H_c + \Delta H_m \qquad (9)$$

The heats of solution can be determined experimentally in these experiments, and the heats of condensation are available in the literature. The heat of mixing can then be calculated with Equation 9. The heat of mixing describes the basic thermodynamic relationship between the polymer and the penetrant.


EXPERIMENTS

Typical Experiment. Figure 3 shows the results of a typical experiment with the co-VDC film. The permeant was ethylbutyrate. The experiment was done at 100°C. The warm temperature caused the experiment to progress quickly. Three ions fragments, m/z = 60, 71, and 88 were monitored by the mass spectrometer. The total ion intensity of these three ions is shown here. The first response in Figure 3 is due to calibration. Here 2.4 ul of a 0.5% solution of ethylbutyrate in acetone was injected into the 1-liter flask and diluted with nitrogen. The resulting gaseous solution was introduced into the mass spectrometer via the interface.

After the baseline was obtained again at 34 minutes into the
run, the permeation experiment began.  Here 2.5 ul of ethylbutyrate
had been injected into the 3-liter flask and diluted with nitrogen
before being circulated past the upstream side of the film.
Breakthrough occurred at about 38 minutes, and steady state was
reached at 90 to 100 min.

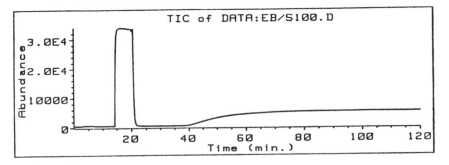

Figure 3.  Permeation Experiment for Ethylbutyrate in a
Vinylidene Chloride Copolymer Film at 100°C

Organized Results.  Many combinations of permeant ester, film, and
temperature were studied.  Figure 4 shows the permeability results
for a single ester-film combination at 5 different temperatures.
Similar graphs were made for the other ester-film combinations.
     Figure 5 shows permeabilities of all the esters at 85°C in the
co-VDC film.  These data were taken from graphs like Figure 4.
Figure 6 shows the diffusion coefficients of all the esters at 85°C
in the co-VDC film.  Figure 7 shows the solubility coefficients of
all the esters in the co-VDC film.  Figure 8 is a re-plot of the
data in Figure 7 with the boiling point of the ester as the
horizontal axis.
     The activation energies for permeation in the co-VDC film were
all positive.  The activation energies for the diffusion coefficient
are shown in Figure 9.  The heats of solution for the esters are
shown in Figure 10.
     Fewer esters were used in experiments with the EVOH because the
trends observed with the co-VDC film were apparent.  Figures 11, 12,
and 13 show the permeabilities, diffusion coefficients, and
solubility coefficients for the esters at 85°C in the EVOH film.
Figures 14 and 15 show the activation energies for diffusion and the
heats of solution.

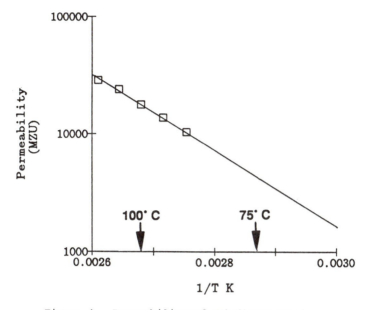

Figure 4. Permeability of Ethylbutyrate in a
Vinylidene Chloride Copolymer Film

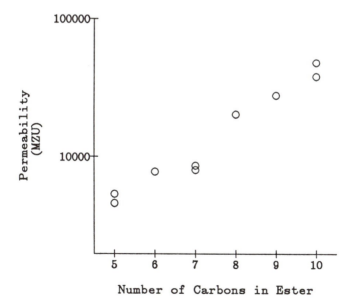

Figure 5. Permeability of Esters at 85°C in a
Vinylidene Chloride Copolymer

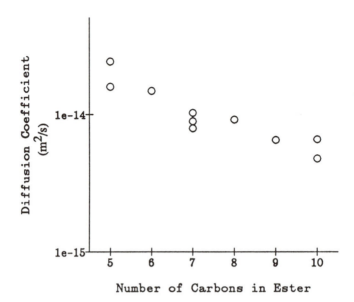

Figure 6.  Diffusion Coefficient at 85°C of Esters
in a Vinylidene Chloride Copolymer Film

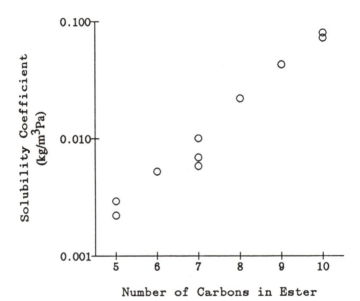

Figure 7.  Solubility Coefficient at 85°C of Esters
in a Vinylidene Chloride Copolymer Film

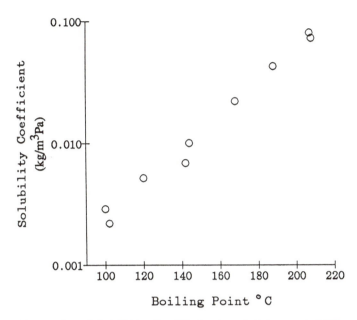

Figure 8. Solubility Coefficients of Esters at 85°C in a Vinylidene Chloride Copolymer Film

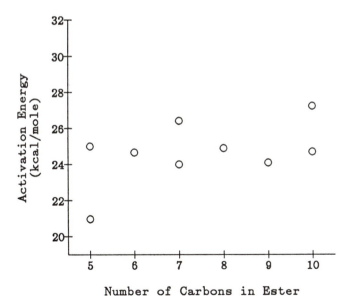

Figure 9. Activation Energy for Diffusion of Esters in a Vinylidene Chloride Copolymer Film

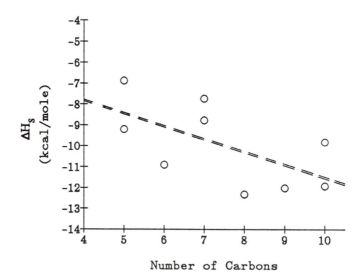

Figure 10.  Heats of Solution for Esters
in a Vinylidene Chloride Copolymer Film

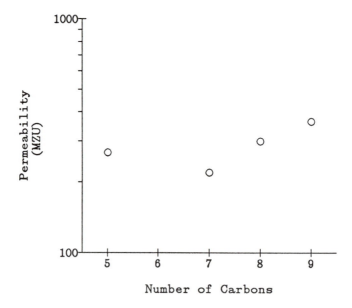

Figure 11.  Permeability Coefficient at 85°C
of Esters Through EVOH

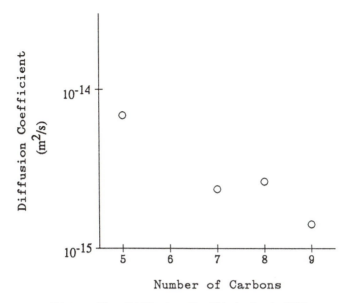

Figure 12. Diffusion Coefficient at 85°C
of Esters Through EVOH

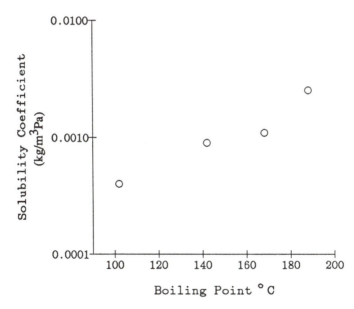

Figure 13. Solubility Coefficient at 85°C vs
Boiling Point of Esters in EVOH

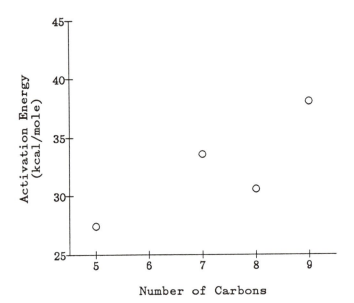

Figure 14.   Activation Energy of Diffusion
             for Esters Through EVOH

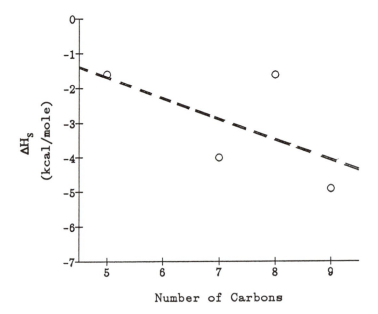

Figure 15.   Heats of Solution for Esters
             in an EVOH Copolymer Film

Figures 16 through 18 show the results for the LDPE film.

Discussion. In each film the permeabilities increase as the size of the ester increases. The data show that this result is driven by the solubility coefficient. The diffusion coefficients decrease as the size of the ester increases; however, the solubility coefficients increase greatly as the size of the ester increases.

The qualitative effect of permeant size on the diffusion coefficient was expected. A recent simple analysis quantified the relationship between D and permeant size for spherical permeants up to $C_6$ in several polymers (10). That analysis is not directly applicable here since these permeants are linear and larger. Another study quantifies the relationship between D and permeant size for linear alkanes in polyethylene (11). The diffusion coefficient at 25°C was observed to decrease from $3.9 \times 10^{-11}$ $m^2/s$ for n-hexane to $2.2 \times 10^{-11}$ $m^2/s$ for n-nonane.

The diffusion coefficients for the esters in the present studies seem to change comparable amounts. At 85°C in co-VDC and EVOH, the diffusion coefficients for the esters decrease by about 3x as the number of atoms (excluding hydrogen) in the backbone increases from 6 to 9 (5 carbons plus one oxygen to 8 carbons plus one oxygen). The difference would be greater at 25°C because the activation energies for diffusion are larger for the larger esters. This is consistant with the results of Landois-Garza and Hotchkiss for an ester series in polyvinyl alcohol (12).

At 30°C in LDPE, the diffusion coefficients decrease by about 30% as the number of atoms in the backbone of the ester increases from 6 to 9. Studies of alkanes in low density polyethylene show similar trends (13).

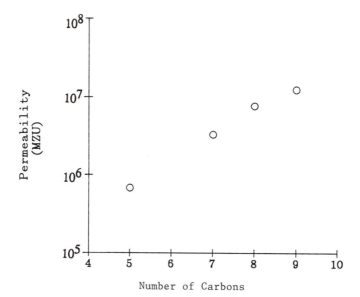

Figure 16. Permeability at 30°C for
Esters in a LDPE Film

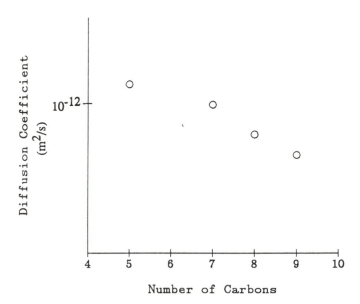

Figure 17.  Diffusion Coefficient at 30°C
of Esters in a LDPE Film

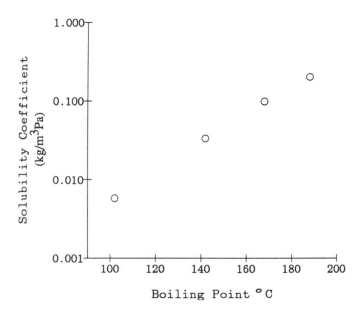

Figure 18.  Solubility Coefficient at 30°C vs.
Boiling Point of Ester in a LDPE Film

The activation energies for diffusion in the co-VDC and EVOH films increase as the size of the ester increases. This is expected since cooperative motion of larger zones of the polymer matrix are necessary to create larger holes for passage. The activation energies for diffusion in LDPE were not determined.

The solubility coefficients make interesting contributions to the permeabilities. The solution process of these gas phase permeants can be separated for analysis into two components - condensation and mixing. Table III contains the heats of condensation for the experimental esters. Table III also contains the heats of solution in the polymer films from linear fits of the data in Figures 10 and 15. Finally, Table III contains the heats of mixing for the esters in the films. The heats of mixing were calculated using equation 9.

Table III Heats of Condensation and Solution for Experimental Esters

| Number of Carbons | $\Delta H_c$ | $\Delta H_s$ co-VDC | $\Delta H_s$ EVOH | $\Delta H_{mix}$ co-VDC | $\Delta H_{mix}$ EVOH |
|---|---|---|---|---|---|
| 5 | $-8.88^1$ | $-8.4$ | $-1.6$ | 0.4 | 7.3 |
| 6 | $-9.47$ | $-9.1$ | --- | 0.4 | --- |
| 7 | $-10.48^1$ | $-9.7$ | $-2.8$ | 0.8 | 7.7 |
| 8 | $-11.20^2$ | $-10.3$ | $-3.5$ | 0.9 | 7.7 |
| 9 | $-11.92$ | $-10.9$ | $-4.1$ | 1.0 | 7.9 |
| 10 | $-12.87^2$ | $-11.5$ | --- | 1.4 | --- |

a) kcal/mole
1) average of two values
2) extrapolated values

The heats of condensation are reliable data: On the other hand, there is considerable experimental uncertainty in the heats of solution. The trends in the heat of solution are more certain; however, the actual values are not so certain. This uncertainty arises from two sources. First, the solubility coefficients are derived from the permeabilities and the diffusion coefficients. Hence, the relative uncertainties of both P and D transfer to S. The heats of solution are derived by fitting equation 7 to only 3 or 4 data. This is not ideal. Future work will expand the data base to test this preliminary analysis.

Nevertheless, the trends for the heats of mixing are probably correct as shown in Table II. As the esters get larger, the molecule becomes more like an alkane, and the polar ester group is diluted. The trend in the heats of mixing for the co-VDC film is interesting. As the esters become larger the heats of mixing become less favorable in the co-VDC film. This is a reflection of the dilution of the ester function. Esters are known to interact well with chlorine containing polymers. The heats of mixing for the esters in the EVOH film are unfavorable. They are more unfavorable than the heats of mixing in the co-VDC film. The heats of mixing in the EVOH film became more unfavorable as the size of the ester increases. This behavior may be predicted by solubility parameters. Future work will include a variety of functional groups to study the specific interactions of polymer and permeant.

SUMMARY

1. The diffusion coefficient decreases modestly as the ester size increases for low density polyethylene, EVOH and co-VDC films.
2. The solubility coefficient undergoes a much greater change with increasing ester size.
3. Thus the permeability increases with increased ester size.
4. Plasticizing the polymer film can result in a dramatic increase in the diffusion coefficient.
5. Important flavor and aroma compounds can be lost to the food package by sorption.
6. The amount of flavor loss to the package is dependent on both diffusion and solubility coefficients.
7. The heats of mixing are dependent on specific interactions between polymer and permeant.

LITERATURE CITED

1. DeLassus, P. T.; Hilker, B. L., Proc. Future-Pak '85 (Ryder), 1985, p. 231.
2. Tou, J. C.; Rulf, D. C.; DeLassus, P. T.; accepted for publiciation in Analytical Chemistry.
3. DeLassus, P. T.; Tou, J. C.; Babinec, M. A.; Rulf, D. C.; Karp, B. K.; Howell, B. A.; ACS Symposium Series 365, Food and Packaging Interactions (J. H. Hotchkiss, ed.) 1988, p. 11.
4. Zobel, M. G. R., Polymer Testing 1989, 5, 153.
5. Lange's Handbook of Chemistry 13th Edition, McGraw-Hill, N.Y., 1985, pp 10-28.
6. DeLassus, P. T.; Strandburg, G., Howell, B. A., TAPPI Journal, 1988, 71(11), 177.
7. Ziegel, K. D.; Frensdorff, H. K.; Blair, D. E., J. Polym. Sci., Part A-2, 1969, 7, 809.
8. Pasternak, R. A.; Schimscheimer, J. R.; Heller, J., J. Polym. Sci., Part A-2, 1970, 8, 467.
9. Zobel, M. G. R., Polymer Testing 1982, 3, 133.
10. Berens, A. R.; Hopfenberg, H. B., J. Membrane Sci., 1982, 10, 283.
11. Peters, H.; Vanderstracten, P.; Verhoeye, L.; J. Chem. Tech. Biotech., 1979, 29, 581.
12. Landois-Garza, J.; Hotchkiss, J. H., Food and Packaging Interactions, ACS Symposium Series 365, 1987, 53.
13. Asfour, A. F. A.; Saleem, M.; DeKee, D., Journal of Applied Polymer Science, 38, 1989, 1503.

RECEIVED January 31, 1990

# Chapter 19

# Sorption and Diffusion of Monocyclic Aromatic Compounds Through Polyurethane Membranes

**U. Shanthamurthy Aithal[1], Tejraj M. Aminabhavi[1], and Patrick E. Cassidy[2]**

[1]Department of Chemistry, Karnatak University, Dharwad, 580003, India
[2]Department of Chemistry, Southwest Texas State University, San Marcos, TX 78666

Sorption and transport of a number of monocyclic aromatics through a polyurethane membrane have been investigated at 25, 44 and 60°C based on an immersion/weight gain method. Activation parameters for the process of diffusion, permeation and sorption are found to follow the conventional wisdom that larger molecules exhibit lower diffusivities and higher activation energies. From a temperature dependence of the sorption constant, the standard enthalpy and entropy have been determined. The molar mass between crosslinks has been determined by using the Flory-Rehner theory. Furthermore, results have been discussed in terms of the thermodynamic interactions between the polymer and penetrants.

Polyurethane elastomers are known to exhibit unique mechanical properties, primarily as a result of two phase morphology (1). These materials are alternating block copolymers made of hard segments of aromatic groups from the diisocyanate/chain extender and soft segments of aliphatic chains from the diol (ether or ester). The hard and soft segments are chemically incompatible and microphase separation of the hard segments into domains dispersed in a matrix of soft segments can occur in varying degrees. In view of the importance of polyurethane as a barrier material in several engineering areas (2,3), it is important to know its transport characteristics with respect to common organic solvents. Thus, knowledge of the transport mechanisms as manifested by sorption, diffusion and permeation of organic liquids (penetrants) in a polyurethane matrix is helpful for establishing the relationships between structures and properties under severe application conditions. Although some previous studies (4-10) have been made on solvent transport through polyurethane, more experimental data are still needed for a better understanding of the thermodynamic interactions between the polymer and solvents. The principal objective of this paper is to follow the transport properties of monocyclic aromatic liquids through a commercially available polyurethane membrane. It is expected that a systematic change in solvent power would lead to results which could be interpreted by considering the possible interactions with soft and hard segments of the polymer. Transport properties viz., diffusivity, D, permeability, P, and sorption, S, have been studied over an interval of temperature from 25 to 60°C, to predict the Arrhenius parameters for each of the transport processes involved. Furthermore, the results have been discussed in terms of thermodynamic interactions between polyurethane and the liquid penetrants. Molar mass, $M_c$, between crosslinks have been estimated by using the Flory-Rehner model (11,12).

EXPERIMENTAL

REAGENTS AND MATERIALS. Polyurethane (PU) used was obtained from PSI, Austin, Texas in sheets of 0.250 cm thickness. The base polymer is a Vibrathane B600 (Uniroyal) which was obtained from the reaction of polypropylene oxide and 2,6-toluene diisocynate (TDI). The base polymer was cured with 4,4'-methylene-bis-(o-chloroaniline) i.e., MOCA, to give the polyurethane. Thus, the two-phase morphology of PU consisted of polyether diol as the soft segment and the aromatic diisocynate acting as the hard segment. The driving force for the phase separation is the incompatibility of the hard and soft segments. A schematical description of the molecular structure of the polyurethane used is given below :

$$\text{\textasciitilde\textasciitilde\textasciitilde}\ \underbrace{AC}_{\text{soft}} - \underbrace{ABABABABABA}_{\text{hard}} - \underbrace{CA}_{\text{soft}} - \underbrace{BABABAB}_{\text{hard}}\text{\textasciitilde\textasciitilde\textasciitilde}$$

where,     A : $-\!\left[O\!-\!\left(CH_2\right)_{\!3}\right]-$     B : $-\!\left[O\!-\!\left(CH_2\right)_{\!3}\right]_n$     C :

        Some representative engineering properties of PU are : tensile strength, 387 kg/sq.cm(5500 psi); maximum percent elongation, 430; modulus for 300% elongation, 155 kg/sq.cm(2200 psi); tear strength, 5 kg/sq.cm(70 psi) (ASTM D- 470) and specific gravity 1.101. The $T_g$ of the polymer was found to be -43.27°C with a heat flow of -1.420 Watts/g as determined by differential scanning calorimetry, duPont model 951. The solvents given in Table I are of reagent grade and double distilled before use.

DIFFUSION EXPERIMENTS :     Polyurethane elastomers were cut into uniform size circular pieces (diameter = 1.9 cm) using a specially designed, sharp-edged, steel die and dried overnight in a desiccator before use. The thickness of the sample was measured at several points using a micrometer (precision ± 0.001 cm) and the mean value was taken as 0.250 cm. Dry weights of the cut samples were taken before immersion into the air tight, metal-capped test bottles containing the liquid. After immersion into the respective liquids, the bottles were placed in a thermostatically controlled oven (± 0.5°C).

At periodic intervals, the samples were removed from the bottles, the wet surfaces were dried between filter paper wraps and weighed immediately to the nearest 0.05 mg by placing it on a covered watch glass within the chamber of the balance. The samples were placed back into test bottles and were transferred to the oven. The experiments were performed at 25, 44 and 60°C.

A possible source of error in this method is that the sample has to be removed from the liquid to allow weighing; if this is done quickly (say within 30-50 secs) compared to the time a sample spent in the liquid in between consecutive weighings, the sample removal exerts a negligible effect. For those liquids whose density is greater than the polymer itself, the sample was kept submerged in the liquid by a glass-plunger attached to the screw cap.

RESULTS AND DISCUSSION

SORPTION KINETICS.   Room temperature sorption curves expressed as percent penetrant uptake, $Q(t)$, versus square root of time, $t^{\frac{1}{2}}$ are displayed in Figures 1 and 2. A perusal of the sorption curves given in Figure 1 suggests a systematic trend in the sorption behavior of methyl substituted benzenes. For instance, the sorption of benzene is higher than other homologues; also, benzene reaches sorption equilibrium more quicker than toluene, p-xylene and mesitylene. This may be due to the presence of more bulkier $-CH_3$ groups rendering a sluggish movement of the solvent within the polymer matrix. The sorption curves of other aromatics, namely, bromobenzene, o-dichlorobenzene, nitrobenzene, chloro-benzene and anisole are presented in Figure 2. Here, bromobenzene exhibits a maximum $Q(t)$ of about 147% (the maximum of all the penetrants) whereas anisole has only about 80%. Thus, it is clear that chloro, bromo, nitro and methoxy substitutions on the benzene moiety tend to increase the extent of sorption with the polyurethane membrane. The maximum sorption of all the penetrants follow the sequence : Bromobenzene > o-dichlorobenzene > chlorobenzene ≃ nitrobenzene > anisole > benzene > toluene > p-xylene > mesitylene.

Sorption data also serve as a guide to study the effect of temperature on the observed transport behavior. Temperature variation of sorption curves for some representative penetrants namely, benzene, mesitylene, anisole, bromobenzene and o-dichloro-benzene are included in Figures 3-7. In almost all cases, the shapes of sorption curves at 25°C are similar to those at the two higher temperatures, although the change in slope is more pronoun-ced between 25 and 44°C than between 44 and 60°C. For all the penetrants except benzene and bromobenzene, the maximum sorption values at 44 and 60°C seem to be more or less identical.

For a Fickian behavior, the plots of $Q(t)$ versus $t^{\frac{1}{2}}$ should increase linearly up to about 50% sorption. Deviations from the Fickian sorption are associated with the time taken by the polymer segments to respond to a swelling stress and rearrange

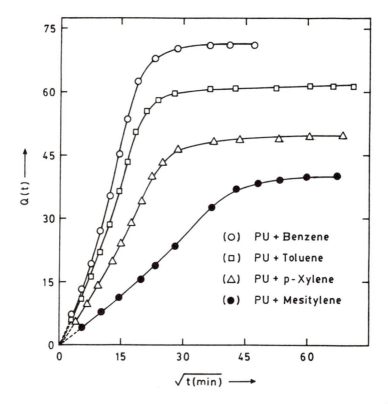

Figure 1  Percentage mass uptake Q(t) of the polymer versus square root of time $t^{1/2}$ for polyurethane (PU) + solvent pairs at 25 ° C.

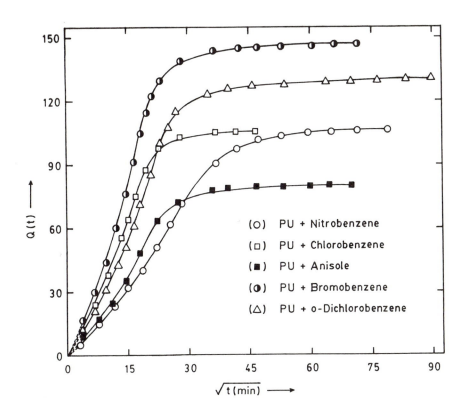

Figure 2  Percentage mass uptake $Q(t)$ of the polymer versus square root of time $t^{1/2}$ for polyurethane (PU) + solvent pairs at 25 ° C.

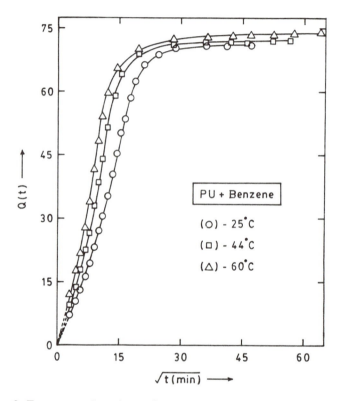

Figure 3 Temperature dependence of percentage mass uptake Q(t) of the polymer versus $t^{1/2}$ for polyurethane + Benzene system.

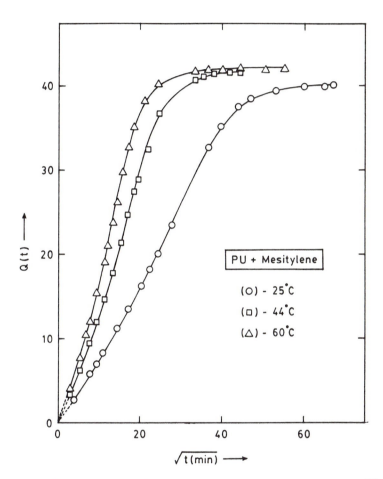

Figure 4 Temperature dependence of percentage mass uptake Q(t) versus $t^{1/2}$ for polyurethane + Mesitylene system.

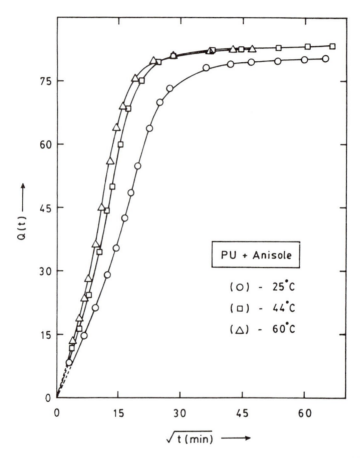

Figure 5  Temperature dependence of percentage mass uptake Q(t) versus $t^{1/2}$ for polyurethane + Anisole system.

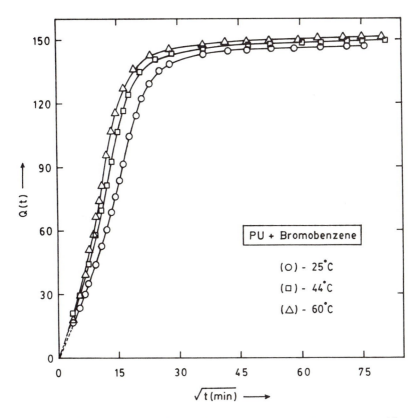

Figure 6 Temperature dependence of percentage mass uptake Q(t) versus $t^{1/2}$ for polyurethane + Bromobenzene system.

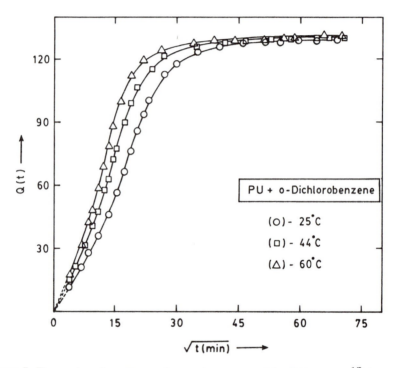

Figure 7 Temperature dependence of percentage mass uptake Q(t) versus $t^{1/2}$ for polyurethane + o-Dichlorobenzene system.

themselves to accommodate the solvent molecules. This usually results in the sigmoidal shapes for the sorption curves: Thus, non-Fickian diffusion involves the tension between swollen (soft segments) and the unswollen (hard segments) parts of polyurethane as the latter tend to resist further swelling. However, during early stages of sorption, the samples may not reach the true equilibrium concentration of the penetrant and thus, the rate of sorption builds up slowly to produce slight curvatures as shown in Figures 1-7. This is indicative of the departure from the Fickian mode and is further confirmed from an analysis of sorption data by using the following equation (13,14) :

$$\log (M_t/M_\infty) = \log k + n \log t \tag{1}$$

Here, the constant $k$ depends on the structural characteristics of the polymer in addition to its interaction with the solvent; $M_t$ and $M_\infty$ are respectively, the mass uptake at time $t$ and at equilibrium, $t_\infty$. The magnitude of $n$ decides the transport mode; for instance, a value of $n = 0.5$ suggests the Fickian mode and for $n = 1$, a non-Fickian diffusion mode is predicted. However, the intermediate values ranging from $n = 0.5$ to unity suggest the presence of anomalous transport mechanism.

From a least-squares analysis, the values of $n$ and $k$ have been estimated and these are included in Table I, and Figure 8 represents a typical plot for benzene and toluene. The average uncertainty in the estimation of $n$ is around ±0.007. The value of $n$ do not indicate any systematic variation with temperature. However, a general variation of $n$ from a minimum value of 0.53 to a maximum of 0.74 indicates that the anomalous type transport mechanism is operative and the diffusion is slightly deviated from the Fickian trend. This fact can be further substantiated from the curvature dependencies of $Q(t)$ vs. $t^{\frac{1}{2}}$ plots shown in Figures 1-7. Such observations are also evident from the work of Nicolais et al. (15) for n-hexane transport in glassy polystyrene.

A temperature dependence of $k$ suggests that it increases with a rise in temperature for all the penetrants except p-xylene and bromobenzene. Furthermore, $k$ appears to depend on structural characteristics of the penetrant molecules i.e., it decreases successively from benzene to mesitylene; this decrease in $k$ parallels the decrease in the values of sorption equilibrium. Similarly, for chlorobenzene to nitrobenzene via o-dichlorobenzene $k$ decreases successively. Thus, it appears that $k$ not only depends on the structural characteristics of the polymer and penetrant molecules, but also on solvent interactions with the polyurethane chains. At any rate, the greater tendency for non-Fickian coefficients in Equation (1) seen for 60°C, may reflect some aspect of swelling of interphase regions between the hard and soft domains.

TRANSPORT COEFFICIENTS: From the slope $\theta$, of the initial linear portion of the sorption curves i.e., $Q(t)$ vs. $t^{\frac{1}{2}}$, the diffusion

Table I
Analysis of Penetrant Transport at Various Temperatures

| Penetrant | Molecular Volume x $10^{23}$ ($cm^3$/molecule) | Temp (°C) | Exponent n (Eq 1) | k (g/g·hr)$10^2$ (Eq 1) | $Q(t)_{max}$ (% of dry sample wt) |
|---|---|---|---|---|---|
| Benzene | 14.85 | 25 | 0.554 | 2.809 | 71.01 |
|  |  | 44 | 0.593 | 3.295 | 72.15 |
|  |  | 60 | 0.599 | 3.975 | 74.14 |
| Toluene | 17.75 | 25 | 0.567 | 2.602 | 60.17 |
|  |  | 44 | 0.600 | 3.207 | 59.09 |
|  |  | 60 | 0.599 | 3.874 | 60.08 |
| p-xylene | 20.48 | 25 | 0.561 | 2.295 | 49.68 |
|  |  | 44 | 0.557 | 3.334 | 49.95 |
|  |  | 60 | 0.595 | 3.021 | 49.31 |
| Mesitylene | 23.19 | 25 | 0.532 | 1.261 | 40.15 |
|  |  | 44 | 0.571 | 2.203 | 41.52 |
|  |  | 60 | 0.602 | 2.444 | 41.99 |
| Anisole | 18.16 | 25 | 0.566 | 2.166 | 80.25 |
|  |  | 44 | 0.585 | 2.765 | 83.89 |
|  |  | 60 | 0.581 | 3.227 | 82.80 |
| Nitrobenzene | 17.14 | 25 | 0.602 | 1.134 | 106.32 |
|  |  | 44 | 0.584 | 1.797 | 107.17 |
|  |  | 60 | 0.618 | 1.839 | 114.81 |
| Chlorobenzene | 16.98 | 25 | 0.588 | 2.371 | 105.52 |
|  |  | 44 | 0.569 | 3.597 | 109.08 |
|  |  | 60 | 0.584 | 3.739 | 108.84 |
| o-Dichlorobenzene | 18.78 | 25 | 0.583 | 1.658 | 131.36 |
|  |  | 44 | 0.565 | 2.441 | 131.17 |
|  |  | 60 | 0.589 | 2.625 | 131.52 |
| Bromobenzene | 17.53 | 25 | 0.575 | 2.300 | 147.50 |
|  |  | 44 | 0.586 | 2.790 | 150.06 |
|  |  | 60 | 0.736 | 1.600 | 151.58 |

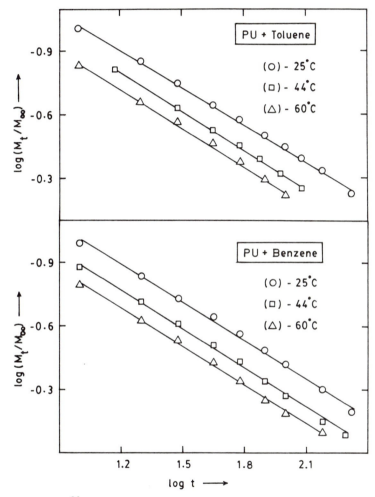

Figure 8 Log $^{M_t}/_{M_\infty}$ versus log t for polyurethane + Benzene and polyurethane + Toluene systems.

coefficients D, have been calculated by using ($\underline{16},\underline{17}$)

$$D = \pi(h\theta/4M_\infty)^2 \qquad (2)$$

Here, $M_\infty$ has the same meaning as before and h is the initial sample thickness; the slope $\theta$, is usually obtained before 50% completion of sorption. The values of D determined in this manner can be regarded as independent of concentration, and are thus, applicable for the Fickian mode of transport. A triplicate evaluation of D from sorption curves gave us D values with an error of ±0.003 units at 25°C and ±0.005 units at 60°C for all polymer-penetrant systems. These uncertainty estimates regarding diffusion coefficients suggest that the half times were very reproducible (to within a few tens of seconds). The values of D are compiled in Table II for each polyurethane-solvent pair. Included in the same table are the values of sorption coefficients, S, as computed from the plateau regions of the sorption curves and permeability coefficient, P as calculated from the simple relation ($\underline{18}$) :

$$P = D.S \qquad (3)$$

In all cases, both permeability and diffusivity of methyl-substituted benzenes vary in an inverse manner with their molecular volumes (as calculated by dividing the molecular weight by density and Avogadro number to yield the volume per molecule). For other penetrants, namely, anisole, nitrobenzene, chlorobenzene, o-dichlorobenzene and bromobenzene, the volume per molecule varies in the range 17-19. However, their diffusion trends are quite different. For instance, though nitrobenzene and chlorobenzene having almost the same molecular volume ($\sim 17 \times 10^{-23}$ $cm^3$/molecule), yield widely different diffusivity and permeability; however, the maximum S values of both the penetrants are almost identical. Similarly, for bromobenzene, the diffusive trends are higher whereas, o-dichlorobenzene exhibits somewhat intermediatory transport behavior between chloro- and bromobenzene. However, anisole exhibits diffusive trends that are in between mesitylene and p-xylene.

When our results are compared with the literature data, a good agreement could be seen. For example, in a study by Schneider and coworkers ($\underline{10}$) for the transport of o-dichloro-benzene in a segmented polyurethane elastomer at 25°C, a value of $D = 0.95 \times 10^{-7}$ $cm^2$/s agrees with our data at 25°C (i.e., $2.01 \times 10^{-7}$ $cm^2$/s). Similarly, for benzene, toluene, and chlorobenzene at 30°C in a polyurethane, Hung ($\underline{5}$) found the values of D as $1.29 \times 10^{-7}$, $1.43 \times 10^{-7}$ and $1.37 \times 10^{-7}$ $cm^2$/s respectively, which agree somewhat with our values, 2.90, 2.60 and $3.42 \times 10^{-7}$ $cm^2$/s respectively, at 25°C.

Table II.   Sorption and Transport Data of
Polyurethane-Penetrant Systems

| Penetrant | Temp (°C) | S (g/g) | $K_S$ ( mmol/g) | $D \times 10^7$ ($cm^2/s$) Eq. (2) | $P \times 10^7$ ($cm^2/s$) Eq. (3) |
|---|---|---|---|---|---|
| Benzene | 25 | 0.710 | 9.09 | 2.90 | 2.06 |
|  | 44 | 0.721 | 9.24 | 5.23 | 3.73 |
|  | 60 | 0.741 | 9.49 | 8.09 | 6.00 |
| Toluene | 25 | 0.602 | 6.53 | 2.60 | 1.56 |
|  | 44 | 0.591 | 6.41 | 5.71 | 3.37 |
|  | 60 | 0.601 | 6.52 | 7.52 | 4.52 |
| p-Xylene | 25 | 0.497 | 4.68 | 2.34 | 1.16 |
|  | 44 | 0.500 | 4.71 | 4.33 | 2.16 |
|  | 60 | 0.493 | 4.64 | 7.02 | 3.46 |
| Mesitylene | 25 | 0.402 | 3.34 | 0.86 | 0.34 |
|  | 44 | 0.415 | 3.45 | 2.24 | 0.93 |
|  | 60 | 0.420 | 3.49 | 3.76 | 1.58 |
| Anisole | 25 | 0.803 | 7.42 | 1.66 | 1.33 |
|  | 44 | 0.839 | 7.76 | 3.57 | 2.99 |
|  | 60 | 0.828 | 7.66 | 4.82 | 3.99 |
| Nitrobenzene | 25 | 1.063 | 8.64 | 0.87 | 0.92 |
|  | 44 | 1.071 | 8.71 | 1.57 | 1.68 |
|  | 60 | 1.148 | 9.23 | 2.47 | 2.83 |
| Chlorobenzene | 25 | 1.055 | 9.38 | 3.42 | 3.61 |
|  | 44 | 1.091 | 9.69 | 5.41 | 5.90 |
|  | 60 | 1.088 | 9.67 | 6.90 | 7.50 |
| o-Dichloro-benzene | 25 | 1.314 | 8.94 | 2.01 | 2.64 |
|  | 44 | 1.312 | 8.92 | 2.87 | 3.77 |
|  | 60 | 1.315 | 8.95 | 4.39 | 5.77 |
| Bromobenzene | 25 | 1.475 | 9.39 | 2.81 | 4.14 |
|  | 44 | 1.501 | 9.56 | 4.56 | 6.85 |
|  | 60 | 1.516 | 9.65 | 7.19 | 10.90 |

The diffusion coefficients as calculated from Equation 2 have been used in Equation 4 to generate the theoretical sorption curves (19-22) :

$$M_t/M_\infty = 1 - (\frac{8}{\pi^2}) \sum_{n=0}^{\infty} \frac{1}{(2n+1)^2} \exp [-D(2n+1)^2 \pi^2 t/h^2] \quad (4)$$

Equation 4 describes the Fickian diffusion mode. The simulated sorption curves are compared in Figures 9 and 10 with the experimental profiles for some representative penetrants. The overall agreement is only fair. During early stages of sorption the agreement is not so good and the experimental curves show slight curvatures; this suggests that the transport is not strictly of Fickian type.

TEMPERATURE EFFECTS. The Arrhenius activation parameters viz., $E_D$ and $E_P$ for the processes of diffusion and permeation have been computed from a consideration of the temperature variation of D and P respectively, by using the relation :

$$\log X = \log X_o - (E_X/2.303 \; RT) \quad (5)$$

where X refers to D or P and $X_o$ is a constant representing $D_o$ and $P_o$; $E_X$ denotes the activation energy for the process under consideration and RT has the conventional meaning. The estimated parameters $E_D$ and $E_P$ are given in Table III. The Arrhenius plots are given in Figure 11.

The $E_D$ and $E_P$ values vary from about 16 to 35 kJ/mol for the penetrants under study. As regards the effect of $CH_3$- substitution on the benzene molecule (i.e., on going from benzene to mesitylene) there is a systematic increase in $E_D$ and $E_P$ values. These results could be explained on the basis of Eyring's hole theory (23), according to which, the energy required to "open a hole" in the polymer matrix to accommodate a diffusing molecule bears a direct relationship with $E_D$. Thus, the larger molecules in a related series will have larger $E_D$ and smaller diffusion coefficients. This is in conformity with the experimental observations reported here.

Attempts have also been made to calculate the equilibrium sorption constants, $K_S$, from considerations on the equilibrium process occurring in the liquid phase at constant temperature and pressure (5). Thus,

$$K_S = \frac{\text{number of moles of penetrant sorbed}}{\text{unit mass of the polymer}}$$

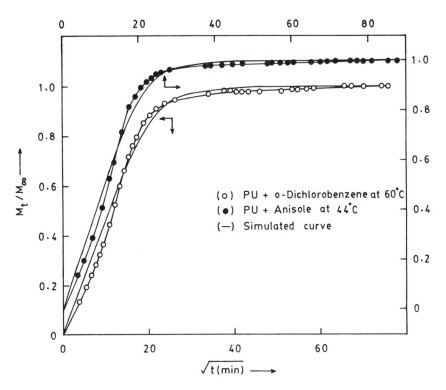

Figure 9 Comparison between experimental and simulated sorption curves for polyurethane + solvent systems.

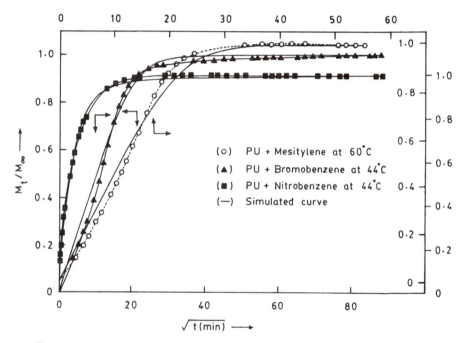

Figure 10  Comparison between experimental and simulated sorption curves for polyurethane + solvent systems.

Table III.   Activation Parameters and Thermodynamic Functions
for Polyurethane-Penetrant Systems

| Penetrant | $E_D$ kJ/mol | $E_P$ kJ/mol | $\Delta S^\circ$ J/mol.K | $\Delta H^\circ$ kJ/mol |
|---|---|---|---|---|
| Benzene | 24.17 | 25.13 | 21.6 | 0.99 |
| Toluene | 25.43 | 25.40 | 15.3 | −0.08 |
| p-Xylene | 25.83 | 25.72 | 12.3 | −0.17 |
| Mesitylene | 34.95 | 35.90 | 13.7 | 1.08 |
| Bromobenzene | 22.04 | 22.67 | 20.8 | 0.64 |
| Anisole | 25.42 | 26.24 | 19.4 | 0.79 |
| Chlorobenzene | 16.64 | 17.40 | 21.2 | 0.76 |
| o-Dichlorobenzene | 18.20 | 18.23 | 18.3 | 0.02 |
| Nitrobenzene | 24.54 | 26.21 | 23.7 | 1.74 |

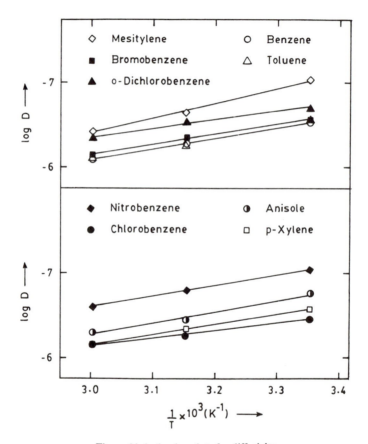

Figure 11 Arrhenius plots for diffusivity.

$$K_S = \frac{m \text{ moles of penetrant}}{g \text{ membrane}} \qquad (6)$$

Figure 12 shows the dependence of $K_S$ on penetrant molecular weights at 25, 44 and 60°C; a systematic decrease in $K_S$ with molecular weight can be seen for benzene to mesitylene suggesting an inverse dependence of $K_S$ on molecular weight of the penetrant. This may be more logical because larger molecules tend to occupy more free volume in the amorphous regions of PU chains than smaller molecules. On the other hand, anisole, nitrobenzene, bromobenzene, chloro- and o-dichlorobenzene show positive deviations (i.e., higher solubility) from linearity. This could be attributed to the structural and polarity similarities of the solvents. Another factor might be the affinity of these solvents towards polyurethane. Following the generalization "like absorbs like", this explanation is consistent with our experimental results. Using the data of $K_S$ we have estimated change in standard enthalpy (i.e., heat of sorption) $\Delta H^O$, and standard entropy of sorption $\Delta S^O$, by using van't Hoff relationship (24) :

$$\log K_S = \frac{\Delta S^O}{2.303R} - \frac{\Delta H^O}{2.303R} (\frac{1}{T}) \qquad (7)$$

The plots of $\log K_S$ versus $1/T$ as shown in Figure 13, are linear within the temperature interval of 25 to 60°C. The estimated values of $\Delta H^O$ and $\Delta S^O$ are also included in Table III. The average estimated error in $\Delta H^O$ is about ±4 J/mol whereas for $\Delta S^O$, it is about ±1 J/mol/K.

For benzene, $\Delta S^O$ is about 22 J/mol.K and it progressively decreases up to p-xylene for which $\Delta S^O$ is about 12 J/mol.K; $\Delta S^O$ for mesitylene is slightly higher than p-xylene. However, for the remaining penetrants we could not observe any systematic trend in $\Delta S^O$ values because these penetrants possess more or less identical sizes and interact differently with the polyurethane segments. This further confirms that the diffusive portion of absorption does not involve polymeric cooperation to a greater extent, but results essentially from the positioning of penetrant molecules during movement within the pre-existing available sites of the polymer matrix (21). The $\Delta H^O$ values are small and positive excepting toluene and p-xylene for which negative values are observed.

THERMODYNAMIC ANALYSIS. For a comprehensive understanding of the structure-property relationships of the elastomeric materials in the presence of a solvent, it is necessary to know the magnitude of polymer-solvent interaction parameter $\chi$, and hence the molar mass between crosslinks, $M_c$. The criterion for swelling equilibrium

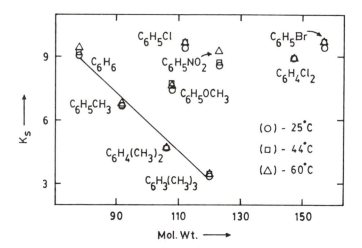

Figure 12  Dependence of sorption constant ($K_s$) on molecular weight of the solvents at (◯) 25 °C; (□) 44 °C; and (△) 60 °C.

was first recognized by Frenkel (25) and was later developed by Flory and Rehner (11,12) into a general theory. For a successful calculation of $M_c$ by this theory we must have in hand reliable data of $\chi$ for the solvent-polymer pair. A number of methods of determining $\chi$ have been suggested in the literature and these have been recently reviewed by Takahashi (26). All these methods are empirical and require the use of solubility parameter of the solvent. Instead, we suggest an alternative phenomenological treatment for the calculation of $\chi$. This approach is based on expressing the Flory-Rehner equation into a derivative of volume fraction of the polymer, $\phi$, in the completely swollen state with respect to temperature. Thus,

$$(\frac{d\phi}{dT}) = \frac{\chi \phi^2 / T}{2 \chi \phi - \frac{\phi}{1-\phi} - [\ln (1-\phi) + \phi + \chi \phi^2] N} \qquad (8)$$

where

$$N = \frac{(\frac{\phi^{2/3}}{3} - \frac{2}{3})}{(\phi^{1/3} - \frac{2\phi}{3})} \qquad (9)$$

so that

$$\chi = \frac{(d\phi/dT) [\{\phi/(1-\phi)\} + N \ln (1-\phi) + N\phi]}{[2\phi (d\phi/dT) - \phi^2 N (d\phi/dT) - \phi^2/T]} \qquad (10)$$

The volume fraction of the swollen polymer is calculated as :

$$\phi = [1 + \frac{\rho_p}{\rho_S} (\frac{M_a}{M_b}) - \frac{\rho_p}{\rho_S}]^{-1} \qquad (11)$$

Here, $M_b$ and $M_a$ are respectively, the mass of polymer before and after swelling, $\rho_S$ is solvent density and $\rho_p$ is density of polyurethane. Computation of the coefficient of volume fraction $(d\phi/dT)$ can be done from a least-squares fit of the $\phi$ data versus temperature, T.

The molar mass between crosslinks can then be obtained from a modification of Flory-Rehner relation as :

$$M_c = \frac{- \rho_p V (\phi)^{1/3}}{[\ln (1-\phi) + \phi + \chi \phi^2]} \qquad (12)$$

where V is molar volume of solvent and the parameter $\chi$ to be used here has been obtained from Equation 10. The estimated quantities are compiled in Table IV. Wide disparity in $\chi$ and $M_c$

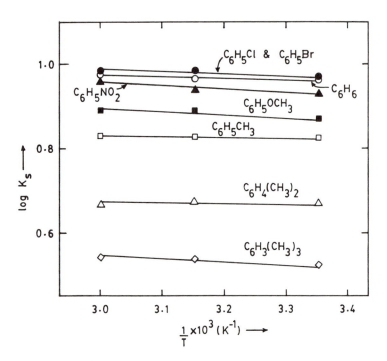

Figure 13  van't Hoff's plots for polyurethane + solvent systems.

Symbols:   (●)  Chloro- and Bromobenzene;
           (O)  Benzene; (■) Anisole; (□) Toluene;
           (△)  p-xylene; and (◇) Mesitylene; (▲) Nitrobenzene.

Table IV.   Results of Flory-Rehner Theory

| Penetrant | $\chi$ | $M_c$ |
|---|---|---|
| Benzene | 0.47 | 860 |
| Toluene | 0.29 | 551 |
| p-Xylene | 0.23 | 469 |
| Mesitylene | 0.37 | 517 |
| Bromobenzene | 0.36 | 1018 |
| Anisole | 0.41 | 886 |
| Chlorobenzene | 0.39 | 1017 |
| o-Dichlorobenzene | 0.24 | 848 |
| Nitrobenzene | 0.58 | 1793 |

values are observed which depend on the nature of the solvents used. For toluene, p-xylene and o-dichlorobenzene, both $\chi$ and $M_c$ respectively, vary from : $\chi$ = 0.23 to 0.29 and $M_c$ = 470 to 850. This suggests that there is a considerable solvent interaction with the polyurethane segments. For benzene $\chi$ = 0.47 and $M_c$ = 860, whereas nitrobenzene exhibits $\chi$ = 0.58 and $M_c$ = 1793, the highest among the liquids considered. On the other hand, mesitylene, anisole, chloro- and bromobenzene exhibit almost identical values of $\chi$ i.e., 0.36 to 0.41; among these, the latter two exhibit higher $M_c$ ( $\sim$ 1020) than either mesitylene or anisole for which $M_c$ are respectively, 517 and 886. Such variations in $M_c$ data indicate the serious limitations of the Flory-Rehner theory to study polymer swelling. However, the complexity of solvent interactions may probably affect in various degrees, the hard and the soft segments of PU. This will alter the overall swelling behavior thereby contributing to the variability of $M_c$ results. This is further indicative of the subtle non-Fickian effects as observed for most of the solvent-polyurethane pairs.

## ACKNOWLEDGMENTS

The authors appreciate the financial support from the Robert A. Welch Foundation (Grant AI-0524); TMA and USA thank the University Grants Commission, New Delhi, India for the award of a teacher fellowship to Mr. Aithal to study at Karnatak University.

## LITERATURE CITED

1. Cooper, S. L.; Tobolsky, A. V. J. Appl. Polym. Sci. 1966, 10, 1837.

2. Gibson, P. E. Properties in Polyurethane Block Copolymers in Developments in Block Copolymers; Goodman, I., Ed.; Elsevier : London, 1982.

3. Trapps, G. In Advances in Polyurethane Technology; Buist, J. M.; Gudgeon, H., Eds.; Interscience : New York, 1968 ; p 63.

4. Hung, G. W. C.; Autian, J. J. Pharm. Sci. 1972, 61, 1094.

5. Hung, G. W. C. Microchem. J. 1974, 19, 130.

6. Hopfenberg, H. B.; Schneider, N. S.; Votta, F. J. Macromol. Sci.-Phys. 1969, B3(4), 751.

7. Nierzwicki, W.; Majewska, Z. J. Appl. Polym. Sci. 1979, 24, 1089.

8. Sefton, M. V.; Mann, J. L. J. Appl. Polym. Sci. 1980, 25, 829.

9.   Yokoyama, T.; Furukawa, M. In International Progress in Urethanes; Ashida, K.; Frisch, K. C., Eds.; Technomic Publishing Co., Inc., 1985; Vol. 4, Chapter 1.

10   Schnieder, N. S.; Illinger, J. L.; Cleaves, M. A. Polym. Eng. Sci. 1986, 26, 1547.

11.  Flory, P. J.; Rehner, Jr, J. J. Chem. Phys. 1943, 11, 521.

12.  Flory, P. J. J. Chem. Phys. 1950, 18, 108.

13.  Lucht, L. M.; Peppas, N. A. J. Appl. Polym. Sci. 1987, 33, 1557.

14.  Chiou, J. S.; Paul, D. R. Polym. Eng. Sci. 1986, 26, 1218.

15.  Nicolais, L.; Drioli, E.; Hopfenberg, H. B.; Apicella, A. Polymer 1979, 20, 459.

16.  Britton, L. N.; Ashman, R. B.; Aminabhavi, T. M.; Cassidy, P. E. J. Chem. Educ. 1988, 65, 368.

17.  Aminabhavi, T. M.; Cassidy, P. E. Polym. Commun. 1986, 27, 254.

18.  Cassidy, P. E.; Aminabhavi, T. M.; Thompson, C. M. Rubb. Chem. Technol., Rubb. Revs. 1983, 56, 594.

19.  Crank, J. The Mathematics of Diffusion; 2nd Ed; Clarendon Press : Oxford, 1975.

20.  Gent, A. N.; Tobias, R. H. J. Polym. Sci. Polym. Phys. Ed. 1982, 20, 2317.

21.  Enscore, D. J.; Hopfenberg, H. B.; Stannett, V. T. Polym. Eng. Sci. 1980, 20, 102.

22.  Garrido, L.; Mark, J. E.; Clarson, S. J.; Semlyen, J. A. Polym. Commun. 1984, 25, 218.

23.  Zwolinski, B. J.; Eyring, H.; Reese, C. E. J. Phys. Colloid Chem. 1949, 53, 1426.

24.  Golden, D. M. J. Chem. Educ. 1971, 48, 235.

25.  Frenkel, J. Rubb. Chem. Technol. 1940, 13, 264.

26.  Takahashi, S. J. Appl. Polym. Sci. 1983, 28, 2847.

RECEIVED December 5, 1989

# Chapter 20

# Toluene Diffusion in Natural Rubber

Lawrence S. Waksman[1,3], Nathaniel S. Schneider[1,4], and Nak-Ho Sung[2]

[1]Polymer Research Branch, SLCMT-EMP, U.S. Army Materials Technology Laboratory, Watertown, MA 02172
[2]Department of Chemical Engineering, Tufts University, Medford, MA 02152

Immersion swelling and incremental vapor sorption experiments were carried out with toluene on a lightly crosslinked natural rubber sample with varying amounts of carbon black. Only small differences in equilibrium swelling were found. The sorption isotherms were superimposable for samples at all carbon black levels up to an activity of 0.9 and could be fitted with the Flory–Rehner relation. Sorption and desorption curves, above 25% toluene, showed slight "S" shaped curvature. Diffusion constants, D, obtained by the half–time method or the Joshi–Astarita analysis of coupled diffusion and relaxation, showed a similar maximum in D with concentration. When converted to solvent mobilities, $D_1$, the values leveled out rather than extrapolating to the self–diffusion coefficient of toluene, $D_1^*$. Application of the Armstrong–Stannett treatment of heating effects during sorption lead to significant corrections in D and to better agreement with an empirical extrapolation to $D_1^*$.

Studies of the diffusion of benzene in natural rubber represent some of the earliest detailed examinations of the interaction of an organic solvent with a polymer. Hayes and Park carried out measurements at low concentrations by the vapor sorption method (1), and at higher concentrations by determining the concentration distribution using an interferometric method (2). Complementary measurements by vapor transmission to determine the diffusion coefficient from time–lag data were carried out at low concentrations by Barrer and Fergusson (3). The main results of these studies have been summarized in Fujita's review (4) of organic vapor diffusion in polymers above the glass transition temperature. However, the problems with these measurements were not referenced. In the work of Hayes and Park, the calculated solvent mobilities extrapolated to a value, at unit solvent volume fraction, which

[3]Current address: United Technologies, Hamilton Standard, 1 Hamilton Standard Road, Windsor Locks, CT 06096
[4]Address correspondence to this author.

0097–6156/90/0423–0377$06.00/0
© 1990 American Chemical Society

exceeded the self-diffusion coefficient of pure benzene by two
orders of magnitude. This lead the authors to question the thermodynamic correction factors used in computing the solvent
mobilities. In Barrer and Fergusson's study, the values of the
diffusion coefficient from steady state were higher than from the
time-lag, leading to the conclusion that the behavior might be
complicated by relaxation effects. Thus, it appears that even in
this classical system there are problems which deserve
consideration.

The goals of the present study were to reexamine vapor sorption
in a lightly crosslinked rubber, both as an unfilled sample and in
samples containing two types of carbon black in varying amounts.
Complications in the vapor sorption-rate curves motivated a more
detailed study of the diffusion problems as the main area of
concern.

EXPERIMENTAL

SAMPLE PREPARATION.  Samples of natural rubber were prepared by
mixing all ingredients in a Haake-Buchler system 40 internal mixer,
using an accelerated sulfur cure and excluding any oil extender or
plasticizer which could leach out in the immersion experiments. Two
types of carbon black were used; N110, a fine particle, high
structure black and N774, a large particle, low structure black. A
two-stage mixing procedure was used to minimize scorch. The carbon
black was incorporated in the first stage, followed by later
addition of the curatives, since the addition and mixing of carbon
black tends to elevate the batch temperature to unacceptable levels.
The formulations and the outline of the procedure are summarized in
Table 1. The compounded rubber was milled to a thickness of 60 mil
and cured at $121^{\circ}$C for 84 minutes in a hydraulic press using a
picture frame shim with a thickness of 20 mil. The low cure
temperature was chosen in order to maximize the scorch time so that
the uncured rubber could flow and fill the frame.

Table 1.  Natural Rubber Formulations and Processing

| Stage 1 | PHR | Stage 2 | PHR |
|---------|-----|---------|-----|
| Natural Rubber | 100 | Sulfur | 2.5 |
| Stearic Acid | 2 | Santocure | 0.8 |
| Zinc Oxide | 4 | | |
| Agerite Resin D | 1 | Stage 1 mix: $80^{\circ}$C, 77 RPM | |
| Carbon Black | 0-50 | mill: T<$119^{\circ}$C | |
| Volume Fraction | 0-20 | Stage 2 mix: $80^{\circ}$C, 50 RPM | |
| | | mill: T< $93^{\circ}$C | |

SORPTION MEASUREMENTS.  Immersion experiments were carried out in
toluene at room temperature using a blot drying technique. Samples
for vapor sorption studies were thoroughly extracted before use,
resulting in a 3 percent weight loss. Vapor sorption measurements
were carried out using a standard vacuum sorption system and quartz

spring balance. The samples were exposed to a series of approximately ten successive vapor pressure increments in the sorption cycle, followed by an equivalent desorption cycle.

RESULTS AND DISCUSSION

LIQUID IMMERSION. The equilibrium liquid sorption data are summarized in Table 2, where the values listed are the average of results for three samples. The large degree of swelling of the unfilled sample is reduced by the addition of carbon black, essentially in proportion to the amount of carbon black added. When the sorption results are normalized to the rubber content, columen 3, it is seen that the increasing carbon black content still reduces the swelling on this basis, but only to a rather small extent. The reduced swelling is expected if the carbon black contributes, in some way to the effective degree of crosslinking, perhaps through formation of bound rubber. However, it is anomalous that the N774 carbon black produces a larger reduction in swelling than the more highly reinforcing N110. This discrepancy is emphasized by comparison, in the next to last column of Table 2, with the predictions of the empirical relation obtained by Kraus (5) for the effect of carbon black content in reducing the extent of solvent swelling. The expression includes a constant that is characteristic of a particular type of carbon black. A possible explanation for the inverted behavior is a change in the rubber during processing. The samples compounded with N110 showed a higher viscosity and larger heat build-up than the rubber with N774. This could have resulted in a lower molecular weight and, hence, a higher degree of swelling. A quantitative estimate of the change in molecular weight appears later.

Table 2. Immersion Sorption Data for Natural Rubber Samples in Toluene

| PHR Black | Wt. % Sorption | Wt. % Corrected | % Change Expt. | Predict | $M_c$ |
|---|---|---|---|---|---|
| N110 | | | | | |
| 0 | 296 | 303 | | | 3220 |
| 10 | 267 | 300 | 1 | 3.4 | 3320 |
| 30 | 206 | 273 | 10 | 10.3 | 2850 |
| 50 | 174 | 265 | 12.5 | 17.2 | 2710 |
| N774 | | | | | |
| 10 | 252 | 293 | 6.5 | 2.7 | 3010 |
| 30 | 206 | 267 | 12.0 | 8.2 | 2730 |
| 60 | 162 | 247 | 19.5 | 13.7 | 2410 |

The immersion sorption-rate curve for the unfilled rubber was highly reproducible for three separate samples and was almost Fickian in appearance. However, the extrapolated linear portion of the curve did not pass exactly through the origin, but had a small positive intercept on the time axis. This slight non-Fickian

character can be interpreted as an indication of some relaxation
contribution to the sorption process. An estimate of the apparent
diffusion coefficient, using the half-time method discussed later,
gave a value of $D = 6.10 \times 10^{-7}$ $cm^2/sec$.

VAPOR SORPTION ISOTHERMS. The sorption isotherm, representing
sorbed concentration as weight percent (g solvent/100g polymer)
versus toluene activity (vapor pressure/saturation pressure, $p/p_0$)
is completely superimposable on the sorption and desorption cycles
for each of the samples, as shown in Figure 1 for the unfilled
rubber. This indicates that there are no permanent changes in the
sulfur crosslink structure or rubber-carbon black interactions as a
result of the solvent exposure. Although the sorption curves for the
different samples reflect the presence of the carbon black, when the
results are normalized to the weight of rubber, the data for all
samples, filled and unfilled appear to follow a single curve up to
an activity $a_1 = 0.9$ and concentration $c = 100\%$, (Figure 2).

The absence of any effect of carbon black on the normalized
sorption isotherms and the small differences in immersion uptake
raised a question about the effectiveness of the carbon black as a
reinforcing filler. Modulus measurements were made by elongation to
4% strain on unfilled and 40 phr N110 samples, both before and after
immersion and drying. Measurements on the toluene swollen samples
were not possible due to the rapid loss of solvent. The carbon black
produced an almost seven-fold increase in modulus, both in the
preswollen samples and following immersion and drying. Assuming
that these results are representative of the modulus differences in
the swollen state, it can be concluded that the mechanism
responsible for the change in modulus behavior has little effect on
swelling.

The sorption isotherms, normalized to the rubber content, were
fitted by the Flory-Rehner (6) relation using a least squares
routine:

$$\ln(a_1) = \ln(1-\Phi_2) + \Phi_2 + \chi\Phi_2^2 + \frac{V_1}{M_c V} \left[1 - \frac{2M_c}{M}\right] (\Phi_2^{1/3} - \Phi_2/2) \qquad (1)$$

where $a_1$ is the solvent activity, $\Phi_2$ is the volume fraction of
polymer and $M_c$ is the molecular weight between crosslinks. The
values used for the various parameters were as follows: 106.1
$cm^3$/mole for $V_1$, the molar volume of the solvent; 1.092 $cm^3$/g for V,
the specific volume of the polymer; 200,000 for the initial
molecular weight, M, of the unfilled rubber and 100,000 for the
molecular weight of the filled rubber. The fit to the data for the
unfilled crosslinked rubber is indicated by the solid curve in
Figure 1. The resulting value of chi = 0.36 is consistent with
literature values (7) and the molecular weight between crosslinks,
$M_c$ = 2970, indicates that this is a lightly crosslinked sample. If
the immersion point is also included, the value of chi is
unaffected, but $M_c$ increases to 3210. The combined data in Figure 2,

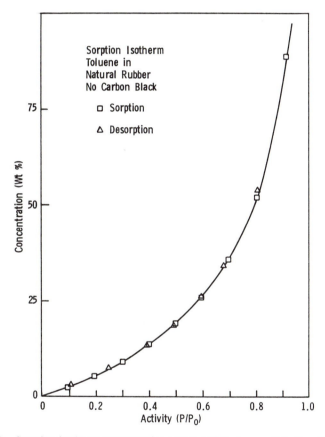

Figure 1. Sorption isotherm; concentration versus activity in crosslinked, unfilled natural rubber.

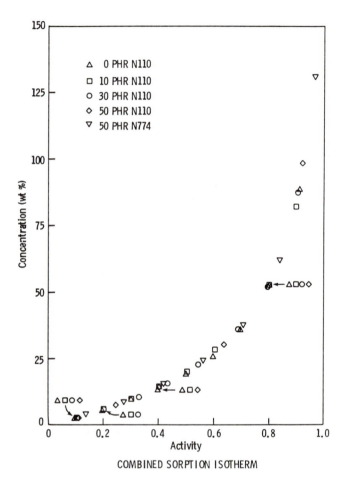

Figure 2. Sorption isotherm for combined data from the crosslinked unfilled and the various carbon black filled samples.

exclusive of the immersion results, could be fitted with a single curve leading to a value of chi = 0.34 and $M_c$ = 2300.

Having shown that the sorption isotherms, exclusive of the swelling under immersion conditions, can be fitted with the Flory Rehner relation, this relation can also be applied to estimate the differences in $M_c$ implied by the small differences in immersion uptake with varying carbon black content. The results of the calculations, using the value of chi = 0.36 are shown as the final column in Table 2. The values decrease progressively with increasing carbon black content, as might be expected from additional crosslinking associated with an increasing amount of bound rubber. It is also possible to obtain an estimate of the decrease in initial molecular weight, M, during mixing, which could account for the greater than expected swelling of the N110 reinforced samples relative to the N774 samples. Considering the two samples with 50 phr carbon black, if the N110 sample were to have the same $M_c$ value as the N774 sample ($M_c$ = 2410), the greater swelling in the N110 sample would correspond to a reduction in M from 100,000 to 30,250.

VAPOR SORPTION-RATE CURVES.  Sorption-rate curves, representing the fractional weight uptake, $M(t)/M(\infty)$, versus square root of time, are shown for the unfilled rubber at three activities in Figure 3. The slope increases from $a_1$ = 0.3 (C=0.2%) to $a_1$ = 0.5 (C=19%) and then decreases at $a_1$ = 0.92 (C=88.5%). Correspondingly, the diffusion coefficient must increase initially and then decrease above some concentration, resulting in a maximum in D versus C. Closer scrutiny of the sorption curves indicates, that, above $a_1$ = 0.5, the extrapolated linear portion of the curve no longer goes through the origin. The curves at first display a subtle "S" shaped behavior, which becomes increasingly marked with increasing concentration, illustrated by the curve at highest activity in Figure 3. The same general behavior is observed on desorption, indicating reversibility, and in all the carbon black filled samples. These anomalies in the sorption-rate curves might be attributed to relaxation effects, which implies that the polymer response to the swelling stress occurs at a slower rate than the diffusion process.

ANALYSIS OF THE SORPTION-RATE CURVES (8).  At low concentrations the sorption-rate curves are Fickian in appearance. The fractional weight increase against square-root time is linear over at least the first 60% of the weight uptake and goes through the origin. The diffusion constant, D, can be calculated from the time to reach one-half the equilibrium weight gain:

$$D = \frac{0.492 \ L^2}{t_{1/2}} \qquad (2)$$

where $t_{1/2}$ is the half-time in seconds, L is the dry sample thickness and D has dimensions of $cm^2/sec$.

The sigmoidal sorption-rate curves were analyzed using the model of Joshi and Astarita (9), which is based on a combination of a

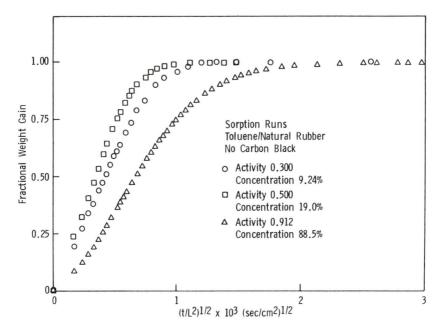

Figure 3. Sorption rate curves for three concentrations of toluene in crosslinked, unfilled natural rubber sample.

Fickian diffusion and first order relaxation rate process. These two processes are linked through a coupling constant, $\Phi^2 = \Theta_D/\Theta_R$. The quantity $\Theta_D = L^2/4D$ is the characteristic time for diffusion. The quantity $\Theta_R = 1/k_R$, where $k_R$ is the first order relaxation rate constant, is the characteristic time for relaxation. The fractional weight uptake is give by the following equation:

$$M(t)/M(\infty) = f(t)_D - m[f_{DR}(t) - (\tan\Phi/\Phi)(1-\exp-(t/\Theta_R))] \qquad (3)$$

Here m is the ratio of uptake due to relaxation relative to the total uptake, $M(\infty)$. The quantities $f_D$ and $f_{DR}$ are, respectively, the series solution for Fick's second law and a related series which includes the coupling constant. For $\Phi \gg 1$, the behavior reduces to classical Fickian diffusion. For $\Phi \ll 1$, that is relaxation very slow compared to diffusion, the equation reduces to the sum of an independent Fickian and a first order relaxation process. This simple model was originally proposed and widely used by Berens and Hopfenberg (10).

The JA model was applied to the sorption-rate curves using a routine in which the three parameters, D, $k_R$ and m, were varied incrementally over a prescribed range and the fit to the experimental data was subjected to a least-squares test to select the best set of values. The experimental curves could be matched with considerable accuracy, as shown by the example in Figure 4. The diffusion coefficients resulting from application of the half-time method, ignoring the initial curvature, are compared with the results from the Joshi-Astarita analysis in Figure 5. The diffusion constant goes through a pronounced maximum with concentration and the results from JA are only slightly higher than the half-time values.

SOLVENT MOBILITIES. One check on the physical significance and the reliability of the data representing the concentration dependence of the diffusion coefficient is to convert these results to solvent mobilities. The values should increase rapidly with increasing concentration and extrapolate to the self-diffusion coefficient for toluene. The procedure for carrying out the calculations was outlined in previous publication (11) and is repeated here in a brief form for convenient reference. The diffusion coefficient obtained directly in the vapor sorption experiment is a polymer, mass-fixed, mean diffusion coefficient, $\bar{D}$, in the sorption interval. Duda et. al. (12) have shown that, if the concentration interval is small, the true diffusion coefficient, D, is simply related to the mean diffusion coefficient at a prescribed intermediate concentration in the interval:

$$\bar{D} = D[C_i + k(C_f-C_i)] \qquad (4)$$

where k = 0.7 if D increases with C and 0.56 if D decreases with C. The values of D are then converted to the mutal diffusion coefficient, $D_{12}$, using the relation due to Crank (8):

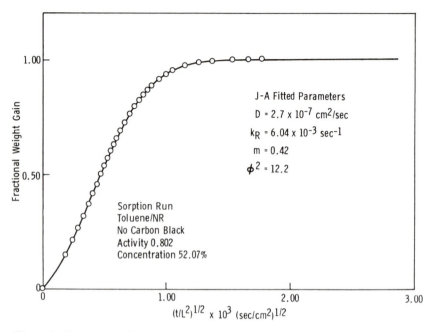

Figure 4. Comparison of experimental data and theoretical curve, solid line, determined from Joshi-Astarita analysis.

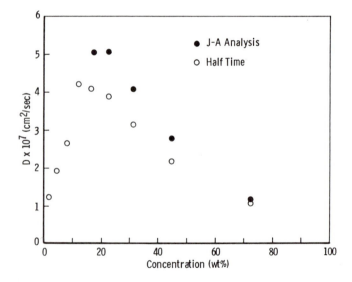

Figure 5. Diffusion constants versus concentration calculated from half-time and from the Joshi-Astarita analysis.

$$D_{12} = D/\Phi_2^2 \tag{5}$$

where $\Phi_2$ is the volume fraction of polymer in the swollen sample.

The concentration dependence of $D_{12}$ can be expressed as the product of two factors, a solvent mobility, $D_1$, and a thermodynamic factor, Q (13);

$$D_{12} = D_1 Q = D_1 \Phi_2 \left(\frac{\partial \ln a_1}{\partial \ln \Phi_1}\right)_{T,P} \tag{6}$$

Since the solvent mobility will generally increase with concentration and the thermodynamic factor decreases with concentration, $D_{12}$ will usually display a maximum when data are obtained over a sufficiently large concentration range. The thermodynamic factor was obtained analytically from the derivative of the Flory–Rehner relation. For consistency, the value was calculated at the adjusted concentration defined in equation 4 above. The resulting values of $D_1$, shown in Figure 6, increase rapidly at low concentration then reach a maximum at a value of 1.0 x $10^{-6}$ cm$^2$/sec. This is a factor of about twenty lower than the estimated self diffusion coefficient for toluene, $1.75 \times 10^{-5}$ cm$^2$/sec (14).

It is possible that the lower than required values of $D_1$ reflect a problem with incorrect values of Q, which if too large would result in smaller values of $D_1$. In an interferometric study of the diffusion of toluene in an uncrosslinked natural rubber sample, Mozisek (15) reported results for the mutual diffusion coefficient which were similar to the results of Hayes and Park. In the absence of thermodynamic data from Mozisek's work, correction factors calculated for the present work were applied to his data. The results are shown in Figure 7, which reproduces Mozisek's data along with the values for $D_1$. The extrapolated value at $\Phi_1 = 1$, would exceed the self diffusion coefficient for toluene by about two orders of magnitude, similar to the discrepancy seen with Hayes and Park's data. This indicates that the fault with the results in the present case is not due to overly high values of the correction factors. Moreover, the method of calculating $D_1$ from $D_{12}$ has been confirmed experimentally by Duda and Vrentas (16) in a comparison of vapor sorption results for toluene diffusion in molten polystyrene with the values of $D_1$ obtained directly using radio–labeled toluene.

HEATING EFFECTS. In searching for an explanation of the problems with the data in the present work, consideration was given to the possible complication of heating effects accompanying vapor sorption. The influence of heating effects on the diffusion coefficent of water vapor in wool fibers and, later, in ethyl cellulose films was treated by Armstrong and Stannett (17,18). They showed that the resulting temperature increases could cause large reductions in the diffusion coefficient determined by the usual analysis. This is not ordinarily considered to be a problem for organic vapors due to the much lower heat of vaporization, eg. 102.5 cal/gm for toluene compared to 582 cal/gm for water.

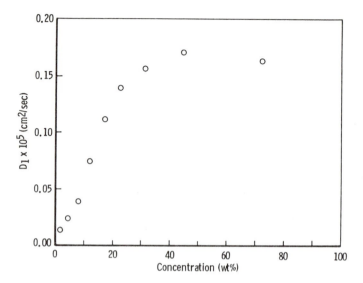

Figure 6. $D_1$ versus concentration in crosslinked, unfilled natural rubber.

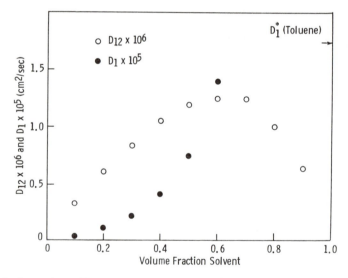

Figure 7. Comparison of $D_1$ versus volume fraction of solvent in natural rubber, calculated from results of Von Mozisek using thermodynamic correction factors from present study, with self-diffusion coefficient for pure toluene. $D_1$ obtained using Q from Flory–Huggins (Chi=0.36).

In the analysis provided by Armstrong and Stannett, the diffusion coefficient is obtained from the limiting, long-time slope of the sorption-rate curve:

$$D = -1/\lambda_1^2 \, dln[(1 - M(t)/M(\infty))]/dt \qquad (7)$$

where $\lambda_1$ is the first eigenvalue in the solution of the differential equations governing the combined heat and mass-transfer. It might be noted that, in the absence of heating effects, this is the usual equation for determining D from the final rate of sorption with $\pi$ replacing $\lambda_1$ (8). The parameter $\lambda_1$ is obtained from the following equation:

$$\frac{1}{H_v \omega} \left[ \frac{H}{b\rho * slope} - C \right] = \frac{tan(\lambda_1 b)}{\lambda_1 b} \qquad (8)$$

where the various parameters are defined below. A figure is provided in their publications to assist in interpolating $\lambda_1$ from the right hand side of the above relation.

In applying the analysis to the present data the following values were used: latent heat of vaporization of toluene, $H_v$ = 102.5 cal/gm; heat capacity, C = 0.437 and 0.357 cal/gm °C for the unfilled and 50 phr filled rubber, respectively; density of rubber, $\rho$ = 0.9126 gm/cm$^3$; half thickness, b = 0.234 and 0.257 cm for the unfilled rubber and 50 phr N111 filled rubber, respectively. The value obtained experimentally by Armstrong and Stannett for the heat transfer coefficient was, H = 1.5 x10$^{-4}$ cal/cm$^2$ sec°C. When this value was used with the present data, the corrections were exceptionally large and values at a concentration of 35% and above were controlled completely by heat transfer. Therefore, a value of H=3.0*10$^{-4}$ was chosen, since it was the smallest value which gave finite corrections over the entire concentration range for the unfilled sample and, therefore, would illustrate the general effect of these corrections. The temperature coefficient of weight gain, $\omega$ (g/g °C) was calculated from the sorption isotherm, rather than being measured directly, using the following relation:

$$\omega = \left( \frac{1}{1-\Phi_1} \right)^2 \frac{H_v}{RT^2} \left( \frac{\partial \Phi_1}{\partial \ln a_1} \right)_{T,P} \qquad (9)$$

The derivation of this relation will be treated in a future publication dealing more extensively with heating corrections to diffusion data. finally, it was found that values of D obtained from the final rate of sorption using (8) with $\pi$ replacing $\lambda_1$, were very close to the half-time or JA values, which varied in a more systematic manner with concentration. Therefore, the final slope was computed from these values of D.

In general, the corrections to the diffusion coefficient increase progressively, with the next to the last point being increased by a factor of 2.4. The resulting trend in $D_1$ versus C, shown in Figure 8, is almost consistent with the continuous, but empirical extrapolation to the self-diffusion coefficient for toluene. The last point falls below this line, possibly due to an error in the initial measurement. However, there is a slight trend for the data to rise too rapidly, in comparison to the dashed line. A comparable analysis of the data for the filled 50 phr N774 sample leads to excessively high values for the solvent mobilities.

CONCLUSIONS

This study of the comparative immersion and vapor sorption behavior of unfilled and carbon black filled rubber samples shows that carbon black, up to 50 phr, has a surprisingly small effect on the behavior, although it produces a major change in the initial tensile modulus. The Flory-Rehner equation can be used to show the modest effect of carbon black loading on the increase in the apparent degree of crosslinking, corresponding to the small reduction in swelling.

Unexpected complications were observed in the vapor sorption-rate curves, which were no longer Fickian above a moderate concentration, but showed slight "S" shaped behavior. These curves can be accurately fitted by the Joshi-Astarita analysis of combined diffusion and relaxation, suggesting that relaxation effects might be involved in the swelling behavior, perhaps in keeping with early observations reported from the studies of benzene permeation by Barrer and Fergusson. Relaxation might arise from slow response of the network structure to the swelling stresses. However, vapor sorption experiments on an uncrosslinked rubber, not described here, showed the same type of non-Fickian sorption-rate curves, making this explanation less likely. In addition, a test of the data, by conversion to solvent mobilities gives results which fail to extrapolate to the self-diffusion coefficient of toluene.

The evaluation of the sorption conditions, using the Armstrong-Stannett treatment of the effect of the heat of sorption on the measured diffusion constant, clearly shows that the sorption process, at even moderate concentrations, is no longer isothermal. To illustrate the effects, corrections to the diffusion coefficient data were made using an empirically chosen value for the heat transfer coefficient. When these results were converted to solvent mobilities, the extrapolation as a function of the volume fraction of solvent appeared to be consistent with the self-diffusion coefficient for toluene. However, this comparison is dependent on the method used to reduce the diffusion data.

There are many details which still need to be considered in assessing the quantitative significance of these heating effects, including the correct value of the heat transfer coefficient and the effect on the shape of the sorption-rate curve. However, there can be no doubt that the corrections to the diffusion constant can be large for highly swelling organic solvents, even though the heat of condensation is significantly lower than that of water vapor.

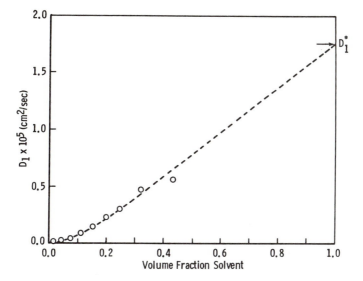

Figure 8. $D_1$ for heat corrected data in unfilled, crosslinked, natural rubber sample, versus volume fraction of solvent, compared with empirical extrapolation (dashed line) to the self-diffusion constant for toluene.

Accordingly, it would seem prudent in any study of vapor sorption kinetics, involving a highly swelling organic liquid, to test whether the conditions are, in fact, isothermal by application of the Armstrong-Stannett analysis.

LITERATURE CITED

1.  Hayes, M.J.; Park, G.S.  Trans. Faraday Soc. 1955, 51, 1134.
2.  Hayes, M.J.; Park, G.S.  Trans. Faraday Soc. 1956, 52, 949.
3.  Barrer, R.M.; Ferguson, R.R.F. Trans. Faraday Soc. 1985, 54, 989.
4.  Fujita, H. In  Diffusion in Polymers; Crank, J.; Park, G.S., Ed; Academic: New York, 1968.
5.  Kraus, G., Ed.; Reinforcement in Elastomers; Wiley Interscience: New York, 1965.
6.  Flory, P. Principles of Polymer Chemistry; Cornell University: Ithaca, N.Y., 1953.
7.  Southern, E.; Thomas, A.F. Rubber Chem. Technol. 1968, 42, 495.
8.  Crank, J.; Park, G.S.  Diffusion in Polymers; Academic: New York, 1968.
9.  Joshi, S.; Astarita, G.  Polymer 1979, 20, 455.
10. Berens,A.R.; Hopfenberg, H.B. J. Polym. Sci.: Polym. Phys. Ed. 1979, 17, 1757.
11. Schneider, N.S.; Mee, C.R.; Goydan, R.; Angelopoulos, A.P. J.Polym. Sci.: Polym. Phys. Ed., 1989, 27, 939.
12. Duda, J.L.; Ni, Y.C.; Vrentas, J.S.  J. Polym. Sci., Polym. Phys. Ed. 1977, 15, 2039.
13. Vrentas, J.S.; Duda, J.L. AICHE Journal 1979, 25, 1.
14. Rimschussel, W.; Hawlicka, E. J. Phys. Chem. 1974, 78, 230.
15. Mozisek, Von M. Macromol. Chemie 197, 136, 87.
16. Duda, J.L.; Ni, Y.C.; Vrentas, J.S. Macromolecules 1979, 12, 459.
17. Armstrong, Jr. A.A.; Stannett, V. Macromol. Chemie 1966, 90, 145.
18. Armstrong, Jr., A.A.; Wellons, J.D.; Stannett, V. Makromol. Chemie 1966, 95, 78.

RECEIVED January 26, 1990

# Author Index

# Affiliation Index

# Subject Index

## O

Orange juice, factors affecting shelf life, 296–297
Organic vapor diffusion in polymers, studies, 377
Organoleptic analysis, description, 187
Orientability of barrier resins
  effect of stretch rate, 243
  effect of temperature, 243
  illustration of cold crystallization behavior, 243,244f
  influencing factors, 243,245
Orientation, molecular
  barrier resins, dependence on morphological nature, 248,249f
  effects on gas transport
    amorphous polymers, 70–71,72f,t
    influencing factors, 69–70
    semicrystalline polymers, 71,73–78
  effects on permeability of competitive oxygen-barrier resins, 245–249
    broadening of melting peak with biaxial orientation, 245,247f,248
    experimental materials, 241
    film preparation, 241
    orientation conditions, 241,242t
    oxygen transmission rates vs. overall draw ratio, 248,249f
  See also Molecular orientation
Oxygen ingression
  measurement, 211,213
  oxygen vs. time, 211,213,214f
Oxygen permeability
  bag in the box packages, 309t
  barrier resins, effect of molecular orientation, 245,246f,248
  calculation of coefficients, 193–194
  commercial polymers, 160,163f
  3,3'-disubstituted bisphenol containing polycarbonates, 164t,165
  effect of diacid structure on polyacrylate permeability, 168,169t,170,171t
  effect of glass transition temperature, 170,172f
  films, 309t
  influencing factors, 194
  measurement procedures, 193
  polymers containing common bisphenol, 165–166,167t
  vs. specific free volume, 170,172f,173,174f
  See also Permeability
Oxygen permeation
  amorphous polyamides
    correlation to specific free volume, 116,118f
    dielectric loss, 116,118f

Oxygen permeation,
  amorphous polyamides—Continued
    effect of amide reversal, 120t
    effect of aromatic ring substituents, 121–122
    effect of chain length, 115–119
    effect of cycloaliphatic groups, 122
    effect of N-methyl groups, 121t
    effect of relative humidity, 122–123,124f
    experimental precautions, 113,115
    influencing factors, 115–120
    structures of monomers, 113,114f
  effect of relative humidity, 111,112f
  in coextruded structures
    blow-molded containers, 233–237
    films containing SELAR OH resins, 231t
    thermoformed containers, 231,232t
  See also Permeation
Oxygen-barrier resins, effect of orientation on permeability, 240–249

## P

Pace and Datyner's model of gas diffusion for rubbery and glassy polymers
  applications, 51
  description, 51
  diffusion coefficient, 52,53f
  energy of activation, 51–52,53f
  microcavity population, 52,54
  problems, 54
Package(s)
  aseptic, See Aseptic packaging
  blood platelet storage requirements, 2,4
  complex barrier responses, 13
  complex barrier structures, 14,15f,16
  function, 3
  gas barrier materials, 7–8
  juices, using polymeric barrier containers, 295–318
  permeation process, 4
  requirements, 2
  search for new barrier materials, 8–13
  use of polymers, 1–2
  walls, flavor scalping, 16
  water permeation requirements, 2,3t
Packaging industry
  application of polymer science, 1–2
  goal, 1
Packaging materials, sorption of flavor compounds, 319–320
Packing efficiency
  definition, 133
  importance, 133
Partial immobilization hypothesis
  description, 41
  search for evidence, 47–48,49f,50

*Production: BethAnn Pratt-Dewey*
*Indexing: Deborah H. Steiner*
*Acquisition: Cheryl Shanks*

*Elements typeset by Hot Type Ltd., Washington, DC*
*Printed and bound by Maple Press, York, PA*

*Paper meets minimum requirements of American National Standard*
*for Information Sciences—Permanence of Paper for Printed Library*
*Materials, ANSI Z39.48–1984*  ∞